Diamond Crystals

Diamond Crystals

Special Issue Editor

Yuri N. Palyanov

MDPI • Basel • Beijing • Wuhan • Barcelona • Belgrade

Special Issue Editor
Yuri N. Palyanov
Novosibirsk State University
Russia

Editorial Office
MDPI
St. Alban-Anlage 66
4052 Basel, Switzerland

This is a reprint of articles from the Special Issue published online in the open access journal *Crystals* (ISSN 2073-4352) from 2017 to 2018 (available at: https://www.mdpi.com/journal/crystals/special_issues/diamond_crystals)

For citation purposes, cite each article independently as indicated on the article page online and as indicated below:

LastName, A.A.; LastName, B.B.; LastName, C.C. Article Title. *Journal Name* **Year**, *Article Number, Page Range*.

ISBN 978-3-03897-630-1 (Pbk)
ISBN 978-3-03897-631-8 (PDF)

Contents

About the Special Issue Editor

Yuri N. Palyanov is a head of Laboratory of Experimental Mineralogy and Crystallogenesis at the V.S. Sobolev Institute of Geology and Mineralogy, Siberian Branch of the Russian Academy of Sciences. He graduated from Novosibirsk State University in 1978 and received a Ph.D. degree in 1983 and a D.Sc. degree in 1997 with a dissertation entitled "Growth and properties of diamond crystals". The main areas of his research activities are concerned with the experimental modeling of mineral-forming processes in the Earth's mantle and growth of diamond crystals at high-pressure high-temperature (HPHT) conditions using the original BARS technique. Dr. Palyanov has authored or co-authored more than 170 scientific articles in peer-reviewed journals. He is a docent at the Geology and Geophysics Department of Novosibirsk State University. Dr. Palyanov was awarded with diplomas from the Russian Academy of Sciences and the Ministry of Education and Science of Russian Federation. He is a laureate of A.E. Fersman award of the Russian Academy of Sciences.

Preface to "Diamond Crystals"

Diamond is perhaps the most remarkable crystalline material created by either Nature or a human being. Being structurally and compositionally very simple, it nevertheless possesses an impressive range of extreme and outstanding properties superior to other materials. For this reason, diamond is frequently referred to as "the ultimate engineering material" and the number of evidence supporting this title has been constantly growing. Besides its importance as the technological material, diamond is the classical model object of fundamental research in solid-state physics, chemistry and engineering. It is difficult to overestimate the significance of diamond in the Earth sciences, where it serves as an invaluable source of information about the Earth's interiors. It is safe to say that diamond is truly an interdisciplinary subject. The editorial and research articles collected in this issue are intendent to highlight recent investigations and developments in diamond research related to the diverse problems of natural diamond genesis, diamond synthesis and growth using CVD and HPHT techniques, and the use of diamond in both traditional applications, such as mechanical machining of materials, and the new recently emerged areas, such as quantum technologies. The results presented in the contributions collected in this special issue clearly demonstrate that diamond occupies a very special place in modern science and technology.

I would like to thank all authors who contributed to this issue for their excellent work, and for their interesting and inspiring articles.

Yuri N. Palyanov
Special Issue Editor

Editorial

The Many Facets of Diamond Crystals

Yuri N. Palyanov [1,2]

[1] Sobolev Institute of Geology and Mineralogy SB RAS, Koptyug ave. 3, 630090 Novosibirsk, Russia;
palyanov@igm.nsc.ru; Tel.: +7-383-330-7501

[2] Department of Geology and Geophysics, Novosibirsk State University, 630090 Novosibirsk, Russia

Received: 24 January 2018; Accepted: 29 January 2018; Published: 31 January 2018

Abstract: This special issue is intended to serve as a multidisciplinary forum covering broad aspects of the science, technology, and application of synthetic and natural diamonds. This special issue contains 12 papers, which highlight recent investigations and developments in diamond research related to the diverse problems of natural diamond genesis, diamond synthesis and growth using CVD and HPHT techniques, and the use of diamond in both traditional applications, such as mechanical machining of materials, and the new recently emerged areas, such as quantum technologies. The results presented in the contributions collected in this special issue clearly demonstrate that diamond occupies a very special place in modern science and technology. After decades of research, this structurally very simple material still poses many intriguing scientific questions and technological challenges. It seems undoubted that diamond will remain the center of attraction for many researchers for many years to come.

Keywords: diamond; high pressure high temperature; chemical vapor deposition; defects and impurities; color centers; carbon isotopes; structural defects; crystal morphology

---

1. Introduction

Diamonds, which possess a remarkable range of extreme and outstanding properties superior to other materials, have been attracting huge interest as a versatile and technologically useful material. Advances in diamond synthesis and growth techniques have paved the way for this unique material to many existent and prospective applications, which now range from optics and electronics to biomedicine and quantum computing. Besides its importance as the strategic future of electronic material, diamond has been the classical model object of fundamental research in solid-state physics, chemistry, and engineering. Diamond occupies a very special place in the Earth sciences, where it serves as an invaluable source of information about the Earth's interior. Of course, being the king of gems, diamond is the key stone for the gem industry and gemological science.

Diamond is an allotrope form of carbon that is thermodynamically stable at high pressures. It has a face centered cubic structure with each carbon atom covalently bonded to its four nearest neighbors in a regular tetrahedron. This structural arrangement coinciding with the sp3 hybrid orbitals of carbon yields a rigid framework, which combined with the strength of the C-C bond gives rise to many outstanding properties of diamond.

At first sight, the idea proposed by the editorial team of *Crystals* to organize a special issue dedicated to diamond crystals seemed somewhat perplexing inasmuch as diamond is truly a multidisciplinary subject and the issue will inevitably be composed of contributions from very different areas of diamond research that may complicate its overall comprehension. On the other hand, as it frequently happens, solutions to some particular issue may lie just a step aside of the subject area and breakthroughs are only possible with a multidisciplinary approach. Diamond is such a case. To structure the editorial introduction to this special issue, I conventionally divide the subjects into what concerns natural diamonds and synthetic diamonds. The former are mainly considered in the Earth sciences and the latter in the material science, physics, and engineering.

2. Some Facets of Natural Diamond Crystals

Considering the importance of diamond in the Earth sciences, I cannot help quoting the famous dictum of the eminent scientist Sir Charles Frank: If "A snowflake is a letter to us from the sky" (Nakaga) then "a diamond is a letter to us from the depths, and a letter more worth reading since we can visit the sky". Indeed, most diamonds are known to form at the depths of 140–200 km and brought to the surface by ancient volcanic activity. Due to the exceptional mechanical strength and chemical inertness of diamond, it can bring to the surface the specimens of the deep Earth interiors in an unchanged form. For decades, diamond research in the Earth sciences has been dominated by the studies of inclusions hosted in natural diamonds as well the specific rocks called xenoliths, which host diamonds. A wealth of information has been gathered from these studies. It has been established that most of natural diamonds can be assigned to the two major parageneses: peridotitic and eclogitic [1–4]. By the use of radiogenic isotopes in the minerals trapped within diamond it is found that diamonds are very old, down to 3.0 Ga, that is about two-thirds of the age of the Earth, and the youngest being around 1.0 Ga and possibly less [5]. These and subsequent studies on mineral and fluid inclusions in diamond have laid the foundation for our understanding of the geochemical and mineralogical environment of diamond formation in the Earth's mantle.

In recent decades there has been clear trend in natural diamond studies consisting in paying more attention to the internal structure and properties of diamond itself rather than to focus just on the entrapped inclusions. The real structure of natural diamond crystals can be considered a storehouse for genetically valuable information. Recently, numerous investigations have been undertaken to extract and decipher this diamond message. In their contribution, Giovanna Agrosi, Gioacchino Tempesta, Giancarlo Della Ventura, Mariangella Cestelli Guidi, Mark Hutchison, Paolo Nimis, and Fabrizio Nestola studied diamond crystals from Sao Luiz (Juina, Brazil) [6]. Diamonds from Juina, Brazil, are well-known examples of superdeep diamond crystals formed under sublithospheric conditions at depths greater than 400 km [7]. Such superdeep diamonds are very rare and their investigations provide paramount information about the physical and chemical conditions in the Earth's regions as deep as the mantle transition zone and the lower mantle [8,9]. Using X-ray diffraction topography and FTIR micro-spectroscopy, the authors show that the studied crystals demonstrate features, which are commonly associated with deformation processes by solid-state diffusion creep under high pressure and high temperature. These observations testify the very deep origin of the Juina diamonds.

Detailed studies of the internal structure of diamond derived from different sources may help to establish possible connections between the primary kimberlitic sources and secondary alluvial placers. Alexey Ragozin, Dmitry Zedgenizov, Konstantin Kuper, and Yuri Palyanov consider in their contribution specific diamonds from the Zarnitsa kimperlite pipe (Yakutia) that show transition morphologies between octahedron and rounded rhombic dodecahedron and have sectorial mosaic-block structure [10]. Such diamonds have not been previously reported for any other known kimberlite pipes of the Yakutian diamondiferous province but, as shown in this paper, demonstrate some similarities in the internal structure with diamonds from the alluvial placers of the Siberian Platform. However, the authors come to an important conclusion that, despite the physical resemblance and the similarity of the internal structure, diamond crystals from the Zarnitsa kimberlite pipe and the rounded diamonds from the alluvial placers were formed due to essentially different processes. Diamonds from the Zarnitsa kimberlite pipe evolve from polycrystalline cores to monocrystals, while an alluvial placer diamond splits up the spherulite-like structure.

The problem of the origin of alluvial diamonds is further considered in the contribution from Alexey Ragozin, Dmitry Zedgenizov, Konstantin Kuper, Viktoria Kalinina, and Alexey Zemnukhov, who studied the internal structure of yellow cuboid diamonds from alluvial placers of the Northeastern Siberian platform [11]. Rich diamond alluvial placer deposits are located in this region and the primary sources of the diamonds in these placers have not been discovered yet. An intriguing question here is that the studied cuboid diamonds constitute approximately 7–7.5% of the alluvial placer diamond collection, while diamonds with similar habit are infrequent in Siberian kimberlites and in amount do not exceed 2% of the total diamonds found in the region. It is therefore thought that the kimberlites discovered in the Siberian platform are unlikely to be the source of these diamond placers. The presented paper, being a part of an extensive study of alluvial diamonds from the Northeastern Siberian platform, provides new information on the internal structure of diamonds from these placers and contributes to solving the challenging puzzle of their origin.

Diamond is composed of carbon, which exists in nature in two stable isotopes ^{12}C and ^{13}C with average abundances of 98.9% and 1.1%, respectively. Natural diamonds are known to demonstrate slight variations of the carbon isotope composition, which is measured in parts per thousand (permil) relative to the standard (PDB, a fossil belemnite). A lot of valuable information has been gained from the isotopic studies on diamond (see reviews [12,13]). For instance, diamonds of the two main parageneses, peridotitic and eclogitic, show clear differences in the carbon isotope compositions, which is thought to be due to different carbon sources for peridotitic and eclogitic diamonds. However, the overall picture is much more complicated, and to explain the heterogeneous distribution of carbon isotopes in natural diamond crystals two major models are considered: (1) changing of the carbon source; and (2) fractionation of a fluid over the course of diamond crystallization. Vadim Reutsky, Piotr Kowalski, Yuri Palyanov, EIMF, and Michael Wiedenbeck in their contribution provided clear evidence of carbon and nitrogen fractionation related to the growing surfaces of a diamond [14]. This work is a vivid example of the importance of the experimental modelling contribution to solving the problems of natural diamond genesis [15–18]. By studying diamonds produced in high pressure high temperature experiments with various growth systems, the authors demonstrate that regardless of the bulk composition of the system, there exists a measurable fractionation of carbon isotopes and nitrogen impurity on the surface of diamond itself. The ab initio calculations of carbon isotope fractionation on different crystallographic faces of diamond also support this conclusion.

3. Some Facets of Synthetic Diamond Crystals

The modern era of diamond as a technological material began in the 1950s after the first reports on the successful synthesis of diamond through the high pressure high temperature (HPHT) technique [19,20]. The subsequent rapid development of the technologies for a mass production of diamond abrasive grids and powders has revolutionized many branches of the manufacturing, mining, and construction industries. As research and developments on synthesis and growth of diamond has further progressed, along with our growing understanding of the unique properties of diamond, new, sometimes quite unexpected, areas of diamond applications have been opening. Today, approaching the end of the second decade of the 21st century, we can see that diamond has entered many important technologies where not only its mechanical properties but its other exceptional properties are utilized. This process is steadily developing. Now, one very promising area to be engaged with diamond is connected with the so-called quantum technologies emerging over the last 10–15 years. It turns that color centers in diamonds represent a suitable platform for realizing solid-state single-photon sources, which are indispensable for the quantum physics technologies [21,22]. The most outstanding example of such color centers in diamond is the negatively charged nitrogen-vacancy (NV) center proving to be a promising system for quantum information processing [23], nanoscale electromagnetic field sensing [24,25], plasmonics [26], and biolabeling [27,28]. A number of impressive examples of using the NV centers in diamond has been demonstrated recently, including experiments on the loophole Bell inequality test [29], and nanoscale imaging magnetometry under ambient conditions [30]. In their contribution, Ettore Bernardi, Richard Nelz, Selda Sonusen, and Elke Neu review recent progress in nanoscale sensing using nitrogen-vacancy centers in single-crystal diamond [31]. After a comprehensive introduction of the nanoscale sensing based on individual nitrogen vacancy centers they consider two key challenges of the field: (1) the creation of highly-coherent, shallow NV centers less than 10 nm below the surface of a diamond crystal; and (2) the fabrication of tip-like photonic nanostructures that enable efficient fluorescence collection and can be used for scanning probe imaging based on color centers with nanoscale resolution. They discuss several approaches for creating optimal, shallow NV centers, and conclude that enhanced sensitivities and resolution in NV-based imaging are feasible in the nearest future.

The NV-based single-photon sources are known to suffer from a disadvantage connected with spectrally broad emission of the NV centers. Therefore, active research has also been undertaken to find out other optical centers in diamond with properties suitable for the novel applications [32]. Recently, silicon-vacancy [33,34], germanium-vacancy [35–37], and tin-vacancy [38,39] centers have been demonstrated to possess characteristics that are promising for single-photon applications. In addition, attempts at creating optical centers related to other impurities in diamond, such as Ni [40], Cr [41], and Eu [42] have been reported. In this context, the contribution by Vladimir Nadolinny, Andrey Komarovskikh, and Yuri Palyanov provides a comprehensive review of point defects in diamond related to the incorporation of large impurity atoms into the diamond crystal lattice [43]. Special emphasis is given to nickel, which is commonly used as a solvent-catalyst in high pressure high temperature (HPHT) diamond synthesis and growth. It is shown that at sufficiently low growth temperatures, nickel atoms occupy a substitutional position in the diamond lattice forming a defect with Td symmetry. The large Ni-C bond length (~2 Å) gives rise to a significant strain around the growth defect; the strain can be relaxed due to the displacement of one of the nearest carbon atoms to the interstitial position at high annealing temperatures. In this case, the nickel atom shifts and forms the so-called double semi-vacancy defect; the observed HFS of ^{13}C and ^{14}N confirms the proposed structure. Besides the nickel, the incorporation of titanium, cobalt, phosphorus, silicon, and germanium impurities into the diamond structure are discussed in the review. These impurity atoms have large atomic radii; a formation of double semi-vacancy defects is energetically favorable for them. The nickel, cobalt, and titanium are catalysts and nitrogen getters in the HPHT synthesis, that is why the resulting impurity defects are of interest. At the same time, doping with phosphorus, silicon, and germanium is promising for high-tech applications. For example, high-power semiconductor devices have already been constructed; silicon- and germanium-vacancy centers are perspective single-photon light sources that can be used in quantum communication systems. I believe that this review can serve as a reference book for the researchers studying defects in diamond.

The progress in the development of new diamond-based technologies is inseparably linked to research and development into the synthesis and growth of this material. At present, bulk diamond crystals are produced using two main methods, chemical vapor deposition (CVD) and growth at high pressure high temperature (HPHT) conditions. HPHT diamond synthesis and growth relies on the catalytic ability of some substances to convert graphite to diamond at the conditions of thermodynamic stability of diamond. These substances are commonly termed to as solvent-catalysts and are typically represented by transition-metal melts, especially Fe, Ni, and Co and their alloys. In recent decades, active research on synthesis and growth of diamond at HPHT conditions has been carried out using a variety of solvent-catalysts, both metallic and non-metallic. The main objectives of these investigations are connected with the fundamental aspects of diamond nucleation and growth, modeling natural diamond-forming processes, and the development of new routes of synthesizing diamond crystals with specific and unusual properties. In their contribution, Yuri Palyanov, Igor Kupriyanov, Yuri Borzdov, Denis Nechaev, and Yuliya Bataleva study HPHT diamond crystallization in the Mg-Si-C system [44]. The Mg-based solvent-catalysts have recently focused considerable attention, which is caused by the following main findings [45–49]: from these catalysts, diamond crystallizes in the kinetically controlled regime with high growth rates, up to 8.5 mm/h; the produced diamond crystals are nitrogen-free type II; effective doping of diamond with silicon and germanium impurities, creating silicon-vacancy and germanium-vacancy color centers, is possible. By studying diamond crystallization in the Mg-Si-C system at 7.5 GPa and 1800 °C with the Mg-Si compositions spanning the range from the Mg-C to Si-C end-systems the effect of the Mg/Si ratio on the graphite-to-diamond conversion degree, diamond crystal morphology, and optical properties is established and discussed [44].

The CVD technology of diamond growth has advanced tremendously to the extent that single-crystal diamond can now be produced with an unprecedentedly low impurity content, down to less than a one-part-per-billion level. One of the challenging problems lying in the focus of the current research is reaching the control over the extended defects, such as dislocations [50]. These defects affect electronic and optical properties of diamond and should be overcome for the most demanding applications. It is frequently mentioned that, in silicon technology, the elimination of dislocations was a major step in microelectronics. It has been established that with careful selection of diamond substrates and applying special surface treatments and tailored growth conditions, the dislocation content of the produced diamonds can be significantly reduced [51]. One of the approaches useful to suppress the dislocation content is the epitaxial lateral overgrowth (ELO), which has been successively applied for other semiconductor materials (GaN, GaAs, etc.) [52]. It consists of filtering dislocations by promoting lateral growth from a substrate that is partly covered by an appropriate mask. Recently, Tallaire et al. [53] have shown that using lateral growth over a macroscopic hole made in the diamond substrate it is possible to produce diamond crystals with significantly reduced dislocation density. Following this general approach, Fengnan Li, Jingwen Zhang, Xiaoliang Wang, Minghui Zhang, and Hongxing Wang describe in their contribution a way to produce low dislocation density single crystalline diamond by two-step epitaxial lateral overgrowth [54]. They show that using substrate processing, such as patterning with inductively coupled plasma (ICP) etching and metallization, it is possible to stimulate diamond lateral growth in the CVD process. The diamond films produced by the two-step ELO show much reduced content of dislocation related etching pits. The proposed method seems very promising for technological applications.

Chemical vapor deposition is a versatile technique allowing diamond growth on various substrates. Since the early stages of CVD diamond growth, it has been recognized that many tools used for material machining would significantly benefit from being covered with a thin film of diamond. Nowadays, diamond coated tools and instruments with increased hardness and durability are omnipresent in the modern industry. However, as described in the contribution by Evgeny Ashkinazi, Roman Khmelnitskii, Vadim Sedov, Andrew Khomich, Alexander Khomich, and Viktor Ralchenko, even in this field which appears well established and widely used in practice, there is still room for further research and development [55]. The functional properties of the cutting tools can further be enhanced if bilayered or multilayered micro- and nanocrystalline diamond coatings are applied. To take all advantages of such coatings, one has to pay special attention to nanocrystalline diamond (NCD) growth on microcrystalline diamond (MCD) film/substrate. The question of how nanocrystalline diamond is deposited on different crystallographic faces of single crystal diamond substrates is considered in this contribution and the results obtained are important for making proper choices for the growth conditions enabling uniform NCD layers on MCD films.

The advancements seen over the last decades in diamond science and technology rely on the detailed knowledge and understanding of the basic properties of diamond, which were intensively investigated in the past and continue to be an area of active research. Most of the extreme material properties of diamond are related to its unique lattice and corresponding phonon spectrum. One of the important tool for probing the vibrational spectrum of diamond is the Raman scattering. The Raman spectrum of diamond bears important information on the optical phonon lifetime and its temperature dependence provides insights into anharmonicity of the lattice vibrations. In addition, both the position and linewidth of the Raman peak are sensitive to the lattice imperfections and commonly used as a measure of crystalline quality of diamond and diamond-related materials. In order to use this property of the diamond Raman spectrum to full extent, it is important to understand the physical mechanisms behind the effect of defects and impurities on the vibrational properties of diamond. This appealing question is considered in the contribution by Nikolay Surovtsev and Igor Kupriyanov, who studied the effect of nitrogen impurities on the Raman line width in diamond [56]. They show that the defect-induced broadening of the diamond Raman line is temperature independent and comes from the optical phonon scattering on the defects.

Diamond is known to be an ideal material for solid-state detectors of ionizing radiation and high-energy particles. Indeed, in this particular niche, diamond has found diverse applications in biomedical sciences [57,58] and high-energy physics [59]. It is sufficient to mention the ATLAS and CMS experiments at the Large Hadron Collider (LHC), both of which include diamond detectors and have been involved, for example, in the search, discovery, and exploration of the elusive Higgs boson. A challenging problem for solid-state detectors is the radiation damage under high-energy loads, resulting in performance degradation. Solving this issue demands clear understanding of the fundamental physics behind the interaction of various high-energy particles with the detector material, diamond in our case. In addition, this question is also important for the process of ion implantation, which is frequently used to modify diamond properties [60]. Yury Belousov in his contribution reports in-depth theoretical calculations of the time evolution of radiation defects induced by negative pions and muons in diamond, silicon, and germanium [61]. Radiation damage of diamond by these particles is not commonly considered, but since they are the products of the interaction of high-energy protons with a target, the corresponding radiation defects may be important for the relevant applications.

As it is already noted, owing to its outstanding mechanical and physicochemical properties, diamond dominates in ultra-precision machining of various metallic and non-metallic materials. There is however one critical limitation. Diamond cannot be used for machining of ferrous metals because of the severe tool wear rate. This hinders many important applications where ultra-precision machining of steel alloys is required. In order to reduce this catastrophic tool wear, certain process modifications have been proposed in the literature, including use of different carbon-saturated atmospheres [62], cryogenic [63], and ultrasonic assisted [64] turning, and others. However, the problem has not been still solved satisfactorily, that stimulates further investigations. In their contribution, Lai Zou, Yun Huang, Ming Zhou, and Guijian Xiao report the results of a series of thermal analysis experiments simulating the wear process of single crystal diamond [65]. The aim of this study is to investigate the thermochemical wear of diamond surface catalyzed by iron at elevated temperatures under different gas atmospheres, leading to a better understanding of the mechanisms of diamond tool wear in machining of ferrous materials.

4. Conclusions

Summarizing, this special issue contains 12 papers which highlight recent investigations and developments in diamond research related to the diverse problems of natural diamond genesis, diamond synthesis and growth using CVD and HPHT techniques, and the use of diamond in both traditional applications, such as mechanical machining of materials, and recently emerged areas, such as quantum technologies. The results presented in the contributions collected in this special issue clearly demonstrate that diamond occupies a very special place in the modern science and technology. After decades of research, this structurally simple material still poses many intriguing scientific questions and technological challenges. It seems undoubted that diamond will remain the center of attention for many researchers for many years to come. It is very tempting to finish this editorial with the famous slogan of the De Beers' advertising line, which has just celebrated its 70th anniversary, "A diamond is forever".

Acknowledgments: I am greatly indebted to my colleagues Igor Kupriyanov, Yuliya Bataleva, and Vladimir Nadolinny for their help during the manuscript preparation. I warmly thank Sweater Shi for her kind editorial assistance during the preparation of this special issue. Support from the Russian Science Foundation (grant no. 14-27-00054) is gratefully acknowledged.

Conflicts of Interest: The author declares no conflict of interest.

References

1. Sobolev, N.V. *The Deep-Seated Inclusions in Kimberlites and the Problem of the Composition of the Upper Mantle*; American Geophysics Union: Washington, DC, USA, 1977.
2. Meyer, H.O.A. Inclusions in diamond. In *Mantle Xenoliths*; Nixon, H.P., Ed.; John Wiley and Sons: New York, NY, USA, 1987; pp. 501–523.

3. Harris, J.W. Diamond geology. In *The Properties of Natural and Synthetic Diamond*; Field, J.E., Ed.; Academic Press: London, UK, 1992; pp. 345–389.
4. Haggerty, S.E. A diamond trilogy: Superplumes, supercontinents, and supernovae. *Science* **1995**, *285*, 851–860. [CrossRef]
5. Richardson, S.H.; Gurney, J.J.; Erlank, A.J.; Harris, J.W. Origin of diamonds in old enriched mantle. *Nature* **1984**, *310*, 198–202. [CrossRef]
6. Agrosì, G.; Tempesta, G.; Della Ventura, G.; Cestelli Guidi, M.; Hutchison, M.; Nimis, P.; Nestola, F. Non-Destructive In Situ Study of Plastic Deformations in Diamonds: X-ray Diffraction Topography and μFTIR Mapping of Two Super Deep Diamond Crystals from São Luiz (Juina, Brazil). *Crystals* **2017**, *7*, 233. [CrossRef]
7. Kaminsky, F.V.; Khachatryan, G.K.; Andreazza, P.; Araujo, D.P.; Griffin, W.L. Super-deep diamonds from kimberlites in the Juina area, Mato Grosso State, Brazil. *Lithos* **2009**, *112*, 833–842. [CrossRef]
8. Stachel, T.; Brey, G.P.; Harris, J.W. Kankan diamonds (Guinea) I: From the lithosphere down to the transition zone. *Contrib. Mineral. Petrol.* **2000**, *140*, 1–15. [CrossRef]
9. Kaminsky, F. Mineralogy of the lower mantle: A review of 'super-deep' mineral inclusions in diamond. *Earth Sci. Rev.* **2012**, *110*, 127–147. [CrossRef]
10. Ragozin, A.; Zedgenizov, D.; Kuper, K.; Palyanov, Y. Specific Internal Structure of Diamonds from Zarnitsa Kimberlite Pipe. *Crystals* **2017**, *7*, 133. [CrossRef]
11. Ragozin, A.; Zedgenizov, D.; Kuper, K.; Kalinina, V.; Zemnukhov, A. The Internal Structure of Yellow Cuboid Diamonds from Alluvial Placers of the Northeastern Siberian Platform. *Crystals* **2017**, *7*, 238. [CrossRef]
12. Cartigny, P.; Palot, M.; Thomassot, E.; Harris, J.W. Diamond formation: A stable isotope perspective. *Ann. Rev. Earth Planet. Sci.* **2014**, *42*, 699–732. [CrossRef]
13. Shirey, S.B.; Cartigny, P.; Frost, D.G.; Keshav, S.; Nestola, F.; Nimis, P.; Pearson, D.G.; Sobolev, N.V.; Walter, M.J. Diamonds and the geology of mantle carbon. *Rev. Mineral. Geochem.* **2013**, *75*, 355–421. [CrossRef]
14. Reutsky, V.N.; Kowalski, P.M.; Palyanov, Y.N.; EIMF; Wiedenbeck, M. Experimental and Theoretical Evidence for Surface-Induced Carbon and Nitrogen Fractionation during Diamond Crystallization at High Temperatures and High Pressures. *Crystals* **2017**, *7*, 190. [CrossRef]
15. Palyanov, Y.N.; Bataleva, Y.V.; Sokol, A.G.; Borzdov, Y.M.; Kupriyanov, I.N.; Reutsky, V.N.; Sobolev, N.V. Mantle–slab interaction and redox mechanism of diamond formation. *Proc. Natl. Acad. Sci. USA* **2013**, *110*, 20408–20413. [CrossRef] [PubMed]
16. Palyanov, Y.N.; Shatsky, V.S.; Sokol, A.G.; Tomilenko, A.A.; Sobolev, N.V. Crystallization of Metamorphic Diamond: An Experimental Modeling. *Dokl. Earth Sci.* **2001**, *381*, 935.
17. Bataleva, Y.V.; Palyanov, Y.N.; Borzdov, Y.M.; Bayukov, O.A.; Sobolev, N.V. Conditions for diamond and graphite formation from iron carbide at the P-T parameters of lithospheric mantle. *Russ. Geol. Geophys.* **2016**, *57*, 176–189. [CrossRef]
18. Palyanov, Y.N.; Sokol, A.G.; Khokhryakov, A.F.; Kruk, A.N. Conditions of diamond crystallization in kimberlite melt: Experimental data. *Russ. Geol. Geophys.* **2015**, *56*, 196–210. [CrossRef]
19. Bundy, F.P.; Hall, H.T.; Strong, H.M.; Wentorf, J.R. Man-made diamonds. *Nature* **1955**, *176*, 51–55. [CrossRef]
20. Bovenkerk, H.P.; Bundy, F.P.; Hall, H.T.; Strong, H.M.; Wentorf, J.R. Preparation of diamond. *Nature* **1959**, *184*, 1094–1098. [CrossRef]
21. Wrachtrup, J.; Jelezko, F.J. Processing quantum information in diamond. *J. Phys. Condens. Matter* **2006**, *18*, S807–S824. [CrossRef]
22. Weber, J.R.; Koehl, W.F.; Varley, J.B.; Janotti, A.; Buckley, B.B.; Van de Walle, C.G.; Awschalom, D.D. Quantum computing with defects. *Proc. Natl. Acad. Sci. USA* **2010**, *107*, 8513–8518. [CrossRef] [PubMed]
23. Prawer, S.; Aharonovich, I. (Eds.) *Quantum Information Processing with Diamond*; Woodhead Publishing: Cambridge, UK, 2014; 330p.
24. Dolde, F.; Fedder, H.; Doherty, M.W.; Nöbauer, T.; Rempp, F.; Balasubramanian, G.; Wolf, T.; Reinhard, F.; Hollenberg, L.C.L.; Jelezko, F.; et al. Electric-field sensing using single diamond spins. *Nat. Phys.* **2011**, *7*, 459–463. [CrossRef]
25. Rondin, L.; Tetienne, J.-P.; Hingant, T.; Roch, J.-F.; Maletinsky, P.; Jacques, V. Magnetometry with nitrogen-vacancy defects in diamond. *Rep. Prog. Phys.* **2014**, *77*, 056503. [CrossRef] [PubMed]

26. Schietinger, S.; Barth, M.; Aichele, T.; Benson, O. Plasmon-Enhanced Single Photon Emission from a Nanoassembled Metal−Diamond Hybrid Structure at Room Temperature. *Nano Lett.* **2009**, *9*, 1694–1698. [CrossRef] [PubMed]

27. Barnard, A.S. Diamond standard in diagnostics: Nanodiamond biolabels make their mark. *Analyst* **2009**, *134*, 1751–1764. [CrossRef] [PubMed]

28. Mohan, N.; Chen, C.S.; Hsieh, H.H.; Wu, Y.C.; Chang, H.C. In Vivo Imaging and Toxicity Assessments of Fluorescent Nanodiamonds in Caenorhabditis elegans. *Nano Lett.* **2010**, *10*, 3692–3699. [CrossRef] [PubMed]

29. Hensen, B.; Bernien, H.; Dréau, A.E.; Reiserer, A.; Kalb, N.; Blok, M.S.; Ruitenberg, J.; Vermeulen, R.F.L.; Schouten, R.N.; Abellán, C.; et al. Loophole-free Bell inequality violation using electron spins separated by 1.3 kilometres. *Nature* **2015**, *526*, 682–686. [CrossRef] [PubMed]

30. Balasubramanian, G.; Chan, I.Y.; Kolesov, R.; Al-Hmoud, M.; Tisler, J.; Shin, C.; Kim, C.; Wojcik, A.; Hemmer, P.R.; Krueger, A.; et al. Nanoscale imaging magnetometry with diamond spins under ambient conditions. *Nature* **2008**, *455*, 648–651. [CrossRef] [PubMed]

31. Bernardi, E.; Nelz, R.; Sonusen, S.; Neu, E. Nanoscale Sensing Using Point Defects in Single-Crystal Diamond: Recent Progress on Nitrogen Vacancy Center-Based Sensors. *Crystals* **2017**, *7*, 124. [CrossRef]

32. Pezzagna, S.; Rogalla, D.; Wildanger, D.; Meijer, J.; Zaitsev, A. Creation and nature of optical centres in diamond for single-photon emission—Overview and critical remarks. *New J. Phys.* **2011**, *13*, 035024. [CrossRef]

33. Müller, T.; Hepp, C.; Pingault, B.; Neu, E.; Gsell, S.; Schreck, M.; Sternschulte, H.; Steinmüller-Nethl, D.; Becher, C.; Atatüre, M. Optical signatures of silicon-vacancy spins in diamond. *Nat. Commun.* **2014**, *5*, 3328. [CrossRef] [PubMed]

34. Green, B.L.; Mottishaw, S.; Breeze, B.G.; Edmonds, A.M.; D'Haenens-Johansson, U.F.S.; Doherty, M.W.; Williams, S.D.; Twitchen, D.J.; Newton, M.E. Neutral Silicon-Vacancy Center in Diamond: Spin Polarization and Lifetimes. *Phys. Rev. Lett.* **2017**, *119*, 096402. [CrossRef] [PubMed]

35. Iwasaki, T.; Ishibashi, F.; Miyamoto, Y.; Doi, Y.; Kobayashi, S.; Miyazaki, T.; Tahara, K.; Jahnke, K.D.; Rogers, L.J.; Naydenov, B.; et al. Germanium-Vacancy Single Color Centers in Diamond. *Sci. Rep.* **2015**, *5*, 12882. [CrossRef] [PubMed]

36. Palyanov, Y.N.; Kupriyanov, I.N.; Borzdov, Y.M.; Surovtsev, N.V. Germanium: A new catalyst for diamond synthesis and a new optically active impurity in diamond. *Sci. Rep.* **2015**, *5*, 14789. [CrossRef] [PubMed]

37. Siyushev, P.; Metsch, M.H.; Ijaz, A.; Binder, J.M.; Bhaskar, M.K.; Sukachev, D.D.; Sipahigil, A.; Evans, R.E.; Nguyen, C.T.; Lukin, M.D.; et al. Optical and microwave control of germanium-vacancy center spins in diamond. *Phys. Rev. B* **2017**, *96*, 081201. [CrossRef]

38. Iwasaki, T.; Miyamoto, Y.; Taniguchi, T.; Siyushev, P.; Metsch, M.H.; Jelezko, F.; Hatano, M. Tin-Vacancy Quantum Emitters in Diamond. *Phys. Rev. Lett.* **2017**, *119*, 253601. [CrossRef] [PubMed]

39. Tchernij, S.D.; Herzig, T.; Forneris, J.; Kupper, J.; Pezzagna, S.; Traina, P.; Moreva, E.; Degiovanni, I.P.; Brida, G.; Skukan, N.; et al. Single-Photon-Emitting Optical Centers in Diamond Fabricated upon Sn Implantation. *ACS Photonics* **2017**, *4*, 2580–2586. [CrossRef]

40. Orwa, J.O.; Greentree, A.D.; Aharonovich, I.; Alves, A.D.C.; Van Donkelaar, J.; Stacey, A.; Prawer, S. Fabrication of single optical centres in diamond—A review. *J. Lumin.* **2010**, *130*, 1646–1654. [CrossRef]

41. Aharonovich, I.; Castelletto, S.; Johnson, B.C.; McCallum, J.C.; Prawer, S. Engineering chromium-related single photon emitters in single crystal diamonds. *New J. Phys.* **2011**, *13*, 045015. [CrossRef]

42. Magyar, A.; Hu, W.; Shanley, T.; Flatté, M.E.; Hu, E.; Aharonovich, I. Synthesis of luminescent europium defects in diamond. *Nat. Commun.* **2014**, *5*, 3523. [CrossRef] [PubMed]

43. Nadolinny, V.; Komarovskikh, A.; Palyanov, Y. Incorporation of Large Impurity Atoms into the Diamond Crystal Lattice: EPR of Split-Vacancy Defects in Diamond. *Crystals* **2017**, *7*, 237. [CrossRef]

44. Palyanov, Y.; Kupriyanov, I.; Borzdov, Y.; Nechaev, D.; Bataleva, Y. HPHT Diamond Crystallization in the Mg-Si-C System: Effect of Mg/Si Composition. *Crystals* **2017**, *7*, 119. [CrossRef]

45. Palyanov, Y.N.; Borzdov, Y.M.; Kupriyanov, I.N.; Khokhryakov, A.F.; Nechaev, D.V. Diamond crystallization from an Mg-C system at high pressure high temperature conditions. *CrystEngComm* **2015**, *17*, 4928–4936. [CrossRef]

46. Palyanov, Y.N.; Kupriyanov, I.N.; Borzdov, Y.M.; Bataleva, Y.V. High-pressure synthesis and characterization of diamond from an Mg–Si–C system. *CrystEngComm* **2015**, *17*, 7323–7331. [CrossRef]

47. Palyanov, Y.N.; Kupriyanov, I.N.; Borzdov, Y.M.; Khokhryakov, A.F.; Surovtsev, N.V. High-pressure synthesis and characterization of Ge-doped single crystal diamond. *Cryst. Growth Des.* **2016**, *16*, 3510–3518. [CrossRef]
48. Khokhryakov, A.F.; Sokol, A.G.; Borzdov, Y.M.; Palyanov, Y.N. Morphology of diamond crystals grown in magnesium-based systems at high temperatures and high pressures. *J. Cryst. Growth* **2015**, *426*, 276–282. [CrossRef]
49. Palyanov, Y.N.; Kupriyanov, I.N.; Khokhryakov, A.F.; Borzdov, Y.M. High-pressure crystallization and properties of diamond from magnesium-based catalysts. *CrystEngComm* **2017**, *19*, 4459–4475. [CrossRef]
50. Tallaire, A.; Achard, J.; Silva, F.; Brinza, O.; Gicquel, A. Growth of large size diamond single crystals by plasma assisted chemical vapour deposition: Recent achievements and remaining challenges. *Comptes Rendus Phys.* **2013**, *14*, 169–184. [CrossRef]
51. Martineau, P.M.; Gaukroger, M.P.; Guy, K.B.; Lawson, S.C.; Twitchen, D.J.; Friel, I.; Hansen, J.O.; Summerton, G.C.; Addison, T.P.G.; Burns, R. High crystalline quality single crystal chemical vapour deposition diamond. *J. Phys. Condens. Matter* **2009**, *21*, 364205. [CrossRef] [PubMed]
52. Zytkiewicz, Z.R. Epitaxial Lateral Overgrowth of Semiconductors. In *Springer Handbook of Crystal Growth*; Dhanaraj, G., Byrappa, K., Prasad, V., Dudley, M., Eds.; Springer-Verlag: Berlin/Heidelberg, Germany, 2010; p. 999.
53. Tallaire, A.; Brinza, O.; Mille, V.; William, L.; Achard, J. Reduction of dislocations in single crystal diamond by lateral growth over a macroscopic hole. *Adv. Mater.* **2017**, *29*, 1604823. [CrossRef] [PubMed]
54. Li, F.; Zhang, J.; Wang, X.; Zhang, M.; Wang, H. Fabrication of Low Dislocation Density, Single-Crystalline Diamond via Two-Step Epitaxial Lateral Overgrowth. *Crystals* **2017**, *7*, 114. [CrossRef]
55. Ashkinazi, E.E.; Khmelnitskii, R.A.; Sedov, V.S.; Khomich, A.A.; Khomich, A.V.; Ralchenko, V.G. Morphology of Diamond Layers Grown on Different Facets of Single Crystal Diamond Substrates by a Microwave Plasma CVD in CH_4-H_2-N_2 Gas Mixtures. *Crystals* **2017**, *7*, 166. [CrossRef]
56. Surovtsev, N.V.; Kupriyanov, I.N. Effect of Nitrogen Impurities on the Raman Line Width in Diamond, Revisited. *Crystals* **2017**, *7*, 239. [CrossRef]
57. Ravichandran, R.; Binukumar, J.P.; Amri, I.A.; Davis, C.A. Diamond detector in absorbed dose measurements in high-energy linear accelerator photon and electron beams. *J. Appl. Clin. Med. Phys.* **2016**, *17*, 291–303. [CrossRef] [PubMed]
58. Moignier, C.; Tromson, D.; de Marzi, L.; Marsolat, F.; Hernández, J.C.G.; Agelou, M.; Pomorski, M.; Woo, R.; Bourbotte, J.-M.; Moignau, F.; et al. Development of a synthetic single crystal diamond dosimeter for dose measurement of clinical proton beams. *Phys. Med. Biol.* **2017**, *62*, 5417. [CrossRef] [PubMed]
59. Trischuk, W. (On behalf of the RD42 Collaboration). Diamond Particle Detectors for High Energy Physics. *Nucl. Part. Phys. Proc.* **2016**, *273–275*, 1023–1028.
60. Prins, J.F. Ion implantation of diamond for electronic applications. *Semicond. Sci. Technol.* **2003**, *18*, S27. [CrossRef]
61. Belousov, Y.M. Evolution in Time of Radiation Defects Induced by Negative Pions and Muons in Crystals with a Diamond Structure. *Crystals* **2017**, *7*, 174. [CrossRef]
62. Casstevens, J.M. Diamond turning of steel in carbon-saturated atmospheres. *Precis. Eng.* **1983**, *5*, 9–15. [CrossRef]
63. Evans, C. Cryogenic diamond turning of stainless steel. *CIRP Ann. Manuf. Technol.* **1991**, *40*, 571–575. [CrossRef]
64. Shamoto, E.; Suzuki, N. Ultrasonic vibration diamond cutting and ultrasonic elliptical vibration cutting. *Compr. Mater. Process.* **2014**, *11*, 405–454.
65. Zou, L.; Huang, Y.; Zhou, M.; Xiao, G. Thermochemical Wear of Single Crystal Diamond Catalyzed by Ferrous Materials at Elevated Temperature. *Crystals* **2017**, *7*, 116. [CrossRef]

Article

Non-Destructive In Situ Study of Plastic Deformations in Diamonds: X-ray Diffraction Topography and μFTIR Mapping of Two Super Deep Diamond Crystals from from São Luiz (Juina, Brazil)

Giovanna Agrosì [1,*], Gioacchino Tempesta [1], Giancarlo Della Ventura [2,3], Mariangela Cestelli Guidi [3], Mark Hutchison [4], Paolo Nimis [5] and Fabrizio Nestola [5]

[1] Dipartimento di Scienze della Terra e Geoambientali, Università degli Studi "Aldo Moro", Via Orabona, 4, 70125 Bari, Italy; gioacchino.tempesta@uniba.it

[2] Dipartimento di Scienze, Università di Roma Tre, I-00146 Rome, Italy; giancarlo.dellaventura@uniroma3.it

[3] INFN-LNF, Via Enrico Fermi 40, I-00044 Frascati (Rome), Italy; mariangela.cestelliguidi@lnf.infn.it

[4] Trigon GeoServices Ltd., 2780 South Jones Blvd, Ste 35-15, Las Vegas, NV 89146, USA; mth@trigon-gs.com

[5] Dipartimento di Geoscienze, Università degli Studi di Padova, Via G. Gradenigo 6, 35131 Padova, Italy; paolo.nimis@unipd.it (P.N.); fabrizio.nestola@unips.it (F.N.)

* Correspondence: giovanna.agrosi@uniba.it; Tel.: +39-0805-442-610

Academic Editors: Yuri N. Palyanov

Received: 29 June 2017; Accepted: 25 July 2017; Published: 28 July 2017

Abstract: Diamonds from Juina, Brazil, are well-known examples of superdeep diamond crystals formed under sublithospheric conditions and evidence would indicate their origins lie as deep as the Earth's mantle transition zone and the Lower Mantle. Detailed characterization of these minerals and of inclusions trapped within them may thus provide precious minero-petrogenetic information on their growth history in these inaccessible environments. With the aim of studying non-destructively the structural defects in the entire crystalline volume, two diamond samples from this locality, labelled JUc4 and BZ270, respectively, were studied in transmission mode by means of X-ray Diffraction Topography (XRDT) and micro Fourier Transform InfraRed Spectroscopy (μFTIR). The combined use of these methods shows a good fit between the mapping of spatial distribution of extended defects observed on the topographic images and the μFTIR maps corresponding to the concentration of N and H point defects. The results obtained show that both samples are affected by plastic deformation. In particular, BZ270 shows a lower content of nitrogen and higher deformation, and actually consists of different, slightly misoriented grains that contain sub-grains with a rounded-elongated shape. These features are commonly associated with deformation processes by solid-state diffusion creep under high pressure and high temperature.

Keywords: super deep diamonds; structural defects; X-ray diffraction topography; infrared spectroscopy; plastic deformation

---

1. Introduction

Diamond shows brittle behaviour at low temperatures, but with increasing temperature it softens considerably and plastic flow arises. Most natural diamonds experience post-growth plastic deformation during their period of residence in the mantle, which can be observed, for example, in birefringence patterns. Several different features of deformation can be found in lithospheric and sub-lithospheric diamonds, depending on both extrinsic and intrinsic factors. Temperature, pressure, time of residence and applied stress are extrinsic factors, whereas the different growth defects such as dislocations, twins, stacking faults, point defects and inclusions represent the intrinsic factors. The major difference between the plastic deformation of sub-lithospheric diamonds compared to that of lithospheric diamonds is that the former grow at much higher pressure and temperature, and in a mantle characterized by convection, whereas lithospheric diamonds grow in a stiffer mantle host that is not undergoing convection. Consequently, the plastic deformation of diamonds originating from the Earth's sub-lithospheric mantle is much more complex and can only be investigated in detail by combining different techniques.

In diamond research, X-ray Diffraction Topography (XRDT) and micro-Fourier Transform InfraRed Spectroscopy (μFTIR) are useful tools for investigating extended and point lattice defects, respectively, in transmission mode [1]. With the aim of contributing to the characterization of plastic deformations in sub-lithospheric diamonds, a multi-methodological and non-destructive approach, combining both XRDT and μFTIR, was used here to study two superdeep diamond crystals from São Luiz (Juina, Brazil) that had solid inclusions still trapped within them. This integrated methodological analysis provides a sort of "diamond mapping", which visualizes the growth and post-growth defects, totally preserving the sample.

XRDT is a non-destructive technique that has been extensively used to obtain images of structural defects with a resolution limit of a few μm. In particular, this method has been employed to control the growth process of synthetic crystals used as electronic devices, to characterize their crystalline quality and their physical properties, and to obtain information on crystal growth mechanisms [2,3]. Recently, XRDT has been successfully employed in Earth Science to study the growth history of tourmalines, garnets, and beryl [4–7]. XRDT in transmission mode (Laue geometry) is particularly suitable for studying structural defects in diamonds, because the low attenuation coefficient of the X-ray beam makes this mineral matrix highly transparent to X-rays. This allows investigation of the structural defects of the whole sample [8,9] instead of just its surface, as is the case with cathodoluminescence. Indeed, this peculiarity allows for imaging of the strain fields associated with defects, without the necessity of cutting the sample into slices. The results obtained in two previous studies by XRDT on diamonds from the Finsch mine, South Africa [10], and from the Udachnaya kimberlite, Siberia [11], provided significant minero-petrogenetic insights into their growth conditions. In the first case, a reconstruction of the growth history of a diamond characterized by the development of sub-individuals (twinned and untwinned) was possible. The data showed how the development of sub-grains could be related to a relaxation phenomenon following the stress caused by the incorporation of large pyrope and orthoenstatite inclusions; a complete discrimination between growth and post-growth defects was also performed. The second study contributed to the debate on the criteria required to establish the genetic nature of inclusions in diamonds, supporting the hypothesis of [12] that diamond-imposed morphology can be caused by a processes of selective partial dissolution at the interface between the diamond and, in this case, its olivine inclusions. To the best of our knowledge, XRDT is used in the present paper for the first time to analyse the structural defects of superdeep diamonds from São Luiz.

FTIR spectra are sensitive to the presence of point defects, in particular nitrogen and its aggregation patterns (N-type defects), that are characterized by different absorption bands in the 1000 to 1500 cm^{-1} wavenumber range [13,14]. FTIR also allows for evaluation of the presence of H defects by using an absorption peak centred at ≈ 3107 cm^{-1}. The distribution of these defects in diamond samples is rarely homogenous, and its zoning patterns depend on multiple episodes under changing crystal growth conditions [15–19]. Finally, FTIR spectroscopy allows for studying, with relatively high-resolution (nominally 5–10 μm when using a bi-dimensional FPA detector), the distribution of a target molecule across a sample [20]. FTIR spectroscopy is a relatively well known analytical tool for diamonds and has been extensively used to characterize impurities and/or inclusions typically present in these materials [21–24]. In the diamond literature, the mapping of single FTIR absorption bands are typically reported [1,25–29] to represent the distribution of the selected defect, obtained by plotting the intensity measured at a specific wavenumber or the integrated peak area. Resulting maps can generate a better understanding of the diamond growth environment.

2. Results

2.1. Samples

The samples studied in this work labelled BZ270 (Figure 1a) and JUc4 (Figure 1b), respectively, are from São Luiz (Juina, Brazil; JUc4 from WGS84 Zone 21S 261000; 8708000 and BZ270 from an unkown alluvial location in the wider area); they are broken crystals up to 6–7 mm in maximum dimension, exhibiting an irregular morphology and a brownish color. This color in diamond is commonly related to vacancy clusters, produced by plastic deformation of the crystal structure [30], as recently confirmed also by spectroscopic studies [31]. Both diamond crystals are partially flattened along the (111) plane. Optical observations revealed anomalous birefringence and the presence of dark and opaque inclusions that, from data published in a previous work [32], are mostly identified as ferro-periclase. These inclusions are very large in size (some hundreds of μm), although the BZ270 sample also exhibits smaller inclusions of as yet uncertain origin.

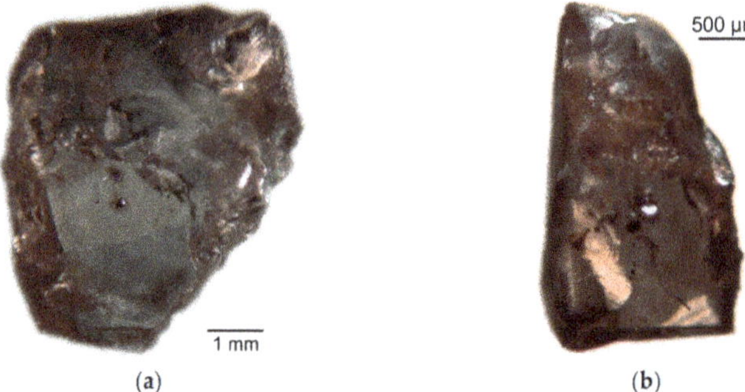

Figure 1. Optical micrographs of diamond crystals: (**a**) BZ270; (**b**) JUc4.

2.2. XRDT

The X-ray topographic images show that both diamonds exhibit an extensive plastic deformation that precludes simultaneous X-ray diffraction of the whole crystal. Nevertheless, despite their common provenance and similarity of inclusions, they show a significantly different type of post-growth deformation.

According to the XRDT data, diamond BZ270 consists of an aggregate of slightly misoriented grains (G) that are not visible simultaneously at the individual angular settings due to the different diffraction vector (Figure 2). Topographs taken with a diffraction vector of type <111> (Figure 2a,b) exhibit a misalignment of about 1° between 2 different portions, irrespective of grains and growth sector. The boundary between these portions (CP in the Figure) belongs to the (−111) plane and represents a typical cleavage plane, interpreted to be due to post-growth brittle deformation (Figure 2).

In the topographs of Figure 2c,d, the diffraction contrast corresponding to two slightly different angular settings of the same diffraction vector (g = 2–20) shows that the diamond region containing the ferro-periclase inclusions is really a grain (G) that is differently oriented with respect to the other diamond regions. In the Figure 2c G is out of contrast, whereas in Figure 2d G is in diffraction contrast and the surrounding grains are out of contrast.

In Figure 2e,f two slightly different angular setting (few seconds of arc) of the same diffraction vector (g = −131) show that the grain labelled G is out of contrast together with other portions corresponding to further grains. It is worth noting that the apparent different size and shape of the topographic images and the different grains are due to the projection effect under the different angular setting of the different topographs. Within the same grains, patchy sub-grains exhibiting a rounded and elongated drop shape are visible (SG).

The diffraction contrast analysis shows that the grains and the subgrains are visible mainly under the reflections **g** = 2–20 and **g** = −131, because of non-simultaneous diffraction conditions for each of them (Figure 2c–f). This finding implies a condition of extinction for some grains and subgrains with respect to the others under these reflections. Applying the extinction criteria, the common direction that satisfies in both cases the extinction corresponds to the (−112) direction.

In Figure 3, an optical micrograph taken under crossed polars of the enlarged detail of the grain (G) containing large inclusions of ferro-periclase is compared with the corresponding topographic image. Comparison reveals that the topographic image exhibits a similar shape of diffraction contrast corresponding to the inclusions (marked by numbers) and the subgrains (SG).

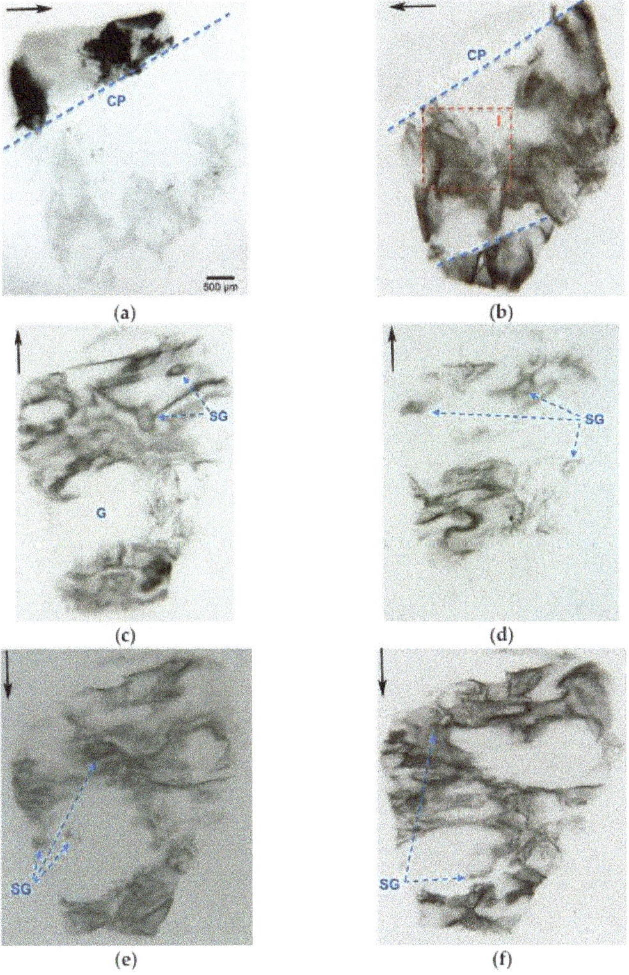

Figure 2. X-ray topographic images (MoKa$_1$) of diamond B7270. The black arrows represent the diffraction vector projection **g**. CP: Cleavage plane; G: misoriented grains; SG: sub-grains indicated by dashed blue arrows; I: area magnified in Figure 3. The directions of the diffraction projection vector are: (**a**) **g** = −1–11; (**b**) **g** = 11–1; (**c**,**d**) **g** = 2–20; (**e**,**f**) **g** = −131. The apparent different size and shape of the topographic images is due to the projection effect of the angular setting for the different reflections.

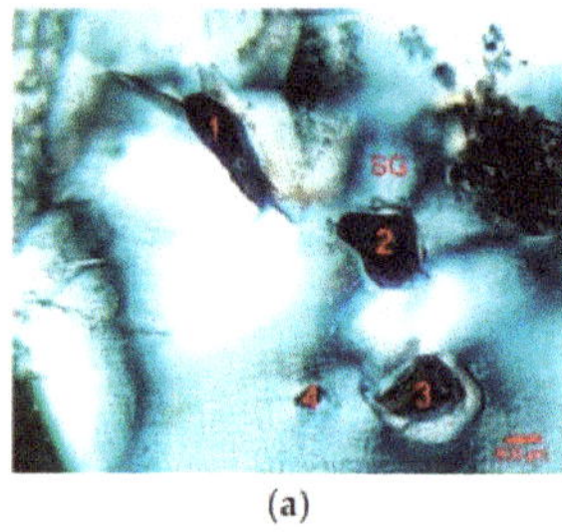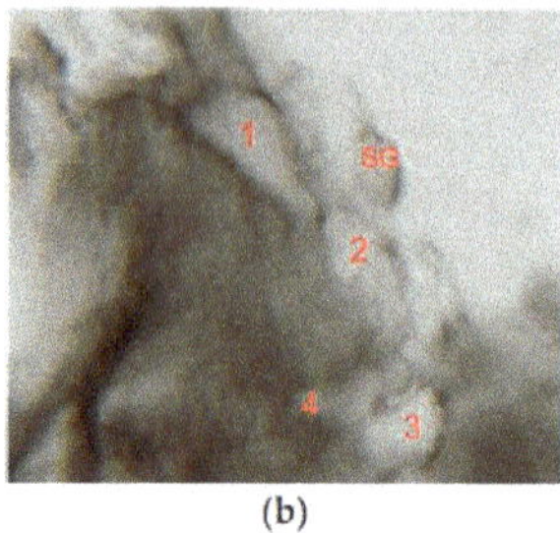

(a)

(b)

Figure 3. Enlarged portion of diamond B7270 surrounded by the red dashed line in Figure 2b. (**a**) Optical micrograph taken under crossed polars; (**b**) X-ray topographic image (MoKα$_1$). The red numbers tag the inclusions and the corresponding contrasts. Note the similar shape of the inclusions and sub-grain (SG).

XRDT images acquired on JUc4 show different plastic deformation patterns: this sample is shown to be a single crystal, extensively plastically deformed, while no subgrains are visible. The diffraction contrast of topographs taken under the reflection with **g** = −220 (Figure 4a,b) exhibits laminations along different slip systems that are out of contrast in the topograph taken with the diffraction vector **g** = −1–11 (Figure 4c). The interaction of the different slip systems and the initial bending of lattice planes determine a resultant elongated strain (Figure 4).

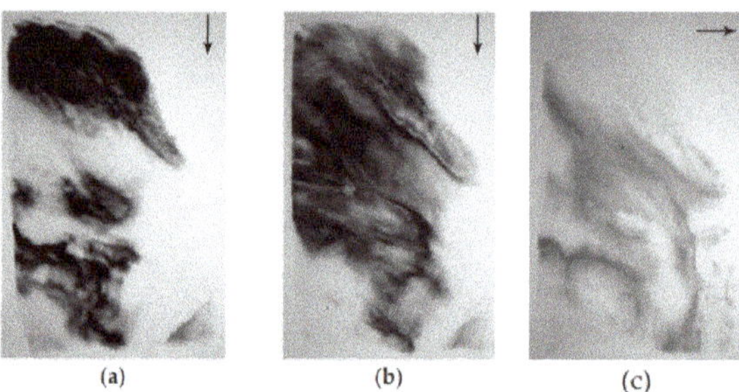

(a)

(b)

(c)

Figure 4. X-ray topographic images (MoKα$_1$) of diamond JUc4. The black arrows represent the diffraction vector projection **g**. (**a**) **g** = −220; (**b**) **g** = −220; (**c**) **g** = −1–11. The apparent different size and shape of the topographic images of the sample are due to the projection effect of the angular setting for the different reflections.

2.3. μFTIR

In Figure 5a two selected single-point spectra (50 μm beam dimension) collected on samples BZ270 and JUc4, respectively, are compared. The broad and convolute absorption extending from 2700 to 1700 cm^{-1} is due to the diamond matrix, while the broad absorption in the range 1400–1000 cm^{-1} is due to the different N defects (for type I diamonds). In more detail, the different peaks in this region allow differentiation of the possible N-related defects and clusters within the structure, i.e., isolated vs. aggregated N impurities [13] (for assistance following the discussion below, the N impurity region for sample JUc4 is enlarged in Figure 5b). In particular, the spectrum of diamond JUc4 shows two very weak peaks at 1405 and 1385 cm^{-1}, respectively, an intense and a sharp peak at 1330 cm^{-1}, and a very broad and intense absorption at 1170 cm^{-1} (Figure 5b). Two broad and relatively weak absorptions can finally be resolved at 1095 and 1010 cm^{-1}, respectively (Figure 5b). The spectra of Figure 5a show that the examined samples have significantly different N contents: while JUc4 has a well resolvable nitrogen absorption, BZ270 has a very low, barely appreciable absorption in the 1400–1000 cm^{-1} range. Based on the N-band nomenclature given in [13] the most intense peaks at 1330 and 1170 cm^{-1} can be assigned to N impurities aggregated in clusters of four atoms and associated structural vacancies (B defects). N impurities aggregated in pairs (A defects) are characterized by an intense band at 1282 cm^{-1}; this band is absent from the spectrum of both examined diamonds. The weak peak at 1385 cm^{-1} can be assigned to B2 aggregates or platelets. This peak is usually found paired to the band of 1170 cm^{-1} corresponding to B aggregates [33]. The broad band at 1095 can be assigned to the overlapping of two characteristic bands: the first at 1090 cm^{-1} corresponding to the A and B defects [34], and the second (1100 cm^{-1}) that could be ascribed to the dislocation loop remaining from the decomposition of the platelets under plastic deformation [35]. Both samples, in addition, show a weak but resolvable peak assigned to the presence of H impurities, centered at 3107 cm^{-1} (Figure 5a). The hydrogen impurity peak is barely appreciable in sample BZ270 (Figure 5a), while it is relatively prominent in sample JUc4. The different intensity associated with the N absorbance in Figure 5a cannot be used for quantitative evaluations, due to the different thickness of the studied crystal sections. Previous direct analytical data carried out destructively on small fragments extracted from both samples (Hutchison, unpublished data) gave different N contents for different fragments from the same sample, suggesting an inhomogeneous N concentration; the average total N for JUc4 is about 64.8 ppm while being 17.8 ppm for BZ270. Based on all available data, sample JUc4 can be classified a type IaB diamond, while based only on its average N content, under 50 ppm, BZ270 could be classified a type IIa diamond, but locally the amount of the detected B defects suggests a classification of a type IaB.

The FTIR maps integrated over the range 1400–1000 cm^{-1} for N defects and over the range 3120–3075 cm^{-1} for H defects are shown in Figures 6 and 7. These maps were superimposed on the optical micrographs of the samples, and care was taken to make the maps transparent. The maps allow a better understanding of the heterogeneous distribution of N in JUc4 and BZ270. Notably, on both samples, the spatial distribution of the intensity associated with H defects is always related to that of N defects. In particular, for BZ270 the maps of total N defects (Figure 6a) and H defects (Figure 6b) show a singular distribution in the micro-domains, which apparently correspond to the grain and subgrains observed in the topographic image (cf. G and SG in Figure 2). In particular, the grain labelled G, which contains the ferro-periclase inclusions, appears completely free of N and H defects. Conversely, the FTIR spectroscopic maps of diamond JUc4 show a clear concentric zoning of total N defects (Figure 7), confirming that this diamond can be interpreted as a single crystal. However, also for this sample, the elongated shape of the zoning pattern reproduces the deformation observed on the topographic images.

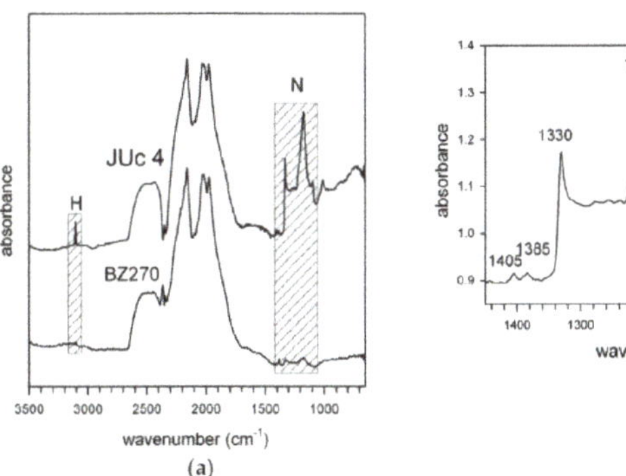

Figure 5. (a) Selected FTIR single-point spectra (50 µm beam size) of the studied samples. The absorbance regions of H and N impurities, respectively, are indicated. Spectra normalized to the diamond absorption. (b) Enlargement of the spectrum of JUc4 in the N impurity region; for explanations, see text.

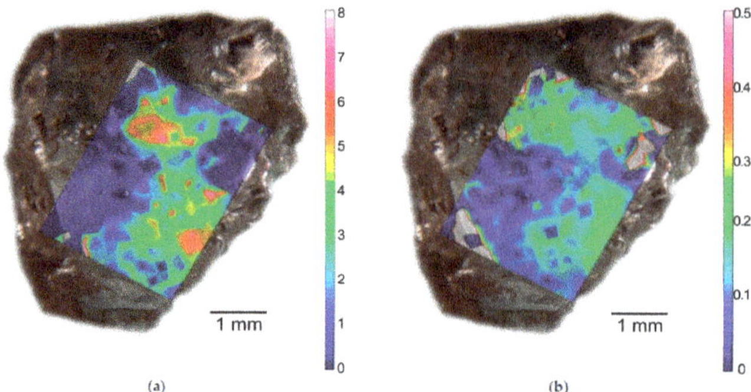

Figure 6. FTIR maps of diamond BZ270. (**a**) N concentration, integration range 1400–1000 cm^{-1}; (**b**) H defects, integration range 3120–3075 cm^{-1}. Both maps were taken with a 50 μm^2 beam at step of 50 μm. The spectroscopic maps are superimposed on the optical micrograph of Figure 1. The color scale is proportional to the measured intensity, from blue (zero intensity) to red/whitish (max intensity).

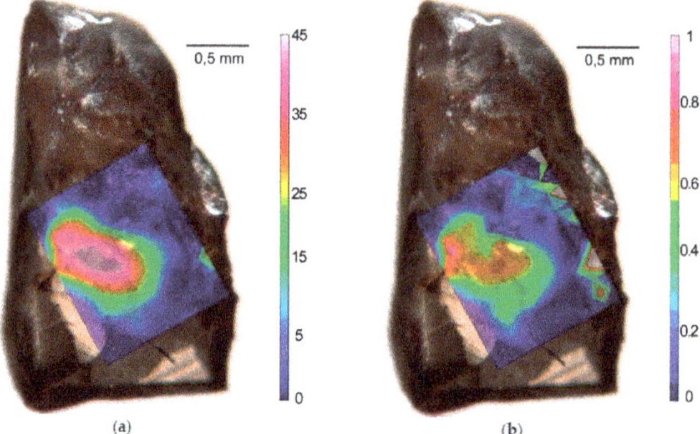

Figure 7. FTIR maps of diamond JUc4. (**a**) N concentration, integration range 1400–1000 cm^{-1}; (**b**) H defects, integration range 3120–3075 cm^{-1}. Both maps were taken with a 50 μm^2 beam at step of 50 μm. The spectroscopic maps are superimposed to the optical micrograph of Figure 1. The color scale is proportional to the measured intensity, from blue (zero intensity) to red/whitish (max intensity).

3. Discussion

The XRDT images are remarkably similar to FTIR maps. The topographs clearly show a strained zone at the boundary separating the domains with different concentrations of N defects, suggesting a migration of point defects during the plastic deformation process. More in detail, the results obtained show that the diamonds studied here experienced different growth and post-growth history.

The sample labelled BZ270 consists of an aggregate of misoriented grains. However, formation of this diamond from multiple simultaneous growth centers can be ruled out. The grain, labelled G in the topographs, containing large inclusions of ferro-periclase and lacking N and H point defects, is different from the others and can be considered as the product of an early stage of growth.

A second step of growth was characterized by the development and aggregation of different grains with the entrapment of a second generation of small inclusions, possibly along some grain boundaries. These grains exhibit a low content of point defects even if their concentration is significantly higher than in the early stage, suggesting modified growth conditions. Episodes of partial dissolution could be invoked to explain the irregular shape of the grains.

After the diamond development, a stage of plastic deformation characterized by the formation of subgrains occurred. This type of plastic deformation has never been reported for diamonds from São Luiz. This could be related to the fact that diamonds from São Luiz have been studied by XRDT in transmission mode for the first time in the present work. Actually, previous studies were performed using mainly cathodoluminescence (surface analyses), and the most common plastic deformation features found were lamination and thus micro-twinning along {111} planes [36]. It is well known that the formation of the twinned micro-lamellae is an important mechanism for deformation accommodation in diamond [37]. Previous high-pressure and high-temperature (HP-HT) experiments carried out on synthetic diamonds suggested that the {111} micro-twins begin to form under a range of P and T conditions typical for the upper mantle [26,38,39]. Howell et al. (2012) [26] showed that further differential stress brings a continuous crystal bending that produces the formation of subgrains. In particular, when multiple slip systems {111} <110> occur simultaneously, a rotation of subgrains arises around <112> axes. However, owing to the quantitative estimation of P and T and/or applied stress responsible for this further deformation, is difficult to estimate for natural diamonds, because the inherent growth defects of the diamond crystal, mechanical properties, and degree of homogeneity of the medium in which the crystal is deformed are unknown [40]. The analysis of diffraction contrasts of this study confirms that the subgrains observed in BZ270 are rotated around the (−112) direction. Commonly, a deformation process by dislocation movement causes a polygonal shape of subgrains [41–43]. The lobate shape of subgrains found in BZ270 suggests a formation mechanism by solid state diffusion creep under HTHP conditions similar to sublithospheric mantle conditions. When a crystal deforms by diffusion creep to accommodate space problems from simultaneous movement of whole grains along grain boundaries, a process of superplastic-like flow occurs [44,45]. This process could also explain the nitrogen aggregation state found by FTIR. The similar shape of diffraction contrast between the inclusions and the subgrains (Figure 3) suggests that the plastic deformation could have also affected the inclusions. An important topic, which can be considered in the future, regards the study of the different behavior of diamond and ferro-periclase under the same stress and the mechanical effects at the interface between diamond and non-diamond inclusions.

Finally, during the ascent of the diamond or during its eruption, the diamond BZ270 experienced brittle deformation with CP formation.

Conversely, diamond JUc4 consists of a single N-zoned crystal with a higher amount of N with respect to the sample BZ270. FTIR maps reveal that also for this sample the concentration of N follows the shape of plastic deformation. No subgrains were found but a polygonised network of different systems of lamination with an elongated shape of deformation can be ascribed to a lower level of plastic deformation compared to BZ270. This difference might be related to the different content of nitrogen and/or to a different density of growth defects, as well as to dislocations. In addition, different types of stress might generate different deformation features.

The plastic deformation found in the studied samples and the differences found between BZ270 and JUc4 are consistent with derivation of these diamonds from a heterogeneous sublithospheric mantle characterized by convection.

4. Materials and Methods

4.1. XRDT

The diamonds were investigated by XRDT in transmission geometry, using a conventional source, at the University of Bari (Italy). The technique, developed by [46], is a non-destructive imaging technique, sensitive to the strain associated with extended defects, that yields the spatial distribution and full characterization of the crystal defects in the whole sample volume. The topographs taken with Laue geometry were carried out using a Rigaku camera (Rigaku International Corporation,

Tokyo, Japan) with monochromatic radiation ($MoK\alpha_1$) and with a micro-focus X-ray tube. XRDT is particularly suitable for the non-destructive investigation of the structural defects in diamond because of its low X-ray attenuation coefficient that allows the optimum kinematic diffraction condition $\mu t \approx 1$ (μ = linear absorption coefficient; t = crystal thickness) to be made, minimizing the X-ray absorption. Spatial resolution using these conditions is about 1–2 μm. The diffraction contrast was recorded on high-resolution photographic films (SR Kodak). Characterization of the structural defects was performed by applying the extinction criteria to their diffraction contrasts, according to kinematic and dynamic X-ray diffraction theories [47]. The detailed XRDT procedures used in this study are given in Agrosì, et al. (2016) [11].

4.2. μFTIR

Micro-FTIR maps were obtained by using a Bruker Hyperion 3000 microscope fitted to a Vertex V70 (both Bruker Optics, Ettlingen, Germany) optical bench, a Globar IR source, and a KBr broadband beamsplitter. The FTIR microscope was equipped with a 15X Schwardzschild objective, a motorized XY stage, and an LN_2 cooled MCT detector. The data were collected at a nominal resolution of 4 cm^{-1}; 128 spectra were averaged for both spectrum and background. The analytical grid covered continuously the mapped area, using a square aperture of 50 μm^2 at step of 50 μm. The diamond sample was initially placed on a ZnSe sample holder, but preliminary tests showed a significant periodic disturbance in the pattern (interference fringes) due to the interference of the IR beam exiting from the sample. This disturbance hindered any reliable integration of the absorbance peaks, therefore to overcome this problem, a thin layer of powdered KBr was inserted between the diamond crystal and the ZnSe sample holder. This shrewdness was effective in eliminating the interference fringes from the patterns. Spectroscopic maps were obtained by integrating the target peak as indicated in the text. Additional information on the method can be found in [20].

5. Conclusions

Although the diamonds studied both derive from the same area of São Luiz and contain the same types of inclusions, they show a heterogeneity of growth and post-growth history that can be explained only if the crystallization and deformation processes occurred within a heterogeneous sublithospheric mantle characterized by convecting movement. This study suggests that the reconstruction of the different stages of growth and plastic deformation of the studied diamonds by means of the analyses of structural defects should be preliminary to the studies on the inclusions in order to obtain robust conclusions on the inclusion–host growth relationships and their minero-petrogenetic implications.

Acknowledgments: The research was supported by ERC Starting Grant INDIMEDEA (grant number 307322) awarded to Fabrizio Nestola, University of Padova (Italy).

Author Contributions: Giovanna Agrosì, Gioacchino Tempesta, and Fabrizio Nestola conceived and designed the experiments; Mark Hutchison contributed the diamond samples; Giovanna Agrosì, Gioacchino Tempesta, Giancarlo Della Ventura, and Mariangela Cestelli Guidi performed the experiments; Giovanna Agrosì, Gioacchino Tempesta, Giancarlo Della Ventura, Mariangela Cestelli Guidi, Mark Hutchison, Paolo Nimis, and Fabrizio Nestola analyzed the data and oversaw the manuscript; Giovanna Agrosì wrote the paper.

Conflicts of Interest: The authors declare no conflict of interest.

References

1. Shiryaev, A.A.; Fisenko, A.V.; Vlasov, I.I.; Semjonova, L.F.; Nagel, P.; Schuppler, S. Spectroscopic study of impurities and associated defects in nanodiamonds from Efremovka (CV3) and Orgueil (CI) meteorites. *Geochim. Cosmochim. Acta* **2011**, *75*, 3155–3166. [CrossRef]
2. Bowen, D.K.; Tanner, B.K. *High Resolution X-ray Diffractometry and Topography*; CRC Press: Bristol, UK, 2005.
3. Agrosì, G.; Tempesta, G.; Capitani, G.C.; Scandale, E.; Siche, D. Multi-analytical study of syntactic coalescence of polytypes in a 6H–SiC sample. *J. Cryst. Growth* **2009**, *311*, 4784–4790. [CrossRef]

4. Agrosì, G.; Bosi, F.; Lucchesi, S.; Melchiorre, G.; Scandale, E. Mn-tourmaline crystals from island of Elba (Italy): Growth history and growth marks. *Am. Mineral.* **2006**, *91*, 944–952. [CrossRef]

5. Agrosì, G.; Scandale, E.; Tempesta, G. Growth marks of titanian-andradite crystals from Colli Albani (Italy). *Period. Di Mineral.* **2011**, *80*, 89–104.

6. Tempesta, G.; Scandale, E.; Agrosì, G. Striations and hollow channels in rounded beryl crystals. *Period. Di Mineral.* **2011**, *79*, 75–87.

7. Pignatelli, I.; Giuliani, G.; Ohnenstetter, D.; Agrosì, G.; Mathieu, S.; Morlot, C.; Branquet, Y. Colombian trapiche emeralds: Recent advances in understanding their formation. *Gems Gemol.* **2015**, *51*, 222–259. [CrossRef]

8. Moore, M. Imaging diamond with X-rays. *J. Phys. Condens. Matter* **2009**, *21*, 364217. [CrossRef] [PubMed]

9. Moore, M.; Nailer, S.G.; Wierzchowski, W.K. Optical and X-ray topographic studies of dislocations, growth-sector boundaries, and stacking faults in synthetic diamonds. *Crystals* **2016**, *6*, 71. [CrossRef]

10. Agrosì, G.; Tempesta, G.; Scandale, E.; Harris, J.W. Growth and post-growth defects of a diamond from Finsch mine (South Africa). *Eur. J. Mineral.* **2013**, *25*, 551–559. [CrossRef]

11. Agrosì, G.; Nestola, F.; Tempesta, G.; Bruno, M.; Scandale, E.; Harris, J.W. X-ray topographic study of a diamond from Udachnaya: Implications for the genetic nature of inclusions. *Lithos* **2016**, *248*, 153–159. [CrossRef]

12. Nestola, F.; Nimis, P.; Angel, R.J.; Milani, S.; Bruno, M.; Prencipe, M.; Harris, J.W. Olivine with diamond-imposed morphology included in diamonds. Syngenesis or protogenesis? *Int. Geol. Rev.* **2014**, *56*, 1658–1667. [CrossRef]

13. Breeding, C.M.; Shigley, J.E. The "type" classification system of diamonds and its importance in gemology. *Gems Gemol.* **2009**, *45*, 96–111. [CrossRef]

14. Kaminsky, F.V.; Khachatryan, G.K. The relationship between the distribution of nitrogen impurity centres in diamond crystals and their internal structure and mechanism of growth. *Lithos* **2004**, *77*, 255–271. [CrossRef]

15. De Vries, D.W.; Pearson, D.G.; Bulanova, G.P.; Smelov, A.P.; Pavlushin, A.D.; Davies, G.R. Re–Os dating of sulphide inclusions zonally distributed in single Yakutian diamonds: Evidence for multiple episodes of Proterozoic formation and protracted timescales of diamond growth. *Geochim. Cosmochim. Acta* **2013**, *120*, 363–394. [CrossRef]

16. Griffin, B.J.; Bulanova, G.P.; Taylor, W.R. CL and FTIR mapping of nitrogen content and hydrogen distribution in diamonds from the Mir pipe—Constraints on growth history. Extended Abstracts of the 6th International Kimberlite Conference, Novosibirsk, Russia, 7–11 August 1995.

17. Palot, M.; Pearson, D.G.; Stachel, T.; Harris, J.W.; Bulanova, G.P.; Chinn, I. Multiple growth episodes or prolonged formation of diamonds? Inferences from infrared absorption data. In Proceedings of the 10th International Kimberlite Conference, New Delhi, India, 6–10 February 2012.

18. Pearson, D.G.; Shirey, S.B.; Bulanova, G.P.; Carlson, R.W.; Milledge, H.J. Re–Os isotope measurements of single sulfide inclusions in a Siberian diamond and its nitrogen aggregation systematics. *Geochim. Cosmochim. Acta* **1999**, *63*, 703–711. [CrossRef]

19. Taylor, W.R.; Bulanova, G.P.; Milledge, H.J. Quantitative nitrogen aggregation study of some Yakutian diamonds: Constraints on the growth, thermal and deformation history of peridotitic and eclogitic diamonds. Extended Abstracts of the 6th International Kimberlite Conference, Novosibirsk, Russia, 7–11 August 1995.

20. Della Ventura, G.; Marcelli, A.; Bellatreccia, F. SR-FTIR Microscopy and FTIR Imaging in the Earth Sciences. *Rev. Mineral. Geochem.* **2014**, *78*, 447–480. [CrossRef]

21. Kaminsky, F.V.; Khachatryan, G.K. Characteristics of nitrogen and other impurity in diamond, as revealed by infrared absorption data. *Can. Mineral.* **2001**, *39*, 1735–1745. [CrossRef]

22. Hainschwang, T.; Notari, F.; Fritsch, E.; Massi, L. Natural, untreated diamonds showing the A, B and C infrared absorptions ("ABC diamonds"), and the H2 absorption. *Diam. Relat. Mater.* **2006**, *15*, 1555–1564. [CrossRef]

23. Gaillou, E.; Post, J.E.; Rost, D.; Butler, J.E. Boron in natural type IIb blue diamonds; chemical and spectroscopic measurements. *Am. Mineral.* **2012**, *97*, 1–18. [CrossRef]

24. Palot, M.; Jacobsen, S.D.; Townsend, J.P.; Nestola, F.; Marquardt, K.; Miyajima, N.; Harris, J.W.; Stachel, T.; McCammon, C.A.; Pearson, D.G. Evidence for H_2O-bearing fluids in the lower mantle from diamond inclusion. *Lithos* **2016**, *265*, 237–243. [CrossRef]

25. Bulanova, G.P.; Pearson, D.G.; Hauri, E.H.; Griffin, B.J. Carbon and nitrogen isotope systematics within a sector-growth diamond from the Mir kimberlite, Yakutia. *Chem. Geol.* **2002**, *188*, 105–123. [CrossRef]

26. Howell, D.; O'Neill, C.J.; Grant, K.J.; Griffin, W.L.; O'Reilly, S.Y.; Pearson, N.J.; Stern, R.A.; Stachel, T. Platelet development in cuboid diamonds: Insights from micro-FTIR mapping. *Contrib. Mineral. Petrol.* **2012**, *164*, 1011–1025. [CrossRef]

27. Lu, T.; Chen, H.; Qiu, Z.; Zhang, J.; Wei, R.; Ke, J.; Sunagawa, I.; Stern, R.; Stachel, T. Multiple core growth structure and nitrogen abundances of diamond crystals from Shandong and Liaoning kimberlite pipes, China. *Eur. J. Mineral.* **2012**, *24*, 651–656. [CrossRef]

28. Spetsius, Z.V.; Bogush, I.N.; Kovalchuk, O.E. FTIR mapping of diamond plates of eclogitic and peridotitic xenoliths from the Nyurbinskaya pipe, Yakutia: Genetic implications. *Russ. Geol. Geophys.* **2015**, *56*, 344–353. [CrossRef]

29. Kohn, S.C.; Speich, L.; Smith, C.B.; Bulanova, G.P. FTIR thermochronometry of natural diamonds: A closer look. *Lithos* **2016**, *265*, 148–158. [CrossRef]

30. Smith, E.M.; Helmstaedt, H.H.; Flemming, R. Survival of Brown Colour in Diamond during Storage in the Subcontinental Lithospheric Mantle. *Can. Mineral.* **2010**, *48*, 571–582. [CrossRef]

31. Yuryeva, O.P.; Rakhmanova, M.I.; Zedgenizov, D.A. Nature of type IaB diamonds from the Mir kimberlite pipe (Yakutia): Evidence from spectroscopic observation. *Phys. Chem. Miner.* **2017**, 1–3. [CrossRef]

32. Mikhail, S.; Howell, D.; Hutchinson, M.; Verchovsky, A.B.; Warburton, P.; Southworth, R.; Milledge, H.J. Constraining the internal variability of carbon and nitrogen isotopes in diamonds. *Chem. Geol.* **2014**, *366*, 14–23. [CrossRef]

33. Brozel, M.R.; Evans, T.; Stephenson, R.F. Partial dissociation of nitrogen aggregates in diamond by high temperature-high pressure treatments. *Proc. R. Soc. Lond. A* **1978**, *361*, 109–127. [CrossRef]

34. Sutherland, G.B.; Blackwell, D.E.; Simeral, W.G. The Problem of the Two Types of Diamond. *Nature* **1954**, *174*, 901–904. [CrossRef]

35. Collins, A.T. *In the Physics of Diamond. Course CXXXV, Proc. Int. School Physics*; IOS Press: Amsterdam, The Netherlands, 1997; p. 195.

36. Hutchinson, M.T.; Cartigny, P.; Harris, J.W. Carbon and nitrogen compositions and physical characteristics of transition zone and lower mantle diamonds from São Luiz, Brazil. In Proceedings of the 7th International Kimberlite Conference, Red Roof Design, Cape Town, South Africa, 11–17 April 1998.

37. Titkov, S.V.; Krivovichev, S.V.; Organova, N.I. Plastic deformation of natural diamonds by twinning: Evidence from X-ray diffraction studies. *Mineral. Mag.* **2012**, *76*, 143–149. [CrossRef]

38. Yu, X.; Raterron, P.; Zhang, J.; Xhijun, L.; Liping, W.; Zhao, Y. Constitutive Law and Flow Mechanism in Diamond Deformation. *Sci. Rep.* **2012**, *2*, 876. [CrossRef] [PubMed]

39. Agrosì, G.; Tempesta, G.; Mele, D.; Allegretta, I.; Terzano, R.; Shirey, S.; Pearson, D.G.; Nestola, F. Multi-analytical approach for non-destructive analyses of a diamond from Udachnaya and its trapped inclusions: the first report of (Fe,Ni)1+xS mackinawite sulfide in diamonds. *Am. Mineral.* **2017**, in press.

40. Shiryaev, A.A.; Frost, D.J.; Langenhorst, F. Impurity diffusion and microstructure in diamonds deformed at high pressures and temperatures. *Diam. Relat. Mater.* **2007**, *16*, 503–511. [CrossRef]

41. Friedel, G. *Dislocations*; [Russian translation]; Mir: Moscow, Russia, 1967.

42. Hirth, J.P.; Lothe, J. *Theory of Dislocation*; [Russian translation]; Atomizdat: Moscow, Russia; McGraw-Hill: New York, NY, USA, 1972.

43. Sumida, N.; Lang, A.R. Cathodoluminescence evidence of dislocation interactions in diamond. *Philos. Mag. A* **1981**, *43*, 1277–1287. [CrossRef]

44. Hiraga, T.; Miyazaki, T.; Tasaka, M.; Yoshida, H. Mantle superplasticity and its self-made demise. *Nature* **2010**, *468*, 1091–1094. [CrossRef] [PubMed]

45. Ohuchi, T.; Kawazoe, T.; Higo, Y.; Funakoshi, K.; Suzuki, A.; Kikegawa, T.; Irifune, T. Dislocation-accommodated grain boundary sliding as the major deformation mechanism of olivine in the Earth's upper mantle. *Sci. Adv.* **2015**, *1*, e1500360. [CrossRef] [PubMed]

46. Lang, A.R. The projection topograph: A new method in X-ray diffraction microradiography. *Acta Cryst.* **1959**, *2*, 249–250. [CrossRef]

47. Authier, A.; Zarka, A. X-ray topographic study of the real structure of minerals. In *Advanced Mineralogy*; Marfunin, A.S., Ed.; Springer-Verlag: Berlin, Germany, 1994; pp. 221–233.

Article

Specific Internal Structure of Diamonds from Zarnitsa Kimberlite Pipe

Alexey Ragozin [1,2,*], Dmitry Zedgenizov [1,2], Konstantin Kuper [3] and Yuri Palyanov [1,2]

[1] V. S. Sobolev Institute of Geology and Mineralogy, Siberian Branch Russian Academy of Sciences, Novosibirsk 630090, Russia; zed@igm.nsc.ru (D.Z.); palyanov@igm.nsc.ru (Y.P.)
[2] Department of Geology and Geophysics, Novosibirsk State University, Novosibirsk 630090, Russia
[3] Budker Institute of Nuclear Physics, Siberian Branch Russian Academy of Sciences, Novosibirsk 630090, Russia; k.e.kuper@inp.nsk.su
* Correspondence: ragoz@igm.nsc.ru; Tel.: +7-383-330-8015

Academic Editor: Helmut Cölfen
Received: 30 March 2017; Accepted: 7 May 2017; Published: 11 May 2017

Abstract: The Zarnitsa kimberlite pipe is one of the largest pipes of the Yakutian diamondiferous province. Currently, some limited published data exists on the diamonds from this deposit. Among the diamond population of this pipe there is a specific series of dark gray to black diamonds with transition morphologies between octahedron and rounded rhombic dodecahedron. These diamonds have specific zonal and sectorial mosaic-block internal structures. The inner parts of these crystals have polycrystalline structure with significant misorientations between sub-individuals. The high consistency of the mechanical admixtures (inclusions) in the diamonds cores can cause a high grid stress of the crystal structure and promote the block (polycrystalline) structure of the core components. These diamond crystals have subsequently been formed due to crystallization of bigger sub-individuals on the polycrystalline cores according to the geometric selection law.

Keywords: diamond; internal structure; diffraction of backscattered electrons; radial mosaic pattern; Zarnitsa kimberlite pipe

---

1. Introduction

The first diamondiferous kimberlite pipe found on the Siberian Platform was named Zarnitsa (Dawn). It was discovered by L.A. Popugaeva in 1954 [1] in the Daldyn-Alakit kimberlite field of the Yakutian diamondiferous province. The Zarnitsa kimberlite pipe is one of the largest pipe in Yakutia, having a surface size of 535 × 480 m. Currently, only limited data has been published on the diamonds excavated from this deposit. According to [2], most of the diamond crystals (96%) in this pipe are octahedral (23%), laminar rhombic dodecahedral (20%), and rounded dodecahedral (29%) with transitional habits between octahedron and dodecahedron (17%). Less common are the grey polycrystalline diamonds (bort), cuboids and coated diamonds.

In this work, we have studied a specific series of individual diamonds and their aggregates which were representing the dark gray- to black-coloured crystals of transitional octahedron and rhombic dodecahedron habits. These diamonds have similar morphological features to the dark grey rounded diamonds which are widespread in the alluvial placers of the northeastern part of the Siberian platform [3–7]. They are classified as variety V with aggregates of variety VII, according to the Yu. Orlov classification [8]. The similarity between the crystals studied in the current work and the rounded diamonds from alluvial deposits can be seen primarily in the presence of many black inclusions, which are unevenly distributed along the samples' volume. The rounded morphology of the diamonds from the Zarnitsa kimberlite pipe, the transition between octahedron and rhombic dodecahedron habits, and the presence of the specific micro-relief surface structure may also be found in the alluvial diamonds of the Siberian Platform [3]. Previous studies demonstrate that rounded alluvial diamonds from these placers are characterized by the radial mosaic internal structure that is unusual for natural diamonds as this structure is formed by the splitting crystal growth mechanism [9]. Until recently, similar diamonds have not been found in kimberlite pipes closer to the alluvial placers nor in any other known kimberlite pipes of the Yakutian diamondiferous province. In the current work, we provide a detailed study of the internal diamond structure found in Zarnitsa kimberlite pipe diamonds. We then compare this structure to the structure of authentic crystals of variety V and VII found in alluvial placers.

2. Results

2.1. Morphology

In the current work, 16 grey- to black-diamonds from Zarnitsa kimberlite pipe were selected for internal structure study (Figure 1). The samples include rounded or half-rounded diamonds, with octahedron and rounded rhombic dodecahedron crystal habits, and the aggregates of such crystals (Figure 2). The yellowish and brownish hue of some diamonds is due to secondary iron oxides/hydroxides deposits that developed on thin cracks in the external parts of the crystals. The dark grey-to-black colour is explained by the presence of numerous black inclusions, within the matrix of the diamond itself which is typically characterized by a colourless and transparent crystal.

Figure 1. Rounded rough diamonds from Zarnitsa kimberlite pipe (optical microphotographs).

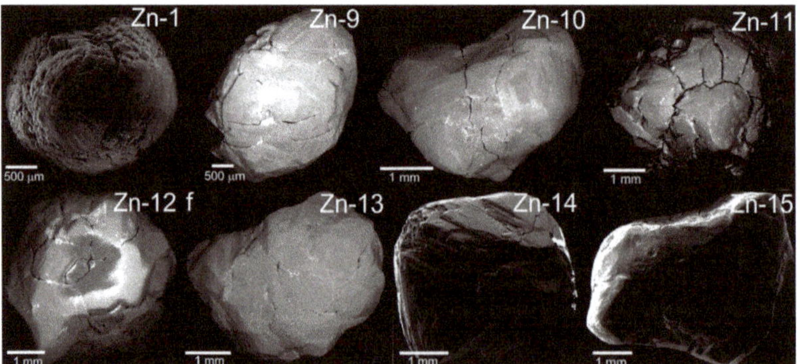

Figure 2. Morphology of diamonds (SEM images).

Morphological features specific to the so-called half-rounded diamonds [10] can be seen in the studied kimberlite pipe diamonds. The crystal morphology includes (1) laminar octahedral faces with clear laminar structure and ditrigonal layers, decreasing in size in a step-by-step pattern towards the periphery; (2) convex rounded surfaces at the octahedral edges with sheaf-like striations produced by the ditrigonal layered-steps of the octahedral faces; and (3) in some crystals, a smooth rounded surface close to the (111) apex is located hypsometrically lower than the combination surfaces and ditrigonal layers. In some cases, the morphology of the diamonds is complicated by: rounded apexes; elongated, drop-shaped hillocks on the dodecahedral crystal faces; etch channels; and a great number of small rounded holes.

The morphology of a Zn-14 diamond is characterized by laminar octahedral edges with clear laminar structure (Figure 2, Sample Zn-1). The vestiges of octahedral edges are also often seen on other crystal types; they are distinguished by ditrigonal layers of varying thickness that diminish step-wise towards the periphery. Triangular pits, which are generally called trigons [11], are a very common and distinctive feature of the octahedral crystal faces of these diamonds. Trigons vary in size up to 300 μm; a single diamond is commonly comprised of many trigons with a wide range of sizes. Even octahedral crystal faces, which visually appear pristine and smooth, can often contain microscopic trigons. Trigons have flat bottoms and are oriented in the direction opposite the underlying crystal face. These trigons are referred to as negative trigons and are typical microrelief features found on the surface reflecting the dissolution processes [12].

In addition to the laminar octahedral faces, the habitus of the crystals includes rounded surfaces. There are usually arched combination surfaces at the octahedral edges, which are sculptured by sheaf-like striation. The striation is a complex of edges found on ditrigonal layers in octahedron faces. These combination surfaces are not true crystal faces (110) but are close enough to the face that the dodecahedral diamond habit is sometimes described as a dodecahedroid. These surfaces also have many microrelief elements. such as drop-shaped hillocks and disc sculptures. The development stage of these combination surfaces varies from sample to sample. Thus, in the Zn-14 sample (Figures 1 and 2), they are weakly developed, which is why the habit of this diamond is mostly octahedral. In some cases (Zn-9 and Zn-10 samples), rounded surfaces prevail in the external morphology (Figures 1 and 2). In these samples, only the remaining laminar octahedral faces are fixed. The Zn-1 sample (Figures 1 and 2) has an almost spherical shape, as in the surface sculpture specific for diamond spherulites (ballas diamonds).

In some crystals (Zn-10, Zn-13, Zn-15 samples), convex rounded surfaces are fixed close to the (111) apex, located hypsometrically below combination surfaces and octahedral ditrigonal layers. On these surfaces, shagreen and round drop-shaped hillocks are occasionally seen. These surfaces have round dodecahedral shape which is the final dissolution form of primarily octahedral crystals. Sometimes the morphology of such crystals is complicated by rounded apexes, rock coatings, deep-etched channels and cracks (see Figure 2c–e). Such features are specific to diamond crystals that have undergone intensive dissolution [12–17].

2.2. Internal Structure

2.2.1. Optical Microscopy

The interference pattern of anomalous birefringence is not expected for crystals of cubic syngony. Diamonds belongs to the cubic crystal system and should therefore exhibit isotropic optical properties. However, many research studies on diamonds often show weak birefringence [18,19]. The appearance of anomalous birefringence may indicate a high degree of crystal structure deformation. High interference colours with cross polarizers indicate that a strong deformation exists within the crystalline structure. Internal stress in diamonds may be caused by (i) dislocations; (ii) lattice parameter deviations (caused by lattice impurities); (iii) inclusions; (iv) cracks; and (v) plastic deformation [20,21].

The optical microscopy-based observations of polished plates reveal zonal and sectorial internal structure in the diamond samples. In the core of the Zarnitsa kimberlite pipe diamond crystals, the highest concentration of black coatings and cracks are usually visible, making them almost non-transparent in most cases (Figure 3). In some cases, high inclusion concentrations are not visible in the central region of the diamond; though, the high-order interference colours (bright colours), from the central diamond region, are visible on the periphery. Thus, in the external zones of a Zn-14 crystal (Figures 3 and 4), only the regions with wavy extinction and birefringence pattern indicates a lack of stress in the crystalline structure. Moreover, diamonds with sectoral inhomogeneity are primarily expressed by the different concentrations of mechanical impurities (inclusions) in the various growth sectors. As seen in Figure 3, some growth sectors contain little three-dimensional defects while other sectors are rich with inclusions. In the Zn-1 sample (Figure 3), black microinclusions are located along the radial fibrous lines and develop from the inner parts of the crystal to its periphery. In addition, birefringence patterns show the various growth sector structures. The areas in the crystals associated with the growth sectors with a small quantity of black inclusions is quantified by fewer deformations compared to the other volume of diamonds. Unlike crystal volumes with abundant inclusions, only the wavy extinction or the specific tatami-like patterns are visible in the crossed polarizers (Figure 3). Such patterns are usually explained by a plastic deformation and fibrous internal structures specific to diamonds with cubic habits [22,23]. In the volume of crystals with abundant inclusions, strong interference is fixed which indicates a high deformation of the crystal structure. Such an interference pattern is a result of the uneven stress distribution within the crystal volume.

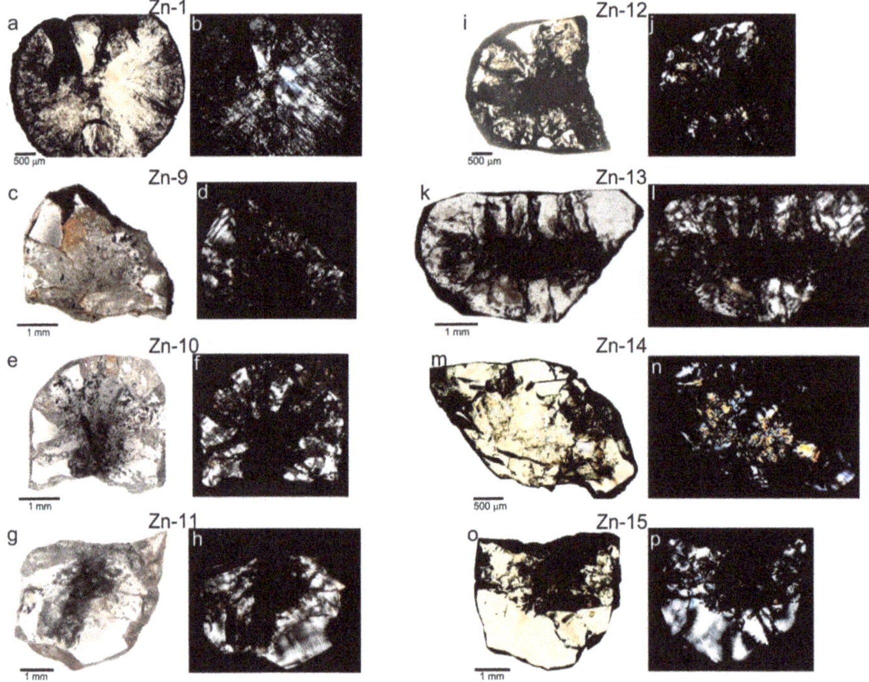

Figure 3. Internal morphology of double-polished diamond plates ((**a,c,e,g,i,k,m,o**)—transmitted polarized light; (**b,d,f,h,j,l,n,p**)—birefringence pattern).

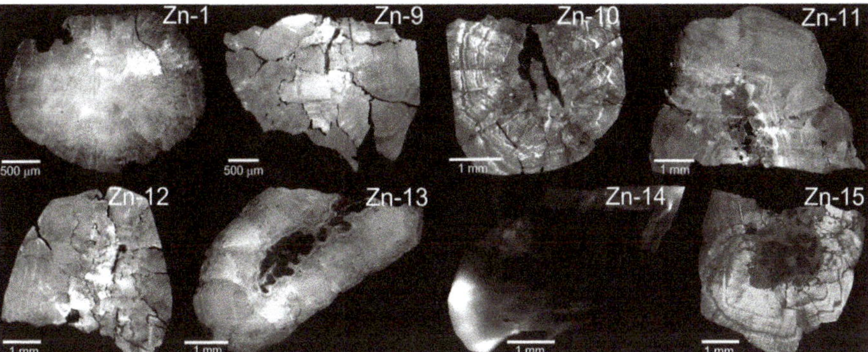

Figure 4. Cathodoluminescence images of polished plates from diamond showing zoning structures, and irregularly shaped polycrystalline cores with dark luminescence.

2.2.2. Cathodoluminescence Imagery

Cathodoluminescence (CL) is one of techniques that allows direct visual observation of crystal growth patterns, zonality and sectorality. The borders of diamond zones and sectors are visible by CL due to the uneven distribution of impurities (isomorphic nitrogen impurities in the form of a defect or microinclusion) found in the crystal volumes [24,25]. Zonal structure is noticed clearly on the CL topographs of diamonds from the Zarnitsa kimberlite pipe (Figure 4). Weak core luminescence is a specific feature of most diamond cores with complicated structures and irregular morphologies. Diamond cores complicate the mosaic structure of the crystal because central parts of the crystals with dark luminescence contain numerous blocks that tightly coalescence with each other, which is specific to polycrystalline bodies (Figure 4, Samples Zn-11, Zn-13, Zn-15). External zones have brighter luminescence on CL topographs. Concentric zonal structures are also seen at the external zones. The shape of growth-zonality close to the diamond centre repeats the contour aggregations in the core (Figure 4, Sample Zn-13). Such zonal structure reflects the primary irregular core morphology. Peripheral parts of the crystal samples also show rounded shapes in some cases; however, in most cases, straight linear octahedral zones are seen.

2.2.3. Electron Backscatter Diffraction (EBSD)

The EBSD technique is capable of precisely determining the textural features of the crystals, expressed as lattice orientation relationships between the macroblocks (subgrains). The EBSD technique reveals that the Zarnitsa kimberlite pipe diamonds have a mosaic block internal structure (Figure 5). The EBSD images show that the diamonds consist of numerous (up to 37) blocks of different size (subindividual), which are disoriented between each other. The disorientation angle of the blocks reach significant values up to 47° (Figure 5). The size of the block increases from the crystal core to the periphery because subindividuals originate at the centre and extend to its surface. Finally, such subindividual systems form the radial structure. At the same time, a bigger quantity of small disoriented blocks is seen in the diamond center. The significant disorientation angle of the block points to elements of polycrystalline structure in the diamond samples. Microscopically, the diamond samples either have the form of single crystals or consist of several (up to 4) monocrystals.

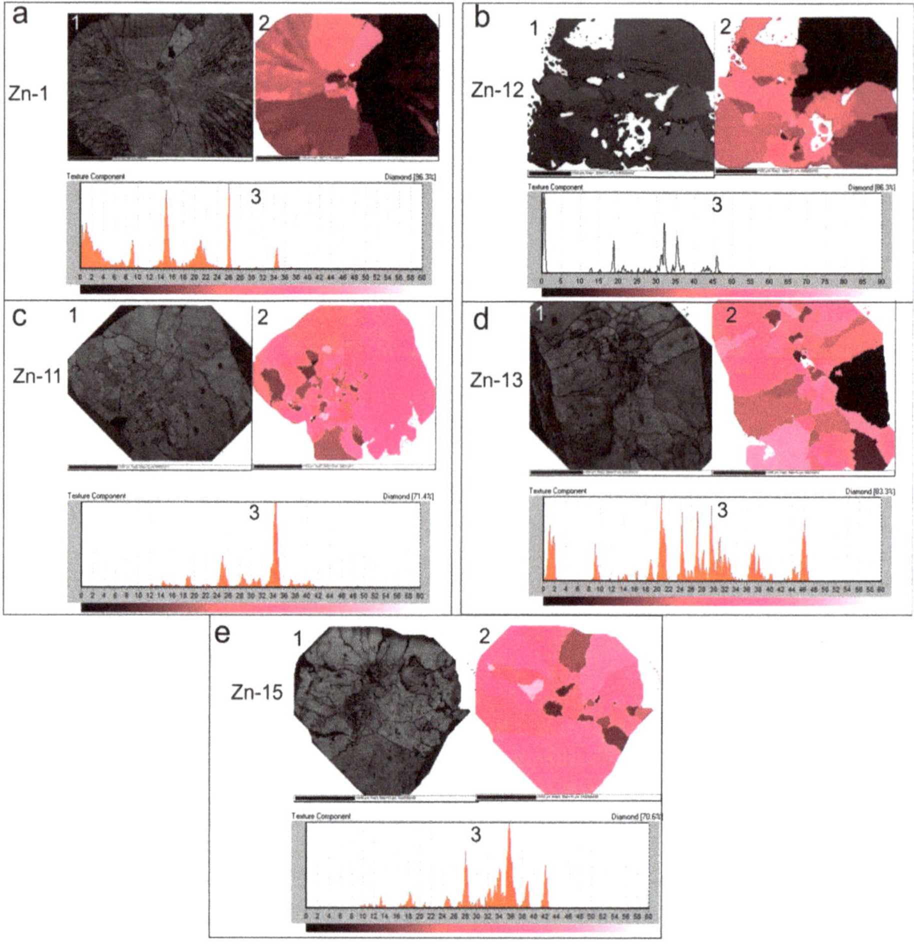

Figure 5. Mosaic-block internal structures of diamonds consisting of subgrains (subindividuals) misorientated relative to each other (1—reflected light, 2—EBSD image, 3—misorientation degrees) (**a**) Sample Zn-1; (**b**) Sample Zn-12; (**c**) Sample Zn-11; (**d**) Sample Zn-13; (**e**) Sample Zn-15.

3. Discussion

The morphological features of diamonds have not been correlated to the geological features of natural diamond formation. This can probably be explained by the fact that the external morphology reflects only characteristics of the latest growth stages which is mainly defined by dissolution and regeneration processes. A large amount of data now shows the deep-seated (mantle) origin of most natural diamonds, while kimberlites and lamproites only bring natural diamonds to the Earth's surface [26–29]. Recently X-ray topography showed that a diamond morphology may be imposed to a full-grown (protogenetic) olivine during their encapsulation, suggesting that the bulk of the inclusion is protogenetic, whereas its more external regions, close to the diamond-inclusion interface, could be syngenetic [30]. Physical and chemical changes in kimberlite melts (changes in pressure, temperature, oxygen fugacity etc.) can lead to the change in the morphology of diamond crystals, which are likely the result of dissolution and magmatic corrosion processes. The high occurrence of dissolution processes has a great influence on many natural diamonds; this is demonstrated by the many micromorphological-based details, such as etch channels, triangular etch pits, dissolution layers, etc. [31,32]. Curved rounded crystal with rhombic dodecahedral habit (dodecahedroid) is the final form of the diamond dissolution [8,13]. Simulated diamond crystal dissolution data, generated by using natural paragenesis system modelling [12–16], shows that the formation of ditrigonal layers in the dissolution process occurs with a pressure range of 2.5–5.7 GPa and a temperature range of 1100–1450 °C. Despite the differences in the simulated data and the natural processes, the main common factors of the morphological evolution of diamonds are likely to be the same. The diamond samples and the partially dissolved simulated diamond crystals have similar complex morphological features. The similarity is seen with the presence of ditrigonal (pseudo-hexagonal) shapes of layers on the (111) faces, negative trigons, deep etch channels, hillocks on rounded surfaces, sheaf-like striations and smooth rounded surfaces near the crystal apexes. Such morphological similarity, combined with the changing crystal habit of each diamond sample (from octahedral to rounded rhombic dodecahedral), results in the development of the various morphological features exhibited in the shape of the diamond samples. This result is likely to be due to the various degrees of dissolution of the primary octahedral crystals and their aggregates. On the other hand, mosaic fibrous internal structures were revealed by X-Ray topograhy in the so-called diamond spherocrystals [33]. Diamond spherocrystals have oriented radial structures: in the cubic sectors elongated subindividuals oriented along the [100] directions and in octahedral sectors along the [111] directions. The habit of diamond spherocrystals was shown to depend on the subindividual growth rate along the [100] and [111] directions; they become close to rhombic dodecahedral or octahedral as the rate increases. A similar structure, in this study, was defined for the Zn-1 diamond. The growth shape of the diamond spherocrystals differs from the dissolution shape via a lack of additional edges. The description "rounded rhombic dodecahedron" is clearly an oversimplification since a rhombus that is divided into two curved surfaces can only be divided by an additional edge that is always the shorter diagonal of the rhombus [34]. Thus, the whole crystal surface is divided into 24 curved surfaces, which in the ideally symmetric specimen would be identical triangular sheets. The polyhedron which best represents this topology is the tetrahexahedron. The additional edges are not seen clearly in the Zarnitsa kimberlite pipe diamond samples, though it can possibly be explained by the rough sculptures of the rounded surfaces.

Morphological features of the diamond samples are similar to V and VII diamond varieties which are wide spread in alluvial deposits of the north-east Siberian Platform [3]. Primary sources of these diamonds have not been discovered yet. The presence of numerous black inclusions, unevenly spread within the crystal volume, in Zarnitsa kimberlite pipe diamond samples makes them similar to rounded diamonds from alluvial placers. At the same time, the maximum concentration of three-dimensional defects are mostly seen in the core of Zarnitsa diamonds, while the inclusions are spread within the whole crystalline volume of the diamonds from alluvial placers, excluding some growth sectors (100) [3]. We noticed sectoral distribution of inclusions in some diamonds from Zarnitsa kimberlite pipe; but in most cases, such inclusions are concentrated in centre region of the crystal. As the birefringence patterns show, the centre crystal regions with the darkest inclusions have significantly more stress inside the crystal structure compared to the external diamond zones. There are likely to be many mechanical admixtures (inclusions) which cause high stress and promote the block (polycrystalline) structure of crystal interiors. External diamond zones have more perfect internal structure compared to their core parts; this is clearly seen in the internal structure found in the Zn-14 sample. Even though this crystal is morphologically represented by a well-faceted octahedron with laminar faces, a high-stress mosaic-block structure was found in the sample.

Thus, the internal structure of some diamonds from the Zarnitsa kimberlite pipe are close to the rounded diamonds from the alluvial placers [3]; despite these similarities, diamonds from the Zarnitsa kimberlite pipe essentially have another mechanism of internal structure formation. Dark grey round diamonds from the Zarnitsa kimberlite pipe have specific zonal and sectorial mosaic-block internal structures visualised by the CD and EBSD techniques. The central regions of the diamond samples have clear polycrystalline structure with significant subindividual disorientation. Large subindividuals develop from the interior and increasing in size outward towards the periphery. Such subindividual systems form the radial structure.

X-Ray diffraction, CL, and birefringence studies have revealed the specific internal structures of rounded V-variety diamonds from alluvial placers of the Northeastern Siberian platform [3]. Although diamonds are single crystals, their volume is separated into slightly misoriented subindividuals, intimately intergrown with each other. The diamonds of the V variety have radial mosaic-block internal structures consisting of slightly misorientated (up 20′) subindividuals. The system of subindividuals forms a radial structure consisting of several larger blocks (subgrains) which are also misoriented to one another by up to 5°; while some of them are split into smaller parts. These internal structures may develop by the splitting crystal growth mechanisms [9]. In this case, crystallography is filled with branched forms. The branching that allows spherulites to fill spherical volumes is called noncrystallographic branching. This branching is distinct from the crystallographic branching of snowflakes, for example, where every branch is in single-crystal register with every other branch [9].

Unlike the rounded V variety diamonds, the internal structure formed by splitting crystal growth mechanisms and the formation of diamonds from Zarnitsa kimberlite pipe, which are in many ways similar, can be explained by different reasons. As shown earlier, diamond centres have clear polycrystalline structure with numerous (>15) subindividuals. According to A.V. Shubnikov's geometric selection law, the enlargement of subindivials, from the centre to the periphery, and the formation of the radial mosaic internal structure are likely to be the result of primarily polycrystalline core development [35]. This law primarily states that single crystals with different orientations grow on the polycrystalline core and crowd over each other when growing before beginning to suppress each other. In other words, the primary development belongs to the crystals (individuals) with the maximum growth rate direction that is perpendicular to the polycrystalline base. It is suggested that the development of spherulitic morphologies may be controlled by growth rate anisotropy and geometrical selection law as well [35]. The geometrical selection law may proceed through competitive or trans-crystalline growth of polycrystalline aggregates [9]. The survival of individual crystals in an aggregate is determined by crystal orientation to the nucleation surface and its relative growth rate. In this case, only radial directions can provide continuous growth of surviving subindividuals.

4. Methods

The morphological features of crystals were studied by optical Zeiss Stemi SV 6 stereo microscope and the scanning electron microscope Leo-1430VP SEM, which has an accelerating voltage of 15 kV, with a working distance of 15 mm. In the past, the 300–500 μm polished plates were made from diamonds. CL diamonds plate patterns were studied using an Oxford Centaurus detector on the Leo-1430VP SEM (accelerating voltages of 12–15 kV and the electron beam current of ~0.5 mA). The deviation from the basic orientation in the diamond grain was mapped by EBSD [36]. The EBSD analysis was performed with the use of the Oxford Instruments HKL detector mounted to a Hitachi S-3400 N scanning electron microscope (SEM). Kikuchi patterns formed by back-scattered electrons were automatically indexed by the Oxford data collection software. The accuracy of determination of the angles of misorientation by this method was 0.5°–1.0°.

Analytical investigations were carried out in the Analytical Center for multi-element and isotope research SB RAS. The part of the work (EBSD) was done using the infrastructure of the Shared-Use Center "Siberian Synchrotron and Terahertz Radiation Center (SSTRC)" based on VEPP-3/VEPP-4M/NovoFEL of BINP SB RAS.

5. Conclusions

Rounded diamonds from Zarnitsa kimberlite pipe have specific zonal and sectorial mosaic-block internal structures. The central sections of the diamond samples have clear polycrystalline structure with significant subindividual disorientation. The high consistency of mechanical admixtures (inclusions) in diamond cores cause high lattice stress of the crystalline structure and promote block (polycrystalline) core structures. Such internal structures in the Zarnitsa kimberlite pipe diamonds are derived from the crystallization of the polycrystalline cores, according to the geometric selection rule. Unlike the mosaic block structure of similar diamonds, described before and after the placers of the north-east of the Siberian platform, which were formed by the splitting crystal growth mechanism. Thus, despite the physical resemblance and the similarity of the internal structure, diamond crystals from the Zarnitsa kimberlite pipe and the rounded diamonds from the alluvial placers were formed due to essentially different processes. Diamonds from the Zarnitsa kimberlite pipe evolve from polycrystalline cores to monocrystals, while an alluvial placer diamond splits up the spherulite-like structure.

Acknowledgments: The research was supported by state assignment project (project No. 0330-2016-0007), and in part by RFBR (project No. 16-05-00614).

Author Contributions: Alexey Ragozin studied samples by optical and scanning electron microscopy, made the polished plates from diamonds; Dmitry Zedgenizov conducted the cathodoluminescence experiments and analyzed the data; Konstantin Kuper performed the electron backscatter diffraction experiments; Yuri Palyanov analyzed the data and discussed the results; Alexey Ragozin and Dmitry Zedgenizov contributed equally by writing the manuscript.

Conflicts of Interest: The authors declare no conflict of interest.

References

1. Sarsadskikh, N.N.; Popugayeva, L.A. New data about manifestation ultramafic-alkaline magmatism within Siberian Platform. *Razvedka I Okhrana Nedr.* **1955**, *5*, 11–20. (In Russian).

2. Kostrovitskiy, S.I.; Spetsius, Z.V.; Yakovlev, D.A.; Fon-der-Flaas, G.S.; Suvorova, L.F.; Bogush, I.N. *Atlas of the Primary Diamond Deposits of the Yakutian Kimberlite Province*; NIGP SC "ALROSA" (PAO): Mirny, Russia, 2015; p. 480. (In Russian)

3. Ragozin, A.L.; Zedgenizov, D.A.; Kuper, K.E.; Shatsky, V.S. Radial mosaic internal structure of rounded diamond crystals from alluvial placers of Siberian Platform. *Mineral. Petrol.* **2016**, *110*, 861–875. [CrossRef]

4. Ragozin, A.L.; Shatsky, V.S.; Rylov, G.M.; Goryainov, S.V. Coesite inclusions in rounded diamonds from placers of the northeastern siberian platform. *Dokl. Earth Sci.* **2002**, *384*, 385–389.

5. Ragozin, A.L.; Shatskii, V.S.; Zedgenizov, D.A. New data on the growth environment of diamonds of the variety V from placers of the northeastern Siberian Platform. *Dokl. Earth Sci.* **2009**, *425*, 436–440. [CrossRef]

6. Afanasyev, V.P.; Agashev, A.M.; Orihashi, Y.; Pokhilenko, N.P.; Sobolev, N.V. Paleozoic U–Pb age of rutile inclusions in diamonds of the V–VII variety from placers of the northeast Siberian Platform. *Dokl. Earth Sci.* **2009**, *428*, 1151–1155. [CrossRef]

7. Smith, E.M.; Kopylova, M.G.; Frezzotti, M.L.; Afanasiev, V.P. Fluid inclusions in Ebelyakh diamonds: Evidence of CO_2 liberation in eclogite and the effect of H_2O on diamond habit. *Lithos* **2015**, *216–217*, 106–117. [CrossRef]

8. Orlov, Y.L. *The Mineralogy of Diamond*; John Wiley: New York, NY, USA, 1977; p. 233.

9. Shtukenberg, A.G.; Punin, Y.O.; Gunn, E.; Kahr, B. Spherulites. *Chem. Rev.* **2012**, *112*, 1805–1838. [CrossRef] [PubMed]

10. Bartoshinskii, Z.V.; Kvasnitsa, V.N. *Crystallomorphology of a Diamond from Kimberlite*; Naukova Dumka: Kiev, Ukraine, 1991; p. 172. (In Russian)

11. Frank, F.C.; Puttick, K.E.; Wilks, E.M. Etch pits and trigons on diamond: I. *Philos. Mag.* **1958**, *3*, 1262–1272. [CrossRef]

12. Khokhryakov, A.F.; Pal'yanov, Y.N. The evolution of diamond morphology in the process of dissolution: Experimental data. *Am. Mineral.* **2007**, *92*, 909–917. [CrossRef]

13. Khokhryakov, A.F.; Pal'yanov, Y.N. The dissolution forms of diamond crystals in $CaCO_3$ melt at 7 GPa. *Russ. Geol. Geophys.* **2000**, *41*, 705–711.

14. Khokhryakov, A.F.; Pal'yanov, Y.N. Evolution of diamond morphology in the processes of mantle dissolution. *Lithos* **2004**, *73*, S57.

15. Khokhryakov, A.F.; Palyanov, Y.N. Effect of crystal defects on diamond morphology during dissolution in the mantle. *Am. Mineral.* **2015**, *100*, 1528–1532. [CrossRef]

16. Khokhryakov, A.F.; Pal'yanov, Y.N.; Sobolev, N.V. Evolution of crystal morphology of natural diamond in dissolution processes: Experimental data. *Dokl. Earth Sci.* **2001**, *381*, 884–888.

17. Khokhryakov, A.F.; Pal'yanov, Y.N. Influence of the fluid composition on diamond dissolution forms in carbonate melts. *Am. Mineral.* **2010**, *95*, 1508–1514. [CrossRef]

18. Howell, D. Strain-induced birefringence in natural diamond: A review. *Eur. J. Mineral.* **2012**, *24*, 575–585. [CrossRef]

19. Howell, D.; Wood, I.G.; Nestola, F.; Nimis, P.; Nasdala, L. Inclusions under remnant pressure in diamond: A multi-technique approach. *Eur. J. Mineral.* **2012**, *24*, 563–573. [CrossRef]

20. Lang, A. Causes of birefringence in diamond. *Nature* **1967**, *213*, 248–251. [CrossRef]

21. Tolansky, S. Birefringence of diamond. *Nature* **1966**, *211*, 158–160. [CrossRef]

22. Frank, F.C. On the x-ray diffraction spikes of diamond. *Proc. R. Soc. Lond. A Math. Phys. Sci.* **1956**, *237*, 168–174. [CrossRef]

23. Moore, M.; Lang, A.R. On the internal structure of natural diamonds of cubic habit. *Philos. Mag.* **1972**, *26*, 1313–1325. [CrossRef]

24. Götze, J.; Kempe, U. Physical principles of cathodoluminescence (CL) and its applications in geosciences. In *Cathodoluminescence and Its Application in the Planetary Sciences*; Gucsik, A., Ed.; Springer: Berlin/Heildelberg, Germany, 2009; pp. 1–22.

25. Lang, A.R. Topographic methods for studying defects in diamonds. *Diam. Relat. Mater.* **1993**, *2*, 106–114. [CrossRef]

26. Sobolev, N.V. *Deep Seated Inclusions in Kimberlites and the Problem of the Composition of the Upper Mantle*; AGU: Washington, DC, USA, 1977; p. 279.

27. Meyer, H.O.A. Genesis of diamond—A mantle saga. *Am. Miner.* **1985**, *70*, 344–355.

28. Meyer, H.O.A. Inclusions in diamond. In *Mantle Xenoliths*; Nixon, P.H., Ed.; Wiley and Sons: New York, NY, USA, 1987; pp. 501–523.

29. Harris, J.W. Diamond geology. In *The Properties of Natural and Synthetic Diamond*; Field, J.E., Ed.; Academic Press: London, UK, 1992; pp. 345–393.

30. Agrosì, G.; Nestola, F.; Tempesta, G.; Bruno, M.; Scandale, E.; Harris, J. X-ray topographic study of a diamond from Udachnaya: Implications for the genetic nature of inclusions. *Lithos* **2016**, *248*, 153–159. [CrossRef]

31. Orlov, Y.L. *Diamond Morphology*; AS USSR: Moscow, Russia, 1963; p. 235.

32. Khokhryakov, A.F.; Pal'yanov, Y.N.; Sobolev, N.V. Crystal morphology as an indicator of redox conditions of natural diamond dissolution at the mantle PT parameters. *Dokl. Earth Sci.* **2002**, *385*, 534–537.

33. Orlov, Y.L.; Bulienkov, N.A.; Martovitsky, V.P. The spherocrystals of diamond—New type of natural single crystals having a fibrous structure. *Dokl. Akad. Nauk SSSR* **1980**, *252*, 703–707.

34. Moore, M.; Lang, A.R. On the origin of the rounded dodecahedral habit of natural diamond. *J. Cryst. Growth* **1974**, *26*, 133–139. [CrossRef]

35. Shubnikov, A.V. About geometric selection law in the formation of a crystalline aggregate. *Dokl. Akad. Nauk SSSR* **1946**, *51*, 679–681.

36. Humphreys, F.J. Review grain and subgrain characterisation by electron backscatter diffraction. *J. Mater. Sci.* **2001**, *36*, 3833–3854. [CrossRef]

 *crystals*

Article

The Internal Structure of Yellow Cuboid Diamonds from Alluvial Placers of the Northeastern Siberian Platform

Alexey Ragozin [1,2,*, **Dmitry Zedgenizov [1,2]**, **Konstantin Kuper [3]**, **Viktoria Kalinina [1,2]** and **Alexey Zemnukhov [4]**

1   V. S. Sobolev Institute of Geology and Mineralogy, Siberian Branch Russian Academy of Sciences, Novosibirsk 630090, Russia; zed@igm.nsc.ru (D.Z.); vika@igm.nsc.ru (V.K.)
2   Department of Geology and Geophysics, Novosibirsk State University, Novosibirsk 630090, Russia
3   Budker Institute of Nuclear Physics, Siberian Branch Russian Academy of Sciences, Novosibirsk 630090, Russia; k.e.kuper@inp.nsk.su
4   JSC Almazy Anabara, Yakutsk 678174, Russia; zemnuchoval@alanab.ru
*   Correspondence: ragoz@igm.nsc.ru; Tel.: +7-383-330-8015

Academic Editors: Yuri N. Palyanov
Received: 30 June 2017; Accepted: 28 July 2017; Published: 31 July 2017

Abstract: Yellow cuboid diamonds are commonly found in diamondiferous alluvial placers of the Northeastern Siberian platform. The internal structure of these diamonds have been studied by optical microscopy, X-Ray topography (XRT) and electron backscatter diffraction (EBSD) techniques. Most of these crystals have typical resorption features and do not preserve primary growth morphology. The resorption leads to an evolution from an originally cubic shape to a rounded tetrahexahedroid. Specific fibrous or columnar internal structure of yellow cuboid diamonds has been revealed. Most of them are strongly deformed. Misorientations of the crystal lattice, found in the samples, may be caused by strains from their fibrous growth or/and post-growth plastic deformation.

Keywords: diamond; internal structure; electron backscatter diffraction; X-ray topography

1. Introduction

Diamond crystals develop diverse morphological and physical properties, which reflect the variation of conditions during diamond formation which largely occurs in the upper mantle [1–9]. Morphology and internal structure of natural diamond crystals reflect the conditions of growth and following post-growth history. The study of mineral inclusion in diamonds and xenoliths of diamondiferous rocks in kimberlites show that diamonds can be crystallized in various upper mantle rocks: peridotites (olivine, orthopyroxene, garnet, and diopside), eclogites (garnet, clinopyroxene) and, rarely, websterites [10–12]. P-T models propose that most diamonds are formed in the diamond's stability field, in temperature ranges (900–1300 °C) and pressures (4.5–6 GPa) that correspond to the depth of formation (140–200 km) [13].

The Siberian platform hosts more than one thousand known kimberlite pipes. Numerous (>100) kimberlite pipes have been discovered within the northeastern part of the Siberian platform, most are barren or very poorly diamondiferous [14]. Despite the very low diamond-bearing capacity of kimberlite pipes in this region, approximately 70% of the diamond alluvial placer deposits of the Siberian platform are located here [15]. The primary sources of the diamonds in these placers have not been discovered yet.

The alluvial deposits of the Northeastern Siberian platform have specific diamond association [16,17]. Previous studies have revealed several diamond populations, which may be related with various primary sources [18,19]. Under the Orlov classification [5], these diamonds can be classified into three groups: (1) typical octahedral and rounded dodecahedral diamonds of variety I; (2) yellow-orange or dark grey cuboids of varieties II and III; and (3) rounded dark crystals of variety V.

Diamonds with cubic crystals (cuboids), forming an incessant colour gradient from yellowish-green to yellow and dark orange, are widespread in alluvial placers of the Northeastern Siberian platform. These diamonds have been categorized as variety II in the mineralogical classification [5]. Diamonds of this variety constitute approximately 7–7.5% of the alluvial placer diamond collection [15]. Diamond crystals of cubic habit are infrequent in Siberian kimberlites and they are no greater in number than 2% of the total diamonds found in the region. Thus, it is unlikely that the kimberlites discovered in the Siberian platform are the source of these diamond placers. In some other regions, e.g., Mbuji-Mayi (Congo), Ekati (Canada), and Jwaneng (Botswana), many cuboid diamonds are found in the kimberlites [8,20,21]. The characteristics of variety II diamonds, i.e., specific morphological features, the low values of $\delta^{13}C$ and high levels of nitrogen in the C-form (single substitutional nitrogen impurity, type Ib—causing yellow coloration), testify to their unusual primary source [19,22]. This study investigates the specific features related to the internal structures of these cuboid diamonds and is part of an extensive study of alluvial diamonds from the Northeastern Siberian platform [23–30].

2. Results

2.1. Morphological Features

An infrequent pattern of natural diamond formation is the cube, which is a rough approximation of the ideal form [31]. Figures 1 and 2 are microphotograph and SEM images of the most typical variety II diamonds from alluvial placers of the Northeastern Siberian platform. These cuboid diamonds are almost cubic or sub-rounded crystals; most studied diamonds are isometric but, in some cases, the crystals are more or less distorted (flattened or elongated at three- or four-fold axes) (Figures 1 and 2). The diamonds of interest in this study are yellow (all diamonds in this study belong to rare type Ib) and some are turbid yellow due to micro-inclusions. Studied diamonds have typical dissolution features, e.g., patterns of numerous few stepped, pyramidal etch pits of rectangular shape on their (100) planes (Figure 2e,f) and rounded surfaces that correspond to tetrahexahedroid corners at cubic edges. There are no crystals with sharp cube edges; all samples have epigenetic dissolution. Rectangular-etched pits on the cubic faces of diamonds are known as tetragons. Tetragonal pits lie on each of the (100) surfaces; each pit is turned around at an angle of 45° with regard to the form of the cubic facets. It was found that convex crystals are usually covered with faces at the maximum speed of dissolution, whereas concave bodies tend to be faceted at a minimal velocity [32]. Rounded convex surfaces produce rounded habits in some crystals with characteristic morphological tetrahexahedroid features of (Figure 2c,h). Experimental data shows that diamond crystals transform their habits during dissolution from cubes to tetrahexahedroids in water-containing systems, when the weight loss is >50% [33]. The surface textures of studied crystals include fine striation along <110> on (100) faces and many elongated hillocks on the rounded convex surfaces of tetrahexahedroid (Figure 2f). Some rounded surfaces of studied crystals contain etched holes (Figure 2c) and/or channels (Figure 2a), which frequently occur at visible cracks extending through the crystals volumes. The relics of cubic faces of studied crystals are not perfectly flat and diverge from precise crystallographic planes (100), i.e., becoming convex or concave.

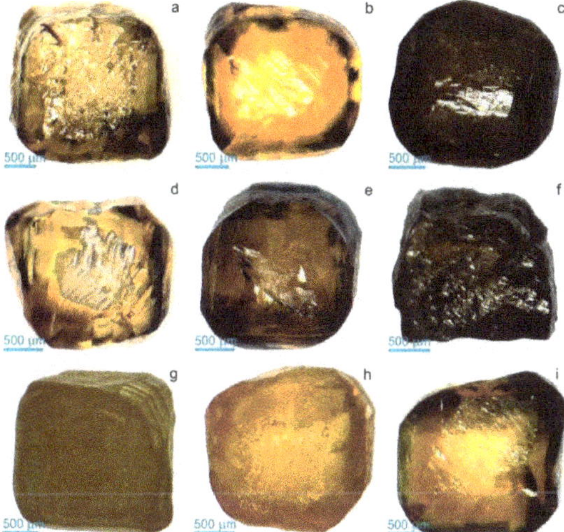

Figure 1. Optical microphotographs of cuboid diamonds from the Northeastern Siberian platform (**a**—Light yellow crystal MP-30 of predominant cubic habit; **b**—Orange-yellow crystal MP-43 of combination form (cube-tetrahexahedroid); **c**—Yellowish-grey diamond rounded crystal MP-56; **d**—Light yellow crystal MP-30 of transition form (cube-tetrahexahedroid); **e**—Dark yellow crystal MP-60 of combinational form (cube-tetrahexahedroid); **f**—Yellowish-grey diamond crystal MP-83 of cubic habit; **g**—Greyish-yellow crystal MP-87 of combination form (cube-tetrahexahedroid); **h**—Yellow crystal MP-98 of transition form (cube-tetrahexahedroid); **i**—Orange-yellow crystal MP-43 of combination form (cube-tetrahexahedroid)).

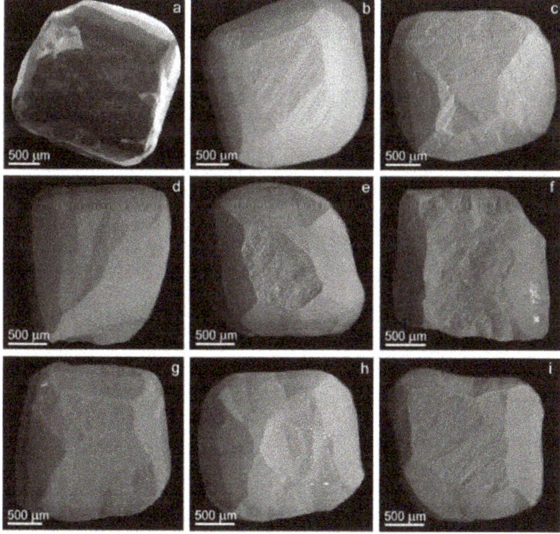

Figure 2. SEM micrographs showing morphological features of cuboid diamonds from the Northeastern Siberian platform (**a**—The crystal MP-30 of predominant cubic habit with small rounded tetrahexahedroid surfaces of at the cube edges; **b**—The crystal MP-43 of combinational form (cube-tetrahexahedroid) with fine striation along <110> on (100) faces; **c**—The diamond crystal

MP-56 of predominantly tetrahexahedroid habit with cube faces which are almost totally replaced by the rounded surfaces, there are only relics of (100) faces with large (~400 μm) negative tetragonal etch pits of rectangular shape and numerous shallow micro-pits with elliptical or irregularly curved outlines (corrosion sculptures) on rounded teraxehedroid surfaces; **d**—The crystal MP-60 of transitional form (cube-tetrahedroid); **e**—The crystal MP-71 of combinational form (cube-tetrahexahedroid), relics of cube face have numerous stepped, flat-bottomed, and pyramidal etch pits of rectangular shape, there are concentric terraces on the rounded surfaces. **f**—The diamond crystal MP-83 of cubic habit with tetragonal etch pits and fine striation along <110>; **g**—The crystal MP-87 of combinational form having flat cubic faces with tetragonal pyramidal etch pits and rounded surfaces of tetrahexahedroid, there are blunted and rounded edges and apices which correspond with intense mechanical abrasion; **h**—The crystal MP-98 of transitional form (cube-tetrahedroid); **i**—The crystal MP-108 of combinational form (cube-tetrahexahedroid) with fine striations along <110> on the (100) faces).

2.2. Internal Structure

2.2.1. Optical Microscopy

The nine diamonds were polished to consist of plates parallel to a crystallographic (110) plane with a 200–400 μm thickness. The observations of polished plates by optical microscopy showed that cuboid diamonds have inhomogeneous internal structures. All diamonds showed concentrical zonation having growth zones with different intensity of colouration (Figure 3). Intensively-coloured zones are regularly sited in the centre of crystals. On the other hand, in some cases (Figure 3d) the centre of the crystals are colourless, while the outer zone is intensely yellow. Moreover, there are more sophisticated patterns with heterogeneous spreading of growth zones with dissimilar coloration (Figure 3b,c,e,i,h). Several crystals showed growth zones with multiple micro-inclusions spreading as a chain along the (111) direction (Figures 3a,f and 4a,b). Such patterns are typical for diamonds of cubic habit and may reflect their fibrous internal structure [25].

In fact a birefringence interference pattern is not supposed for crystals of cubic symmetry which includes diamonds; its appearance suggest strong deformation of the diamond's crystal structure [4,34–37]. In other words, irregular birefringence reflects an accidental imperfection of the crystal structure. The straining of the crystal structure may result from (1) dislocations; (2) lattice parameter deviations (caused by lattice impurities); (3) physical impurities (mineral and fluid inclusions; (4) cracks; and (5) plastic deformation [36]. The fibrous internal structure of diamonds with cubic habits, which consist of sub-parallel fibres diverging in lines from a common centre (crystals core), has been previously recognized by X-ray topography [31,38]. The misorientations between subindividuals (fibres) in studied diamonds produce a deformation of crystals, which results in the patterns of anomalous birefringence (Figure 5). The interference of highest degree is detected at the boundaries between various growth zones and/or sectors (Figure 5a,e,f). The crystal samples exhibit zonal and sectorial birefringence patterns. They often have brindled interference patterns, and in some parts of the crystals patterns include thin lines intersecting each other in different <111> directions. The latter is called "tatami" pattern and explained by the influence of post-growth plastic deformations by the dislocation gliding mechanism [39].

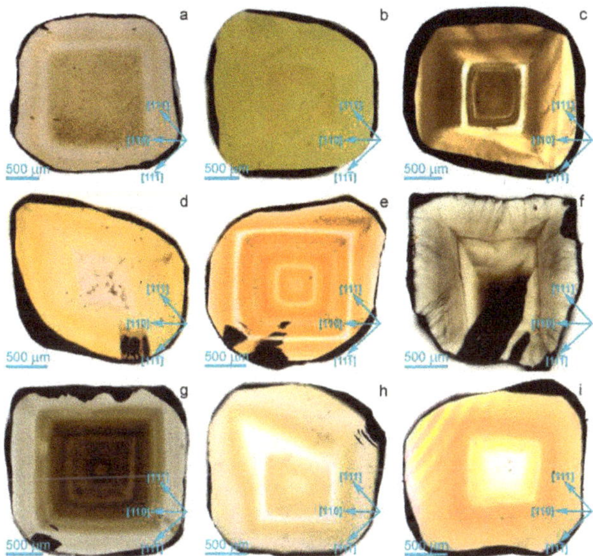

Figure 3. Internal structure of yellow cuboid diamonds from the Northeastern Siberian platform (optical microphotographs of double-polished plates parallel to (110), transmitted polarized light) (**a**—The crystal MP-30 having zonal structure, cube-shaped core with microinclusions located by chains that defined radial distribution; **b**—The diamond MP-43 having cubic zones of slightly different intensities of yellow coloration; **c**—The crystal MP-56 with zonal distribution of numerous microinclusions, there are inclusion-free zones and zones with different densities of microinclusions; **d**—The crystal MP-60 having light yellow inner core and intense yellow outer zones; **e**—The diamond MP-71 having numerous cubic zones with different intensity of yellow coloration; **f**—The diamond MP-83 with clear zonal and sectorial texture; there are radial chains of microinclusions in outer zones; **g**—The crystal MP-87 having zonal structure having muddy yellow core zone with high microinclusion density and an inclusion-free rim. In core there are several zones with different inclusions density; **h**—The diamond having distorted cubic zones with different intensities of yellow colour; **i**—The diamond having colourless inner core and numerous cubic zones with different intensity of yellow colour in outer rim).

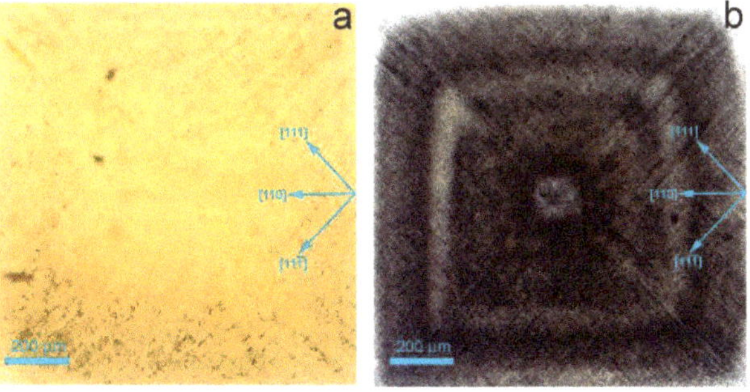

Figure 4. Core regions of diamonds (**a**—M-30, **b**—MP-87) with abundant microinclusions.

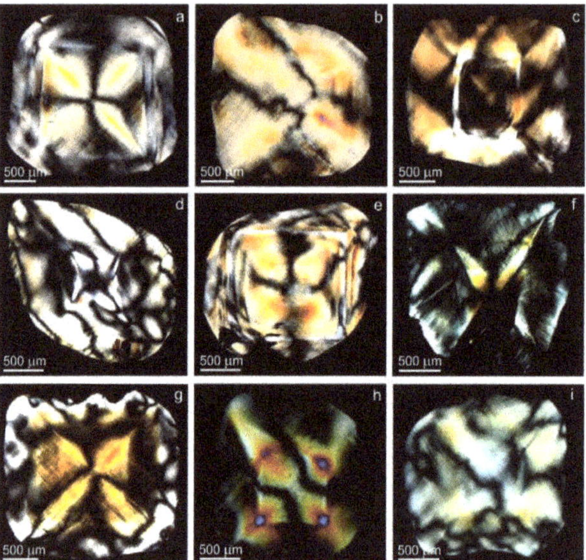

Figure 5. Anomalous birefringence patterns of double-polished plates of yellow cuboid diamonds from the Northeastern Siberian platform (photos taken between crossed polarizers) (**a**—MP-30; **b**—MP-43; **c**—MP-56; **d**—MP-60; **e**—MP-71; **f**—MP-83; **g**—MP-87; **h**—MP-98; **i**—MP-108).

2.2.2. X-ray Topography (XRT)

Figure 6 shows X-ray projection topography images of the cuboid diamond samples. These diamonds normally display a diffraction contrast due to dislocations [40–42]. The topographical structures (Figure 6b,c,g) are like those found in [31,32,38]. The crystal structures found in the XRT study can be characterized as "fibrous" or "columnar" [31]. This is due to the subdivision of the crystal (monocrystalline) volume into slightly mutually misorientated subindividuals (columns), the axial directions of which were <111>. The column's diameter ranged from about 10 μm downwards to the resolution limits of the technique (approx. 1 μm). The X-ray topographs show that the development of the volume of the growing crystal is associated with repeated branching in the equivalent <111> directions. Therefore, this data shows that cuboid diamonds of variety II have been crystallized by fibrous growth in the <111> direction with branching and equal velocities in the equivalent directions. In some X-ray topographs, clear sectorial and zoning structures of the crystals were observed (Figure 6a,c–e). Zoning structures visible on X-ray topographs are due to deformation of the crystal lattice (probably caused by lattice impurities) between growth zones, which often have different colouring in transmitted light (see Figure 3).

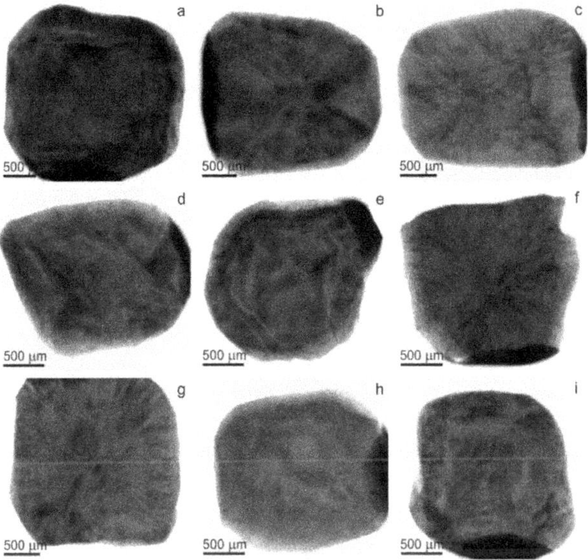

Figure 6. X-ray projection topographs of yellow cuboid diamonds from the Northeastern Siberian platform (the Burgers vector is directed horizontally to the left, (220) reflection) (**a**—MP-30; **b**—MP-43; **c**—MP-56; **d**—MP-60; **e**—MP-71; **f**—MP-83; **g**—MP-87; **h**—MP-98; **i**—MP-108).

2.2.3. Electron Backscatter Diffraction (EBSD)

The electron backscatter diffraction (EBSD) technique allows accurate identification of the structural characters of crystals, manifested as lattice orientation relationships between the columns or as a deformation lamination. The EBSD images (Figure 7) show that the diamond samples consist of domains, which are misoriented to each other. Blue to red coloration on the images correspond to misorientations of up to 2°. The higher misorientations have been observed in the outer parts of the various growth sectors of the cuboids. These patterns support the suggestion that fibrous growth of sample cuboid diamonds are the repeated branching of fibres in equivalent <111> directions. This branching causes the stronger strains of the outer crystal zones. Some diamonds exhibit strain patterns expressed as a lamination in two crossed directions (Figure 7d,e,h). These patterns, as in birefringence, could be attributed to post-growth plastic deformations.

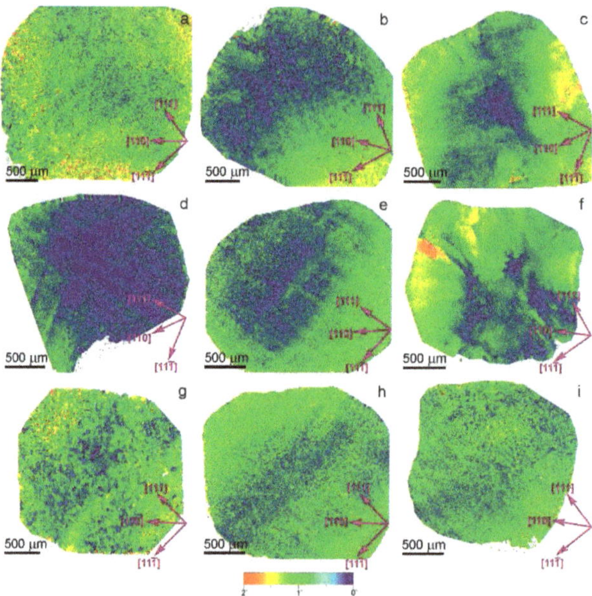

Figure 7. Misorientation patterns of yellow cuboid diamonds from the Northeastern Siberian platform. EBSD maps over a large parts of the polished plates showing a change in orientation up to 2° from blue to red (**a**—MP-30; **b**—MP-43; **c**—MP-56; **d**—MP-60; **e**—MP-71; **f**—MP-83; **g**—MP-87; **h**—MP-98; **i**—MP-108).

3. Discussion

Previous studies [30,39] of yellow cuboid diamonds from alluvial placers of the Northeastern Siberian platform revealed the overall presence of C centres (single substitution-based nitrogen defects). The presence of these centres is usually attributed either to relatively cool conditions of the storage of diamonds in the mantle or short mantle residence time prior to eruption. These cuboid diamonds show a wide range of carbon isotope compositions, from mantle-like values towards lighter values [30]. Mineral inclusions found in these diamonds testify to their formation in eclogitic environments, which is believed to correspond to deeply-subducted protoliths of the former oceanic crust.

The morphology of diamonds may be caused by the different physical and chemical conditions of growth and post-growth alteration. The morphological features of diamonds often reflect only the last stages of their evolution, connected with resorption and/or regeneration. Observed rounded morphology of most yellow cuboid diamonds indicates their post-growth resorption, rather than represents primary growth morphologies. The resorption of cuboid diamond typically result in transformation into rounded dodecahedral morphology through gradual inflation of crystal surfaces. These morphological features are generally similar to those observed in partially resorbed diamond crystals of cubic habit in experiments [33,43].

It was shown that cuboid diamonds have fibrous internal structure [31,32,38,44]. The X-ray topography studies demonstrate that fibres represent less than 20 μm sub-individuals oriented along the equivalent <111> directions direction. Growth of diamonds with such internal structures are not accompanied by a thickening of these sub-individuals. Splitting the fibres in the equivalent equivalent <111> directions directions led to total space filling of the cuboid sectors [31]. The growth mechanism is reflected in the X-ray topography and EBSD images.

According to [44], cuboid diamonds with specific fibrous internal structures could be crystallized under high supersaturation. Fibrous internal structures of diamonds likely resulted from fast abnormal growth at high carbon supersaturation operating as a driving force of this process. This concept has additionally been proposed by theoretical models [45]. The fibrous internal structure identified in the studied cuboid diamonds can also be described by the position of the abnormal growth.

Cuboid diamond crystals may have suffered post-growth annealing and plastic deformation. The plastic deformation of the diamond samples are reflected in internal structure features. These structures appear as a "tatami" pattern (two crossed directions of strain lamination) in EBSD images and birefringence patterns. Moreover, the deformation of a crystal lattice in the sample diamonds also took place during the fibrous or columnar crystal growth. The rounded shape, negative etch pits, and other dissolution features of studied samples are suggested to be the result of resorption. The resorption has resulted in a gradual transformation of crystals with nearly cubic shape into rounded tetrahexahedrons (tetrahexahedroid) [33].

4. Methods

The cuboid diamonds were first examined with optical microscopy using a Zeiss Stemi SV 6 stereo microscope (Carl Zeiss Microscopy GmbH, Göttingen, Germany). The morphology of the crystals has been studied using a scanning electron microscope (JEOL JSM-6510LV (20 kV), Japanese Electron Optics Laboratory Ltd., Tokyo, Japan) installed in the Analytical Centre for Multi-Element and Isotope Research SB RAS, Novosibirsk, Russia.

The internal structure of crystals has been visualized through the method of X-ray projection topography (XRT). The X-ray topograms were obtained over the whole samples synchronously translated with the detector under the beam [42,46]. The images were collected from superposition of the section images in the Bragg reflection geometry. The spatial resolution of the detector with a factor of 20 has been increased using a mirror magnification system. A double-crystal Si monochromator was tuned to the (333) reflection to avoid the influence of the high-energy harmonics.

The misorientations of the crystal structure in cuboid diamonds was determined from the EBSD mapping of polished plates [47]. The EBSD data were collected on a Hitachi S-3400 N scanning electron microscope equipped with an Oxford Instruments HKL detector with an accuracy of misorientations of 0.5–1.0° (Oxford Instruments plc, Abingdon, UK). The Kikuchi pattern of each individual point were automatically indexed by the Oxford data collection software. Patterns were acquired on rectangular grids by moving the electron beam at a regular step size of 1 μm.

The XRT and EBSD study was performed using the infrastructure of the Shared-Use Centre at the Siberian Synchrotron and Terahertz Radiation Centre (SSTRC) based on the VEPP-3/VEPP-4M/NovoFEL at the Budker Institute of Nuclear Physics SB RAS, Novosibirsk, Russia.

5. Conclusions

The internal structure of yellow cuboid diamonds found in alluvial placers of the Northeastern Siberian platform have been investigated via optical microscopy, XRT, and EBSD techniques. We have demonstrated that these cuboid diamonds have a specific fibrous or columnar internal structure. Most are strongly deformed and have a brindled internal structure. Misorientations of the crystal lattice, found in the samples, may be caused by strains from their fibrous growth or/and post-growth plastic deformation. Most of the crystals have typical resorption features and do not preserve primary growth morphology. The resorption leads to an evolution from an originally cubic shape to a rounded tetrahexahedroid.

Acknowledgments: The research was supported by state assignment project (project no. 0330-2016-0007).

Author Contributions: Alexey Ragozin studied samples by scanning electron microscopy and analysed the data; Dmitry Zedgenizov studied samples by optical microscopy and analysed the data; Viktoria Kalinina made the polished plates from diamonds and discussed the results; Konstantin Kuper performed the electron backscatter diffraction and X-ray topography experiments; Alexey Zemnukhov analysed the data and discussed the results; and Alexey Ragozin and Dmitry Zedgenizov contributed equally by writing the manuscript.

Conflicts of Interest: The authors declare no conflict of interest.

References

1. Harrison, E.R.; Tolansky, S. Growth history of a natural octahedral diamond. *Proc. R. Soc. Lond. A* **1964**, *279*, 490–496. [CrossRef]
2. Suzuki, S.; Lang, A.R. Occurrences of facetted re-entrants on rounded growth surfaces of natural diamonds. *J. Cryst. Growth* **1976**, *34*, 29–37. [CrossRef]
3. Lang, A.R. Glimpses into the growth history of natural diamonds. *J. Cryst. Growth* **1974**, *24*, 108–115. [CrossRef]
4. Varshavsky, A.V. *Anomalous Birefringence and Internal Morphology of Diamond*; Nauka: Moscow, Russia, 1968; p. 92. (In Russian)
5. Orlov, Y.L. *The Mineralogy of Diamond*; John Wiley: New York, NY, USA, 1977; p. 233.
6. Gurney, J.J.; Helmstaedt, H.H.; Richardson, S.H.; Shirey, S.B. Diamonds through time. *Econ. Geol.* **2010**, *105*, 689–712. [CrossRef]
7. Harris, J.W. Diamond geology. In *The Properties of Natural and Synthetic Diamond*; Field, J.E., Ed.; Academic Press: London, UK, 1992; pp. 345–393.
8. Welbourn, C.M.; Rooney, M.-L.T.; Evans, D.J.F. A study of diamonds of cube and cube-related shape from the jwaneng mine. *J. Cryst. Growth* **1989**, *94*, 229–252. [CrossRef]
9. Howell, D.; Griffin, W.L.; Piazolo, S.; Say, J.M.; Stern, R.A.; Stachel, T.; Nasdala, L.; Rabeau, J.R.; Pearson, N.J.; O'Reilly, S.Y. A spectroscopic and carbon-isotope study of mixed-habit diamonds: Impurity characteristics and growth environment. *Am. Miner.* **2013**, *98*, 66–77. [CrossRef]
10. Sobolev, N.V. *Deep Seated Inclusions in Kimberlites and the Problem of the Composition of the Upper Mantle*; AGU: Washington, DC, USA, 1977; p. 279.
11. Meyer, H.O.A. Genesis of diamond—A mantle saga. *Am. Miner.* **1985**, *70*, 344–355.
12. Meyer, H.O.A. Inclusions in diamond. In *Mantle Xenoliths*; Nixon, P.H., Ed.; Wiley: New York, NY, USA, 1987; pp. 501–522.
13. Stachel, T.; Harris, J.W. The origin of cratonic diamonds—Constraints from mineral inclusions. *Ore Geol. Rev.* **2008**, *34*, 5–32. [CrossRef]
14. Parfenov, L.M.; Kuzmin, M.I. *Tectonics, Geodynamics and Metallogeny of the Territory of the Republic of Sakha (Yakutia)*; Nauka/Interperiodika: Moscow, Russia, 2001; p. 571. (In Russian)
15. Grakhanov, S.A.; Shatalov, V.I.; Shtyrov, V.A.; Kychkin, V.R.; Suleimanov, A.M. *Diamond Placers of Russia*; Akademicheskoe Izd."Geo": Novosibirsk, Russia, 2007; p. 454. (In Russian)
16. Zinchuk, N.N.; Koptil, V.I. *Typomorphism of Diamonds in the Siberian Craton*; Nedra: Moscow, Russia, 2003; p. 603. (In Russian)
17. Zinchuk, N.N.; Koptil, V.I.; Boris, E.I.; Lipashova, A.N. Typomorphic features of placer diamonds from the Siberian craton: A principle guide in diamond exploration. *Rudy Met.* **1999**, *3*, 18–30.
18. Afanas'ev, V.P.; Zinchuk, N.N.; Koptil, V.I. Polygenesis of diamonds in connection with the problem of northeastern primary sources of Siberian platform. *Dokl. Akad. Nauk* **1998**, *361*, 366–369.
19. Afanas'ev, V.P.; Lobanov, S.S.; Pokhilenko, N.P.; Koptil', V.I.; Mityukhin, S.I.; Gerasimchuk, A.V.; Pomazanskii, B.S.; Gorev, N.I. Polygenesis of diamonds in the Siberian platform. *Russ. Geol. Geophys.* **2011**, *52*, 259–274. [CrossRef]
20. Gurney, J.J.; Hildebrand, P.R.; Carlson, J.A.; Fedortchouk, Y.; Dyck, D.R. The morphological characteristics of diamonds from the Ekati property, Northwest Territories, Canada. *Lithos* **2004**, *77*, 21–38. [CrossRef]
21. Posukhova, T.; Kolume, F. Diamonds from placers in Western and Central Africa: A problem of primary sources. *Mosc. Univ. Geol. Bull.* **2009**, *64*, 177–186. [CrossRef]

22. Galimov, E. The relation between formation conditions and variations in isotope composition of diamonds. *Geochem. Int.* **1985**, *22*, 118–141.
23. Shatsky, V.S.; Zedgenizov, D.A.; Ragozin, A.L.; Kalinina, V.V. Diamondiferous subcontinental lithospheric mantle of the northeastern siberian craton: Evidence from mineral inclusions in alluvial diamonds. *Gondwana Res.* **2015**, *28*, 106–120. [CrossRef]
24. Shatsky, V.S.; Zedgenizov, D.A.; Ragozin, A.L.; Kalinina, V.V. Carbon isotopes and nitrogen contents in placer diamonds from the ne siberian craton: Implications for diamond origins. *Eur. J. Miner.* **2014**, *26*, 41–52. [CrossRef]
25. Zedgenizov, D.A.; Ragozin, A.L.; Shatsky, V.S.; Araujo, D.; Griffin, W.L. Fibrous diamonds from the placers of the northeastern Siberian platform: Carbonate and silicate crystallization media. *Russ. Geol. Geophys.* **2011**, *52*, 1298–1309. [CrossRef]
26. Zedgenizov, D.; Rubatto, D.; Shatsky, V.; Ragozin, A.; Kalinina, V. Eclogitic diamonds from variable crustal protoliths in the northeastern siberian craton: Trace elements and coupled $\delta^{13}C$–$\delta^{18}O$ signatures in diamonds and garnet inclusions. *Chem. Geol.* **2016**, *422*, 46–59. [CrossRef]
27. Ragozin, A.L.; Zedgenizov, D.A.; Kuper, K.E.; Shatsky, V.S. Radial mosaic internal structure of rounded diamond crystals from alluvial placers of Siberian platform. *Miner. Petrol.* **2016**, *110*, 861–875. [CrossRef]
28. Ragozin, A.L.; Shatskii, V.S.; Zedgenizov, D.A. New data on the growth environment of diamonds of the variety V from placers of the northeastern Siberian platform. *Dokl. Earth Sci.* **2009**, *425*, 436. [CrossRef]
29. Ragozin, A.L.; Shatsky, V.S.; Rylov, G.M.; Goryainov, S.V. Coesite inclusions in rounded diamonds from placers of the northeastern Siberian platform. *Dokl. Earth Sci.* **2002**, *384*, 385–389.
30. Zedgenizov, D.A.; Kalinina, V.V.; Reutsky, V.N.; Yuryeva, O.P.; Rakhmanova, M.I. Regular cuboid diamonds from placers on the northeastern Siberian platform. *Lithos* **2016**, *265*, 125–137. [CrossRef]
31. Lang, A.R. Space-filling by branching columnar single-crystal growth: An example from crystallisation of diamond. *J. Cryst. Growth* **1974**, *23*, 151–153. [CrossRef]
32. Orlov, Y.L.; Bulienkov, N.A.; Martovitsky, V.P. A study of the internal structure of variety III diamonds by x-ray section topography. *Phys. Chem. Miner.* **1982**, *8*, 105–111. [CrossRef]
33. Khokhryakov, A.F.; Pal'yanov, Y.N. The evolution of diamond morphology in the process of dissolution: Experimental data. *Am. Miner.* **2007**, *92*, 909–917. [CrossRef]
34. Howell, D. Strain-induced birefringence in natural diamond: A review. *Eur. J. Miner.* **2012**, *24*, 575–585. [CrossRef]
35. Howell, D.; Piazolo, S.; Dobson, D.; Wood, I.; Jones, A.; Walte, N.; Frost, D.; Fisher, D.; Griffin, W. Quantitative characterization of plastic deformation of single diamond crystals: A high pressure high temperature (HPHT) experimental deformation study combined with electron backscatter diffraction (EBSD). *Diam. Relat. Mater.* **2012**, *30*, 20–30. [CrossRef]
36. Lang, A.R. Causes of birefringence in diamond. *Nature* **1967**, *213*, 248–251. [CrossRef]
37. Tolansky, S. Birefringence of diamond. *Nature* **1966**, *211*, 158–160. [CrossRef]
38. Moore, M.; Lang, A.R. On the internal structure of natural diamonds of cubic habit. *Philos. Mag.* **1972**, *26*, 1313–1325. [CrossRef]
39. Titkov, S.V.; Shiryaev, A.A.; Zudina, N.N.; Zudin, N.G.; Solodova, Y.P. Defects in cubic diamonds from the placers in the northeastern Siberian platform: Results of ir microspectrometry. *Russ. Geol. Geophys.* **2015**, *56*, 354–362. [CrossRef]
40. Authier, A. Contrast of images in X-Ray topography. In *Diffraction and Imaging Techniques in Materials Science*; Amelinckx, S., Gevers, R., Van Landuyt, J., Eds.; North-Holland: Amsterdam, The Netherlands, 1978; Volume 2, pp. 715–757.
41. Lang, A.R. Techiques and interpretation in X-ray topography. In *Diffraction and Imaging Techniques in Materials Science*; Amelinckx, S., Gevers, R., Van Landuyt, J., Eds.; North-Holland: Amsterdam, The Netherlands, 1978; Volume 2, pp. 623–714.
42. Lang, A.R. The projection topograph: A new method in X-ray diffraction microradiography. *Acta Crystallogr.* **1959**, *12*, 249–250. [CrossRef]
43. Khokhryakov, A.F.; Pal'yanov, Y.N. Evolution of diamond morphology in the processes of mantle dissolution. *Lithos* **2004**, *73*, S57.
44. Sunagawa, I. Growth and morphology of diamond crystals under stable and metastable contitions. *J. Cryst. Growth* **1990**, *99*, 1156–1161. [CrossRef]

45. Chernov, A.A. Stability of faceted shapes. *J. Cryst. Growth* **1974**, *24–25*, 11–31. [CrossRef]

46. Kuper, K.E.; Zedgenizov, D.A.; Ragozin, A.L.; Shatsky, V.S. X-ray topography of natural diamonds on the VEPP-3 SR beam. *Nucl. Insrum. Meth. A* **2009**, *603*, 170–173. [CrossRef]

47. Humphreys, F.J. Review grain and subgrain characterisation by electron backscatter diffraction. *J. Mater. Sci.* **2001**, *36*, 3833–3854. [CrossRef]

 *crystals*

Article

Experimental and Theoretical Evidence for Surface-Induced Carbon and Nitrogen Fractionation during Diamond Crystallization at High Temperatures and High Pressures

Vadim N. Reutsky [1,*], Piotr M. Kowalski [2], Yury N. Palyanov [1,3], EIMF [4] and Michael Wiedenbeck [5]

[1] Sobolev Institute of Geology and Mineralogy SB RAS, Novosibirsk 630090, Russia; palyanov@igm.nsc.ru
[2] Institute of Energy and Climate Research (IEK-6), Forschungszentrum Juelich, 52425 Juelich, Germany; p.kowalski@fz-juelich.de
[3] Department of Geology and Geophysics, Novosibirsk State University, Novosibirsk 630090, Russia
[4] Edinburgh Ion Microprobe Facility, Grant Institute of Earth Sciences, School of GeoSciences, University of Edinburgh, Edinburgh EH9 3JW, UK; jcraven@staffmail.ed.ac.uk
[5] Deutsches GeoForschungZentrum, 14473 Potsdam, Germany; michael.wiedenbeck@gfz-potsdam.de
* Correspondence: reutsky@igm.nsc.ru; Tel.: +7-383-3306531

Academic Editor: Helmut Cölfen
Received: 31 March 2017; Accepted: 22 June 2017; Published: 26 June 2017

Abstract: Isotopic and trace element variations within single diamond crystals are widely known from both natural stones and synthetic crystals. A number of processes can produce variations in carbon isotope composition and nitrogen abundance in the course of diamond crystallization. Here, we present evidence of carbon and nitrogen fractionation related to the growing surfaces of a diamond. We document that difference in the carbon isotope composition between cubic and octahedral growth sectors is solvent-dependent and varies from 0.7‰ in a carbonate system to 0.4‰ in a metal-carbon system. Ab initio calculations suggest up to 4‰ instantaneous ^{13}C depletion of cubic faces in comparison to octahedral faces when grown simultaneously. Cubic growth sectors always have lower nitrogen abundance in comparison to octahedral sectors within synthetic diamond crystals in both carbonate and metal-carbon systems. The stability of any particular growth faces of a diamond crystal depends upon the degree of carbon association in the solution. Octahedron is the dominant form in a high-associated solution while the cube is the dominant form in a low-associated solution. Fine-scale data from natural crystals potentially can provide information on the form of carbon, which was present in the growth media.

Keywords: mixed-habit diamond crystallization; carbon isotopes; nitrogen impurity; fractionation; experiment; high pressure; high temperature; crystal chemistry; surface structure; SIMS

---

1. Introduction

Compositional inhomogeneity of both natural and synthetic diamond crystals is a well-established phenomenon. Two models for explaining a heterogeneous distribution of carbon isotopes and key impurities within diamond crystals have been debated: changing of the carbon source and fractionation of a fluid over the course of diamond crystallization. Both of these models require marked changes in fluid composition as the driving factor for generating shifts in the carbon isotope ratio and/or the abundances of impurities, the most important of which being nitrogen. A comparison of synthetic and natural crystals reveals some key features that can be obvious for synthetic crystals but which are extremely rare in natural diamonds. First, a strong sectorial distribution pattern for nitrogen both in terms of concentration and isotopic composition has been observed in high temperature and high pressure (HTHP) synthetic diamonds [1–5]. It is also known that the growth of diamonds along the [100] direction during chemical vapour deposition (CVD) synthesis leads to lowering the impurities level in the crystal in comparison with material grown along [111] direction [6]. Less dramatic but consistent sectorial carbon isotope distribution between octahedra and cube growth sectors has also been documented in HTHP synthetic diamonds [2,5], whereas most attempts to observe such a compositional pattern in natural diamond crystals have been without success [7–9]. Sectorial variations in nitrogen abundance and carbon isotope ratios in natural mixed-habit diamond crystals were reported by Zedgenizov and Harte [10] for a diamond crystal from Udachnaya pipe and by Howell et al. [11,12] for three crystals from unknown locations. These data suggest that some compositional heterogeneity can occur in diamonds with no changing of fluid composition and at roughly equal kinetic conditions because simultaneously grown faces of the same crystal certainly feed from the same fluid. This particular case is the focus of the present research.

Until now, most investigated synthetic crystals come from metal-carbon systems that are highly efficient at diamond production [1,13]. Taking into account the general opinion of crystallization of natural diamonds from silicate/carbonate fluids [14], the difference in composition of the fluid/melt from which diamond grows in experiments and in the nature has brought into question the relevance of such experimental results for geochemistry. However, methods for producing synthetic diamonds from non-metallic systems have now become advanced enough to make such material available for investigation. Here, we present the results of a compositional investigation of diamond material obtained from carbonate system in high-pressure and high-temperature experiments. Together with more detailed data for crystals from metal-carbon systems, our results point to a measurable fractionation of carbon isotopes and nitrogen impurity on the surface of diamond itself, regardless of the bulk composition of the system. This conclusion is supported by ab initio calculations of carbon isotope fractionation on different crystallographic faces of diamond.

Further in the text, we will use the term "octahedral growth" to indicate diamond material grown by faces of octahedra and term "cubic growth" for sectors of growth of cube faces.

2. Results

2.1. Cathodolumenesence

As described in the "Samples and Methods" section, samples were cut parallel to the [110] direction to reveal internal structure, including growth zones and sectors, and, especially, to reveal the seed-to-rim relationship. The contrast in cathodolumenesence (CL) response between natural seeds and synthetic rims is clearly visible in Figure 1a,b. Sample 268-1 (octahedral to dodecahedral seed) grew by octahedral faces only (Figure 1a,c). The rim has a dark and relatively uniform CL image in comparison to the seed crystal. In the case of sample 268-2 (dodecahedral seed), octahedral sectors appear simultaneously with cubic sectors (Figure 1b,d). Growth sectors of faces {100} show brighter CL images in comparison with both the natural seed and growth sectors of faces {111} (Figure 1a,b). All the seed-rim boundaries are sharp and linear, which means that no appreciable resorption of the seed crystals took place prior to overgrowth crystallization. As expected, seeds retained their shape and size through the experiments.

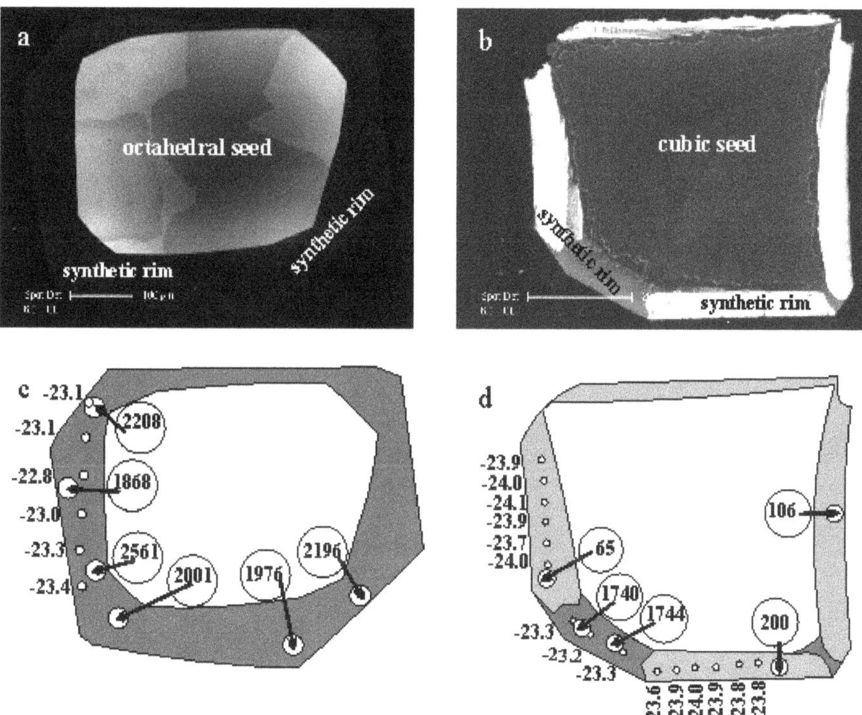

Figure 1. Cathodoluminescence images of investigated samples 268-1 (**a**) and 268-2 (**b**) and schematic pictures of synthetic rims (**c**,**d**). The cores represent seed crystals: 268-1—octahedral to dodecahedral seed crystal, 268-2—cubic seed crystal. Rims grown from a single carbonate melt. On schemas, light-grey parts related to cubic growth and dark grey parts related to octahedral growth. Small circles indicate positions of carbon isotope analysis. Bigger circles correspond to positions of nitrogen analysis.

2.2. Carbon Isotope Ratios by Secondary Ion Mass Spectrometry (SIMS)

We have collected a comprehensive carbon isotope data from all the observed segments of the crystals, including seeds, sectors of octahedral and of the cube (Table S1). The seed crystals have average carbon isotope compositions $-5.4‰$ for sample 268-1 and $-2.3‰$ for sample 268-2, which are normal for natural diamonds. Despite quite complex internal structures within the seed crystals that are revealed with CL, it was no reason for detailed investigation of these parts of the samples. Limited data were collected in order to evaluate the contrast of natural and synthetic material. Carbon isotope composition of synthetic diamond rims is distinctly different from that of the seeds. The rim in sample 268-1 grew by octahedral growth only and this material shows $\delta^{13}C$ values from -22.5 to -23.4 (mean $-23.03‰$, sd = 0.29, n = 15). The synthetic rim in sample 268-2 consists of sectors of both octahedral and cubic growth. The octahedral growth rim material shows $\delta^{13}C$ values from -23.19 to -23.33 (mean $-23.27‰$, sd = 0.07, n = 3), whereas the $\delta^{13}C$ values obtained for cubic growth of sample 268-2 vary from -23.6 to -24.1 (mean $-23.88‰$, sd = 0.14, n = 14). The carbon isotope composition of octahedral growth diamond in both samples is equal. The cubic growth diamond shows statistically more negative $\delta^{13}C$ value in comparison with octahedral growth (Table 1) and the difference of $0.8‰$ exceeds the analytical uncertainties ($0.24‰$).

Table 1. Summarized SIMS data of δ^{13}C and nitrogen abundances measured within synthetic rims of diamonds grown from carbonate fluid. Statistics account for all the data available for particular crystallography in both 268-1 and 268-1 samples.

Sample	Octahedral Growth		Cubic Growth	
	δ^{13}C (‰)	N (at. ppm)	δ^{13}C (‰)	N (at. ppm)
mean	−23.07	1536	−23.88	70
st.dev.	0.28	375	0.14	41
n	18	27	14	18

2.3. Ab Initio Calculations of Carbon Isotope Fractionation

Considering the role of growing surfaces for carbon isotope fractionation during diamond crystallization, we perform an ab initio calculation of carbon isotope equilibrium. Uncertainty concerning the carbon position in a solution makes it impossible to calculate directly the fractionation between a solution and a crystal. Instead, we calculated carbon isotope fractionation between atomic positions within the crystal lattice and selected faces of a diamond. This approach makes possible a comparison of particular faces of a diamond crystal in terms of carbon isotope affinity regardless of solution composition. Results from these computations are shown in Figure 2. Remarkably, the surface of a cube is the most favorable for ^{12}C capture. In the first layer, the instantaneous difference in δ^{13}C between cube and octahedra is expected to be 3.75‰ at 1400 K, and this decreases to 1.85‰ at 2000 K. This difference disappears completely once one models seven or more layers below the crystal surface. Octahedron and rhombododecahedron surfaces display complicated relationships with a changeover in the affinity of ^{12}C isotope arising at the fifth layer inside the crystal lattice (Figure 2). However, a clear qualitative difference of certain isotope affinity between faces of cubes and of octahedra can be traced at least down to the fifth layer.

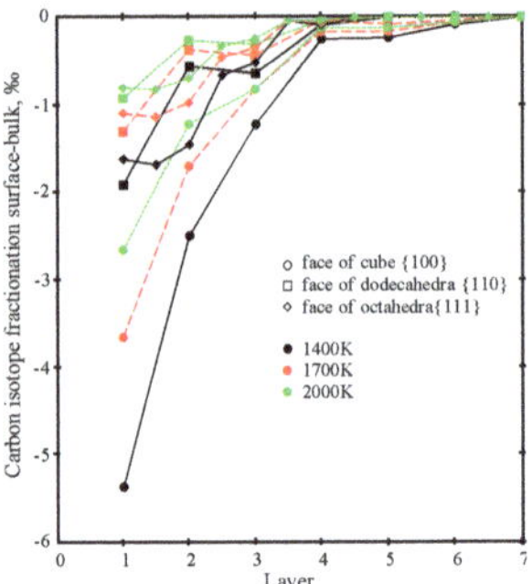

Figure 2. Results of ab initio calculation of isotope equilibrium of carbon positions on particular crystallographic face of diamond crystal in comparison with deeply sited positions in diamond lattice (bulk) at different temperatures.

2.4. Nitrogen Abundances

All the determined nitrogen results are given in Table S1. Nitrogen abundance in the seed crystals varies from 824 to 1516 ppm in sample 268-1 and from 1163 to 1370 ppm in sample 268-2. The nitrogen content in the synthetic layer of sample 268-1 ranges from 1116 to 2561 ppm (mean 1575 ppm, sd = 395, *n* = 21). Growth sectors of octahedra in sample 268-2 are rather small, but the nitrogen content in these sectors range from 1112 to 1744 ppm (mean 1401 ppm, sd = 281, *n* = 6), which is very similar to those obtained from the synthetic layer of sample 268-1. SIMS analysis points within cube growth sector in sample 268-2 all yielded low nitrogen concentrations ranging from 16 to 200 ppm (mean 70 ppm, sd = 41, n = 18). The difference in nitrogen abundance can clearly be seen between octahedral and cubic growth sectors (Table 1). Similar to carbon, nitrogen characteristics of octahedral grown diamond in both samples are equal, but cubic grown diamond is significantly depleted in nitrogen overall.

3. Discussion

Nitrogen distribution shows a strong sector dependence in synthetic diamond crystals [1–5,15]. Sectors of octahedra were always enriched in nitrogen in comparison with sectors of other forms, including cube, trapezohedra and rhombododecahedra. As mentioned, previously published works considered diamond crystals grown in metal-carbon systems only. In the present study, we report first data collected from diamonds grown in a carbonate system.

A marked difference of nitrogen content reaching over a factor of 10 is clearly seen between cubic and octahedral growth diamond from carbonate system. The value for this difference cannot be precisely established due to the limited dataset for cubic growth, but the overall pattern is the same for diamond crystals from metal-carbon and carbonate systems (Table 2). This difference is also confirmed by the data from different SIMS laboratories, which fit each other well in both carbon isotopes and nitrogen abundances.

Table 2. Summary of available data on compositional difference between sectors of growth of octahedra and other crystallographic faces within synthetic diamond single crystals. Methods of data collecting are emphasized: IR—infra-red spectroscopy; MS—convenient bulk combustion with mass-spectrometry; SIMS—secondary ion mass-spectrometry.

System	$\delta^{13}C_{111}$-$\delta^{13}C_{100}$	N_{111}/N_{100}	Source
Carbonate	0.8	~20	Present work (SIMS)
Metal-carbon	0.4	2.58–4.0	[5] (SIMS)
	about 0	2.87	[2] (MS)
	-	2.0	[3] (IR)
Natural reduced conditions	0	5.3–8.8	[12] (SIMS)

Like nitrogen, the carbon isotope composition also shows a measurable difference between cube and octahedra growth sectors. The use of a single carbon source under identical growth conditions for both investigated samples makes it reasonable to combine the data from octahedral growth in sample 268-2 and from octahedral growth in sample 268-1 to construct a single dataset. This approach makes for a more statistically robust comparison between octahedral and cubic growth. Our dataset shows that cubic growth diamonds are depleted in ^{13}C isotope in comparison to octahedral growth diamonds (Table 2, Figure 1). This difference is rather small but exceeds the analytical uncertainty of $\delta^{13}C$ measurements, so it should be consider significant. A similar relationship also can be recognized for crystals grown in metal-carbon system [5]. In order to illustrate this, a detailed dataset for synthetic diamond crystals synthesized in a Fe-Ni-C system and where both octahedra and cube sectors grew simultaneously is shown in Table 3. The cube sectors show systematic depletion by about 0.4‰ in heavy carbon isotope in comparison with an adjacent octahedral sector. The size of ^{13}C depletion of cubic sector for the carbonate system is about 0.7‰. The difference between the metal and carbonate systems is negligible and could be solvent dependent or growth rate dependent. What is important is that the direction and magnitude of the effect are the same in both metal-dominated and carbonate-dominated environments (Table 2).

The obvious similarities between the isotopic patterns of synthetic diamonds from both experimental systems provide evidence of no significant importance of solvent composition for distribution of nitrogen and carbon isotopes between octahedral and cubic growth diamond. By concluding that these effects are structurally defined, a generalized model describing the atomic structures of these particular faces on diamond crystals can be developed.

Table 3. SIMS data of δ^{13}C values measured in adjusted sectors of faces of octahedra and cube in synthetic diamond crystals from Fe-Ni-C system. See [5] for details.

Sample	Octahedra		Cube		Δ, ‰
	Distance *, µm	δ^{13}C, ‰	Distance *, µm	δ^{13}C, ‰	
	1849	−27.1	1825	−27.4	0.4
	2371	−27.3	2380	−27.4	0.1
150/3/5	2907	−26.9	2934	−27.3	0.4
	3422	−26.9	3405	−27.2	0.4
	3839	−26.7	3828	−27.4	0.7
	285	−25.6	286	−26.0	0.4
	524	−25.9	528	−26.4	0.5
	607	−26.1	595	−26.4	0.3
140/4/6	695	−25.9	705	−26.6	0.7
	2120	−26.5	2203	−26.8	0.3
	2492	−26.4	2446	−26.7	0.3
	2865	−26.2	2828	−26.7	0.5

* The distance is given as a projection of actual position of a SIMS spot to a line starting from the seed position and going perpendicular to octahedral growth zones. The diameter of SIMS spots is 20 µm.

3.1. Crystal Structure Characteristics and General Regularities of Carbon Isotopes and Impurities Incorporation into a Crystal

A crystal surface can grow either via the incorporation of single atoms or by attaching coordinated groups of atoms. In the first case, the structure and perfectness of the growing surface determine the fractionation of elements and isotopes. Overall, the fractionation is more pronounced upon lowering a linear growth rate of the surface [16]. In the second case, where carbon atoms have formed a group in a melt/fluid, only minor fractionation would be expected either between a solution and the groups of atoms or between the groups of atoms and the crystal. This process would provide a means for incorporating impurities into the growing crystal while avoiding any significant carbon isotope fractionation. Faces {111}, {110} and {100} differ significantly in their arrangement of atoms and any fractionation of atoms or groups of atoms on attaching to these surfaces should not be equal. Therefore, we consider each of these surfaces separately.

The face of the cube {100} has a single-layer atomic plane (Figure 3), where the surface consists of regularly distributed equal units. Atoms on a cubic face have no direct bonds with neighbours within the same plane but joined with two atoms of the previous plane and two atoms of the next plane. Thus, every unit cell on the growing face of the cube has two open bonds. Every subsequently produced atomic plane of {100} will be identical to the previous one.

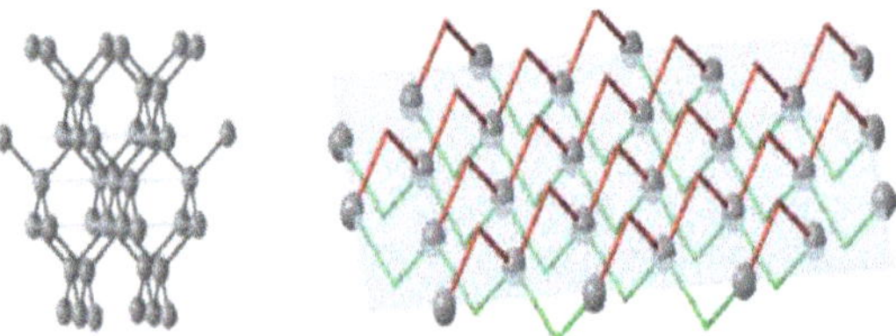

Figure 3. Atomic plane of cubic face.

The face of rhombododecahedra {110} has a single-layer atomic plane that consists of zigzag-like chains of atoms (Figure 4). These chains are not bonded to each other, but every atom in the plane has one link to a previous atomic plane and one link to the next atomic plane. Thus, a growing rhombododecahedra face will consist of single-bond units where every newly formed atomic plane will be the same as the previous one.

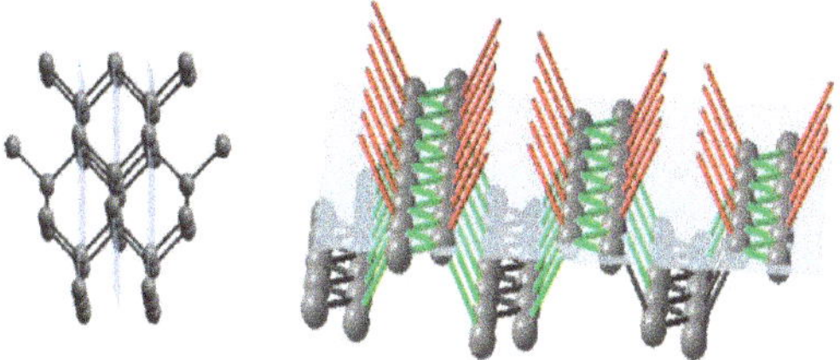

Figure 4. Atomic plane of rhombododecahedral face.

The face of octahedra {111} has a double-layer atomic plane (Figure 5). Every single atom in this plane makes three joins with neighbour atoms of the same plane and one bond connected to the next atomic layer. It means that the growing surface of octahedra consists of a layer of triple-bond units followed by a layer of single-bond units.

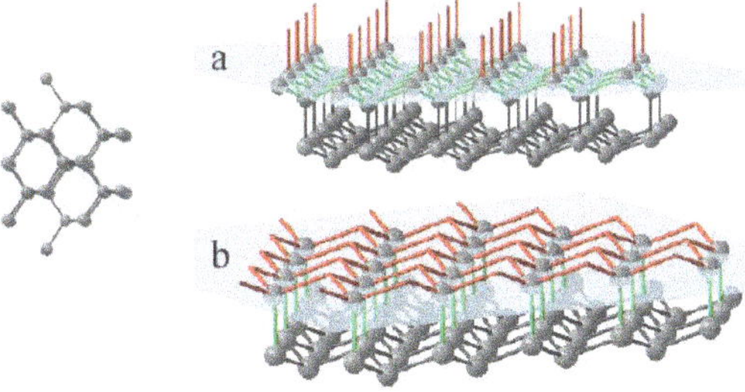

Figure 5. Atomic plane of octahedral face. (**a**)—layer of single-bond units; (**b**)—layer of triple-bond units.

In general, any crystallization from a solution will result from a supersaturation of a solvent by a component with respect to a certain crystal structure. Coordinated groups of atoms of a component can become stable in a solution under certain conditions. Such coordinated groups will have characteristics favouring their incorporation in a particular crystal structure: Composition of the solvent and solvent impurities will also influence the stability of particular coordinated groups of atoms. For example, an increasing abundance of nitrogen impurity in a diamond producing metal carbon melt will favour to crystallization of graphite instead of diamond [17].

A detailed investigation of crystallization of semiconductors with diamond-like structures reveals the importance of atomic clusters in solution [18]. The stability of specific coordinated groups of atoms is required to produce a perfect crystal. A few types of coordinated groups have been recognized in the case of diamond: one-dimension chains; two-dimension nets and three-dimension skeletons. These groups are rather stable in the solution, but can modify or rearrange until they are ultimately incorporated into the crystal lattice. Obviously, atoms from the solvent and other impurities could also take part in these groups. However, the presence of impurity atoms within such clusters leads to changes in both morphology and electron structure of the group, making it more difficult for such impurity-bearing groups to be incorporated onto crystal surface. Therefore, the availability of a perfect and regular structure on a growth surface will suppress the incorporation of groups of atoms holding impurities.

According to [18], when a matrix solution contains no associated carbon atoms, octahedral faces of the diamond will grow the fastest. This, in turn, leads to the slower growing cubic faces to dominate in such a low-associated solution. If pre-existing atomic clusters of carbon are common in a solution, then the octahedra form becomes dominant due to its slow growth rate in comparison with other directions. On the other hand, the highly ordered cube faces provide favorable conditions for the fractionation of atoms and groups of atoms [19]. In contrast, alternation of triple-bond and single-bond units on the growing surface of octahedra could make fractionation of atoms and groups of atoms less likely. In support, it has been shown that octahedra faces have less perfect structures as compared to cube faces in CVD diamonds [6]. Considering data for HPHT diamonds, growth sectors of cube, trapezohedra and rhombododecahedra are always depleted in impurities, including nitrogen, metals and other elements in comparison with growth sectors of octahedra [1–5].

3.2. Ab Initio Calculations of Carbon Isotope Fractionation

The results from our ab initio calculations provide strong support for ^{12}C isotope affinity to cubic growth diamond relative to octahedral crystal faces. The difference of 3.75‰ is expected within the first crystal layer at 1400 K (Figure 2). Instantaneous depletion of cubic face in heavy carbon isotope at about 1100 °C must be close to 4‰ in comparison to simultaneously growing octahedral faces. Again, it should be emphasized that the direction of fractionation is more important than the actual degree of isotope fractionation. It is well known from calculations [20–22] and from experiments [23] that diamond is enriched in heavy carbon isotope in a metal-carbon system but depleted in carbonate system in comparison with parent solution/fluid. However, the surface-induced relative fractionation is the same for both systems.

Theoretically, carbon isotope self-diffusion in the crystal lattice could erase this pattern since no energy difference between atom positions exists below the 6th atomic layer (Figure 2). However, self-diffusion of carbon isotopes in diamond is extremely slow [24], and, once generated, this pattern should remain for a long time even at mantle conditions. Such patterns have been reported in some natural diamonds [10,11]. Galimov [25] shows ~3‰ depletion in ^{13}C of cuboid diamonds in comparison with octahedral crystals from Yakutian kimberlites, which might, at least in part, have resulted from the surface-induced crystallochemical fractionation discussed here. However, the major challenge in documenting considered feature within the natural samples is the relatively high carbon isotope heterogeneity typically found in natural diamond crystals. Furthermore, the very limited dataset reported at the nanogram sampling size with high spatial resolution make it difficult to make firm conclusions about such behavior in natural environments.

3.3. Fractionation of Nitrogen

As shown above, within a single crystal of synthetic diamond sectors of cube are significantly depleted in nitrogen in comparison with growth sectors of octahedra. Detailed SIMS traverses across growth sectors in synthetic diamonds from a metal-carbon system reveal the same distribution of nitrogen between {111} and {100} sectors [5]. Furthermore, sectors of octahedra show nitrogen abundances, which are close to that of the starting graphite [5,17]. Significant nitrogen depletion in cube sectors in comparison to nitrogen content in initial graphite documents the incompatible behavior of nitrogen in the 100 sector of diamond. These observations fit well with the incorporation of coordinated groups of carbon atoms containing nitrogen onto octahedral face with no efficient fractionation of nitrogen impurity, whereas the opposite is true for cube sectors. As mentioned above, the cube faces are dominant in a low-associated solution. Atom by atom incorporation of both carbon and nitrogen into well-ordered regular surfaces of cube provides the strongest discrimination against nitrogen incorporation. This pattern holds true for all synthetic samples investigated until now, including our current data for diamonds from the carbonate system. There are still no direct data reported on the actual nitrogen isotope fractionation between particular faces of diamond and the crystallization medium. The most important result from the above data is that surface-induced nitrogen fractionation is very much the same in metal-carbon and carbonate systems and the compositional difference between sectors of growth of different crystal faces has the same direction in both substrates.

There is a growing body of data documenting coupled variation of nitrogen concentration and carbon isotope fractionation during diamond growth [26–30]. It has been suggested that nitrogen behavior is influenced by both diamond growth rate [31] and fO_2 [22]. According to "a limit sector" model [26], nitrogen must be generally incompatible in diamond at kinetic fractionation, since nitrogen content in diamonds suggested going lower with lowering diamond growth rate. Stachel et al. [27] argued that, in case of equilibrium fractionation, nitrogen can be moderately compatible (K_N = 2) in diamond under reduced conditions and strongly compatible (K_N = 4), when diamond formed from an oxidized fluid. Mentioned conceptions of nitrogen partitioning at diamond crystallization based on model thermodynamic calculations and on empirical modeling of general distribution of nitrogen concentrations in mantle diamonds vs. their carbon isotopic composition. There is no doubt that this approach can be valid for modeling the distribution of nitrogen between bulk fluid and bulk diamond. Our data cannot provide any evidence of fractionation between bulk fluid and bulk diamond since there is no data about nitrogen concentrations in the fluid is available in our experiments. However, it is clear from our experiments that the nitrogen is strongly incompatible to diamond when atomic-flat faces of cube are growing. Most cuboid sectors reported in natural samples from kimberlites, lamproites and associated placers are, in fact, atomic-rough. Therefore, it is rather difficult to expect an exact fit of compositional difference resulted from the surface-induced fractionation in natural and synthetic crystals.

The first reported natural diamonds with flat cube faces found in Tibetan ophiolites [32] reveal the same nitrogen distribution pattern between {111} and {100} growth sectors as has been seen in synthetic crystals. Three natural crystals gave $N_{(111)}/N_{(100)}$ ratios from 5.3 to 8.8 [32]. The paragenesis of the studied ophiolite, including native elements, alloys, carbides and nitrides, pointed strongly towards highly reducing conditions.

A number of specific conditions must be met in order for such a fractionation mechanism to be active for both carbon isotopes and nitrogen concentration during diamond growth. First, crystallization should take place from a carbon rich solution. Secondly, growth must occur predominantly on atomic-flat surfaces of different crystallographic faces. Diamond crystallization is generally associated with low degree of melting of a silicate-dominated matrix within the Earth's mantle. However, this melting must be advanced enough to provide crystallization of euhedral macrocrystals, and it would be reasonable to expect a significant concentration of carbon clusters in coordinated groups in these melts. As mentioned, high association of carbon in a solution leads to the suppression of growth on all crystal faces other than octahedra. Such a situation would explain the overwhelming dominance of octahedra during the crystallization of natural diamonds.

Another important parameter that affects fractionation is growth rate: diamond growth rate has a significant influence on the carbon isotope fractionation between diamond and the carbon solution in metal melt [16]. Presumably, the ability of surface-induced fractionation to be recorded within a crystal depends on the ratio of isotope diffusion rate over crystal growth rate. Because the growth rate is considered as one of the major parameters determining diamond crystal habit [33], this parameter could be of interest for further investigations.

Some natural environments should be favorable for diamond crystallization associated with a relatively low degree of association of carbon atoms in coordinated groups. One such example would be a "magma ocean" setting, such as that which dominated during Earth's early history [34], where the temperature regime and compositional gradients were able to provide conditions for atom-by-atom crystallization. A second environment favoring a low degree of association would be a metal rich environment of lower mantle, where diamond could precipitate from carbon dissolved in metal melts. As can be seen from experiments, metal-carbon systems are most conducive for a stable growth of cube faces and surfaces other than octahedra. A third such environment would be melts from subducted carbonates containing minor quantities of silicates and relatively high amount of biogenic carbon, such as those generating carbonaceous diapirs [35] or during interaction of carbonates with reduced mantle material [36]. Such setting would have the potential to provide conditions that would form diamond crystals with the surface-induced compositional inhomogeneity, such as we observed in our experiments including the carbonate system.

4. Samples and Methods

Two samples of synthetic diamond were prepared in the Na_2CO_3-CO_2-C system. The composition of the system was chosen to provide diamond crystallization near the Carbon-Carbon oxide (CCO) buffer from a single carbon source rather than modelling some specific geological situation. However, sodium carbonates were found in fresh kimberlites [37,38] and even as inclusions in diamonds [38,39] and in olivine recovered from kimberlites [40].

Natural diamond crystals with cubic and octahedral to dodecahedral habits and near 0.5 mm in size were used as seed crystals in our experiments. Both morphological types of seeds were placed together at the bottom of a platinum ampoule and the ampoule was filled with 99.99% pure sodium oxalate ($Na_2C_2O_4$). Finally, the ampoule was hermetically welded using a carbon-free technique. High-pressure experiments used a "split-sphere" type (BARS) multianvil apparatus at 1400 °C and 6 GPa in the V.S. Sobolev Institute of Geology and Mineralogy SB RAS [41]. Sodium oxalate decomposes at experimental conditions producing sodium carbonate fluid (CO_2-containing Na_2CO_3 melt) with about 4.5 wt % excess of carbon. This excess of carbon in the carbonate fluid prevents diamond seeds from dissolution, a phenomenon often seen in metal-carbon systems and leads to the seeds being overgrown by synthetic diamond material with maximum thickness of 120 μm Diamond crystals recovered from the ampoules were inspected for growth morphology as described in [42], then cut in a parallel [110] direction, polished and examined using a Philips XL30CP scanning electron microscope (Philips, Germany) with attached CL detector at the Edinburgh University, thereby revealing internal structure within the samples.

Secondary Ion Mass-Spectrometry data of carbon and nitrogen isotopes and nitrogen abundance were collected using the CAMECA 1270 ion microprobe (AMETEK Inc., France) at the NERC/Edinburgh University Ion Microprobe Facility (EIMF), in 2005–2006 and the CAMECA 1280 HR at the GFZ Potsdam SIMS laboratory in 2015–2016.

The analytical conditions for the EIMF measurements where reported earlier [5]. In all cases, the unknowns and reference material were placed adjacent to one another in the same indium mount. The surface of the 1-inch sample mounts were then cleaned using high-purity ethanol prior to being argon sputter coated with high-purity gold. Magnet calibration and centring of the field aperture and slits was done prior to each point analysis.

At the GFZ Potsdam SIMS facility, carbon and nitrogen data were collected in June 2015 and March 2016, respectively. As for earlier EIMF work, for the GFZ measurements we employed $^{133}Cs+$ primary beam with a total impact energy of 20 keV. Our analyses used a 250 pA probe current for nitrogen and 2.5 nA current for carbon. The beam was focused to a ~4–8 μm diameter with a Gaussian distribution at the sample surface. Secondary ions were extracted using a -10 kV potential applied to the sample holder. Charge compensation involved circa 250 pA low energy electron cloud provide by a normal incidence electron flood gun. The instrument was operated with a circa 80 μm field-of-view, with a 50 eV wide energy window. In the case of the carbon isotope measurements, we used a static multi-collection mode operating at a mass resolution of M/dM $\approx$ 3200, which fully resolves the ^{13}C mass station from the nearby $^{12}C^1H$ molecular isobar. For determining nitrogen contents and nitrogen isotope ratios, our mass spectrometer was operated in mono-collection mode at a mass resolution of M/dM $\approx$ 6900 using the peak-stepping sequence $^{12}C_2$, $^{12}C^{14}N$ and $^{12}C^{15}N$ signals.

Each carbon isotope analysis was preceded by a 60 s presputter using a 25 μm raster, so as to remove the 35 nm tick gold coat and also to implant Cs+ in order to establish equilibrium sputtering conditions. An analysis used a 15 μm raster, together with the tool's dynamic transfer capability, thereby generating a square, flat-bottom sputter crater. Prior to initiating data collection, we conducted automatic beam centering routines in both x and y for the field aperture. A single carbon analysis involved 20 cycles of 4 s integration each and the data were filtered at the 3 sd level. Thus, a single analysis required around 3 min of analysis time, including the presputtering. All data were collected in fully automated data acquisition mode.

Prior to initiating nitrogen data collection on a selected target, the region was sputter cleaned for 80 s using a 3 nA Cs+ primary beam rastered over a 20×20 μm area. Prior to initiating data collection, the primary beam current was reduced to the 250 pA probe current and the rastered area was reduced to 10×10 μm, which was compensated for using the dynamic transfer function of tool. Automatic centering routines were then conducted for x and y for the field aperture. Our data acquisition involved the peak-stepping sequence: $^{12}C_2$ (2 s integration time per cycle), $^{12}C^{14}N$ (4 s) and $^{12}C^{15}N$ (15 s). A single nitrogen analysis consisted of 50 cycles of this peak-stepping sequence, leading to a total analysis time of ~20 min per acquisition.

For the EIMF measurements, the "synAT" standard (a synthetic "A" diamond crystal, partial slice "T") was used [43]. The characteristics of this reference material are: 230.4 ppm N (wt) and $\delta^{13}C_{PDB}$ of $-23.92‰$ as determined by combustion analysis using a gas source mass spectrometer (analysed by S.R. Boyd, at the Laboratoire de Geochimie des Isotopes Stables, Université de Paris VII). The GFZ machine calibration used two diamond reference materials contained in the same indium-based sample mount. The samples were cut out from "steady state" parts of up-octahedral sectors in synthetic crystals 140/4 and 150/3 [5]. The synthetic crystals had been characterized for their $\delta^{13}C_{VPDB}$ values and nitrogen contents refer to "SynAT" reference material during the EIMF sessions in 2005 and 2006. Minor pieces of the specific chips used in this study were additionally checked for their $\delta^{13}C$, $\delta^{15}N$ and nitrogen content using static vacuum gas source mass-spectrometry by Dr. A. Verchovsky at Open University (Milton Keynes, UK). The established characteristics are as follows: Crystal 140/4 with $\delta^{13}C_{VPDB} = -26.5‰$, 460 μg/g nitrogen and $\delta^{15}N_{AIR} = 0‰$; Crystal 150/3 with $\delta^{13}C_{VPDB} = -24.5‰$, 215 μg/g nitrogen and $\delta^{15}N_{AIR} = -5‰$.

Our carbon isotope raw data were tested for the presence of a time dependent linear drift, which was corrected for when needed. Such drifts, which we attribute to changes in sensitivity of the amplifiers in our detector system, were 0.01‰ per hour or less. After correcting for such drift, the repeatability of our $^{13}C/^{12}C$ determinations on the two reference materials was less than 0.24‰ (1 sd), which is our best estimate for the random component that might be present in our carbon isotope data. Hence, we can conclude that the overall reliability of our $\delta^{13}C$ values is better than $\pm0.24‰$ (1 sd) for GFZ measurements. The uncertainty of individual N abundance data points reported for the studied samples are better than $\pm10\%$. These uncertainties represent one standard error of the mean based upon multiple analyses of restricted areas within the reference crystals.

Ab initio calculations of C isotope fractionation factors were performed following the procedure of Kowalski et al. [44,45] for calculation of isotope fractionation factors at high temperatures. This method utilizes the Bigeleisen and Mayer [46] approximation, in which only the force constants acting on the fractionating atom of interest are computed and used in derivation of relevant fractionation factors. This approach has been successfully applied in calculation of Li and B isotope fractionation between complex minerals and fluids [44,45]. For the quantum mechanical calculations of the force constants, we used plane-wave DFT code CPMD [47] with BLYP generalized gradient approximation [48,49] and energy cutoff of 140 Ryd. The norm conserving Goedecker pseudopotentials were applied to describe the core electrons [50]. Beta factors, describing isotope fractionation between species and an ideal monoatomic gas, were computed for three diamond surfaces (100), (110) and (111) with the bulk lattice parameter of face-centered cubic Bravais lattice of 3.56669 Å [51]. The surfaces were modelled with thick 2×2 multi-layer slabs separated by a thick layer of vacuum in the z-direction (at least 12.9 Å thick) and the atoms of the bottom layer were fixed to the bulk positions. The modelled slabs contained 17, 13 and 10 layers for (100), (110) and (111) surfaces, respectively.

5. Conclusions

Our results univocally evidence that compositional heterogeneity can occur in diamonds because of crystallochemically controlled fractionation. Difference in atom's arrangements on atomic-flat faces of octahedra, cube and other faces provide measurable contrast in fractionation of carbon isotopes and nitrogen impurity between crystal and growth media on the certain crystal faces. This effect is equally occurs at both reduced and oxidized conditions and the cubic grown diamond have up to 4‰ lighter carbon isotope composition in comparison with the octahedral grown diamond. At the same time, the cubic grown diamond contain 2 to 20 times less nitrogen content pointing generally incompatible behavior of the nitrogen in diamond. Due to limited diffusion in diamond lattice, the difference in composition of sectors of cube and octahedra will survive through the geological timescale even at mantle temperatures and can be seen in some natural diamond crystals.

Presumably high degree of carbon association in natural mantle fluids makes atomic-flat faces of octahedra dominant at diamond crystallization. As a result, natural diamonds are growing mainly by atomic-flat faces of octahedra or atomic-rough faces of cube. Such conditions reducing the significance of surface-induced carbon and nitrogen fractionation for natural diamonds. However, identification of the compositional pattern examined here within natural diamonds can provide a key for evaluation of degree of carbon association in the mantle fluid.

Supplementary Materials: Supplementary materials can be found at www.mdpi.com/2073-4352/7/7/190/s1. Table S1 contains all the SIMS data of $\delta^{13}C$ and nitrogen abundances measured within natural seeds and synthetic rims of particular samples. Statistical accounts of all the data are available for particular crystallography. Errors (1σ) are 0.2‰ for $\delta^{13}C$ values and 10% for nitrogen abundances. Regular numbers were measured in the Edinburgh microprobe facility. Bold numbers were measured in the GFZ Potsdam SIMS laboratory.

Acknowledgments: Members of the EIMF (Edinburgh ion Microprobe Facility) particularly supporting this work were: Ben Harte (provided the opportunity for initial steps of the research), John Craven (tuning and maintenance of the Cameca IMS 1270 instrument), Nicola Cayzer (collection of CL images by SEM). The help of Sasha Verchovsky (Open University) in obtaining and characterizing standards is much appreciated. The authors are grateful for research support from the United Kingdom NERC (grant NER/A/S/2003/00368) in part for allowing use of the EIMF facility in 2005–2006. We thank Frederic Couffignal for his excellent support during SIMS data acquisition at GFZ Potsdam. The latter part of study was supported by a grant from the Russian Scientific Foundation (project No. 14-27-00054).

Author Contributions: Vadim N. Reutsky conceived the idea of the research, managed opportunities, summarized and interpreted the data, and prepared the manuscript, Piotr M. Kowalski performed ab initio calculations, Yury N. Palyanov designed and performed the experiments, and EIMF and Michael Wiedenbeck contributed SIMS instruments and infrastructure for high spatial resolution isotope analysis.

Conflicts of Interest: The authors declare no conflict of interest. The founding sponsors had no role in the design of the study; in the collection, analyses, or interpretation of data; in the writing of the manuscript, and in the decision to publish the results.

References

1. Strong, H.M.; Chrenko, R.M. Properties of Laboratory-Made Diamond. *J. Phys. Chem.* **1971**, *76*, 1838–1843. [CrossRef]

2. Boyd, S.R.; Pillinger, C.T.; Milledge, H.J.; Mendelssohn, M.J.; Seal, M. Fractionation of nitrogen isotopes in a synthetic diamond of mixed crystal habit. *Nature* **1988**, *133*, 604–607. [CrossRef]

3. Burns, R.C.; Cvetkovic, V.; Dodge, C.N.; Evans, D.J.F.; Rooney, M.T.; Spear, P.M.; Welbourn, C.M. Growth-sector dependence of optical features in large synthetic diamonds. *J. Cryst. Growth* **1990**, *104*, 257–279. [CrossRef]

4. Watt, G.A.; Newton, M.E.; Baker, J.M. EPR and optical imaging of the growth-sector dependence of radiation-damage defect production in synthetic diamond. *Diam. Relat. Mater.* **2001**, *10*, 1681–1683. [CrossRef]

5. Reutsky, V.N.; Harte, B.; EIMF; Borzdov, Y.M.; Palyanov, Y.N. Monitoring diamond crystal growth, a combined experimental and SIMS study. *Eur. J. Mineral.* **2008**, *20*, 365–374. [CrossRef]

6. Schwarz, S.; Rottmair, C.; Hirmke, J.; Rosiwal, S.; Singer, R.F. CVD-diamond single-crystal growth. *J. Cryst. Growth* **2004**, *271*, 425–434. [CrossRef]

7. Bulanova, G.P.; Pearson, D.G.; Hauri, E.H.; Griffin, B.J. Carbon and nitrogen isotope systematics within a sector-growth diamond from the Mir kimberlite, Yakutia. *Chem. Geol.* **2002**, *188*, 105–123. [CrossRef]

8. Cartigny, P.; Harris, J.W.; Taylor, A.; Davies, R.; Javoy, M. On the possibility of a kinetic fractionation of nitrogen stable isotopes during natural diamond growth. *Geochim. Cosmochim. Acta* **2003**, *67*, 1571–1576. [CrossRef]

9.   Reutsky, V.N.; Zedgenizov, D.A. Some specific features of genesis of microdiamonds of octahedral and cubic habit from kimberlites of the Udachnaya pipe (Yakutia) inferred from carbon isotopes and main impurity defects. *Russ. Geol. Geophys.* **2007**, *48*, 299–304. [CrossRef]

10.   Zedgenizov, D.A.; Harte, B. Microscale variations of δ^{13}C and N content in diamonds with mixed-habit growth. *Chem. Geol.* **2004**, *205*, 169–175. [CrossRef]

11.   Howell, D.; Griffin, W.L.; Piazolo, S.; Say, J.M.; Stern, R.A.; Stachel, T.; Nasdala, L.; Rabeau, J.R.; Pearson, N.J.; O'Reilly, S.Y. A spectroscopic and carbon-isotope study of mixed-habit diamonds: Impurity characteristics and growth environment. *Am. Miner.* **2013**, *98*, 66–77. [CrossRef]

12.   Howell, D.; Stern, R.A.; Griffin, W.L.; Southworth, R.; Mikhail, S.; Stachel, T. Nitrogen isotope systematics and origins of mixed-habit diamonds. *Geochim. Cosmochim. Acta* **2015**. [CrossRef]

13.   Pal'yanov, Y.N.; Khokhryakov, A.F.; Borzdov, Y.M.; Sokol, A.G.; Gusev, V.A.; Rylov, G.M.; Sobolev, N.V. Growth conditions and the real structure of synthetic diamond crystals. *Geol. Geofiz.* **1997**, *38*, 882–906.

14.   Jablon, B.M.; Navon, O. Most diamonds were created equal. *Earth Planet. Sci. Lett.* **2016**, *443*, 41–47. [CrossRef]

15.   Babich, Y.V.; Feigelson, B.N. Spatial distribution of the nitrogen defects in synthetic diamond monocrystals: Data of IR mapping. *Geochem. Int.* **2009**, *47*, 94–98. [CrossRef]

16.   Reutsky, V.N.; Borzdov, Y.M.; Palyanov, Y.N. Effect of diamond growth rate on carbon isotope fractionation in Fe–Ni–C system. *Diam. Relat. Mater.* **2012**, *21*, 7–10. [CrossRef]

17.   Palyanov, Y.N.; Borzdov, Y.M.; Khokhryakov, A.F.; Kupriyanov, I.N.; Sokol, A.G. Effect of nitrogen impurity on diamond crystal growth processes. *Cryst. Growth Des.* **2010**, *10*, 3169–3175. [CrossRef]

18.   Stroitelev, S.A. *Crystal Chemical Aspect of Semiconductor Technology*; Nauka: Novosibirsk, Russia, 1976; 192p. (In Russian)

19.   Chernov, A.A. Stability of faceted shapes. *J. Cryst. Growth* **1971**, *24–25*, 11–31. [CrossRef]

20.   Bottinga, Y. Carbon isotope fractionation between graphite, diamond and carbon dioxide. *Earth Planet. Sci. Lett.* **1969**, *5*, 301–307. [CrossRef]

21.   Richet, P.; Bottinga, Y.; Javoy, M. A review of hydrogen, carbon, nitrogen, oxygen, sulphur, and chlorine stable isotope fractionation among gaseous molecules. *Annu. Rev. Earth Planet. Sci.* **1977**, *5*, 65–110. [CrossRef]

22.   Deines, P. The carbon isotopic composition of diamonds: Relationship to diamond shape, color, occurrence and vapor composition. *Geochim. Cosmochim. Acta* **1980**, *44*, 943–961. [CrossRef]

23.   Reutsky, V.N.; Palyanov, Y.N.; Borzdov, Y.M.; Sokol, A.G. Isotope fractionation of carbon during diamond crystallization in model systems. *Russ. Geol. Geophys.* **2015**, *56*, 239–244. [CrossRef]

24.   Koga, K.T.; Van Orman, J.A.; Walter, M.J. Diffusive relaxation of carbon and nitrogen isotope heterogeneity in diamond: New thermochronometer. *Phys. Earth Planet. Inter.* **2003**, *139*, 35–43. [CrossRef]

25.   Galimov, E.M. Isotope fractionation related to kimberlite magmatism and diamond formation. *Geochim. Cosmochim. Acta* **1991**, *55*, 1697–1708. [CrossRef]

26.   Cartigny, P.; Harris, J.W.; Javoy, M. Diamond genesis, mantle fractionations and mantle nitrogen content: A study of δ^{13}C–N concentrations in diamonds. *Earth Planet. Sci. Lett.* **2001**, *185*, 85–98. [CrossRef]

27.   Stachel, T.; Harris, J.W.; Muehlenbachs, K. Sources of carbon in inclusion bearing diamonds. *Lithos* **2009**, *112*, 625–637. [CrossRef]

28.   Smart, K.A.; Chacko, T.; Stachel, T.; Muehlenbachs, K.; Stern, R.A.; Heaman, L.M. Diamond growth from oxidized carbon sources beneath the Northern Slave Craton, Canada: A δ^{13}C–N study of eclogite-hosted diamonds from the Jericho kimberlite. *Geochim. Cosmochim. Acta* **2011**, *75*, 6027–6047. [CrossRef]

29.   Palot, M.; Pearson, D.G.; Stern, R.A.; Stachel, T.; Harris, J.W. Isotopic constraints on the nature and circulation of deep mantle C–H–O–N fluids: Carbon and nitrogen systematics within ultra-deep diamonds from Kankan (Guinea). *Geochim. Cosmochim. Acta* **2014**, *139*, 26–46. [CrossRef]

30.   Petts, D.C.; Chacko, T.; Stachel, T.; Stern, R.A.; Heaman, L.M. A nitrogen isotope fractionation factor between diamond and its parental fluid derived from detailed SIMS analysis of a gem diamond and theoretical calculations. *Chem. Geol.* **2015**, *410*, 188–200. [CrossRef]

31.   Boyd, S.R.; Pineau, F.; Javoy, M. Modelling the growth of natural diamonds. *Chem. Geol.* **1994**, *116*, 29–42. [CrossRef]

32.   Howell, D.; Griffin, W.L.; Yang, J.; Gain, S.; Stern, R.A.; Huang, J.-X.; Jacob, D.E.; Xu, X.; Stokes, A.J.; O'Reilly, S.Y.; et al. Diamonds in ophiolites: Contamination or a new diamond growth environment? *Earth Planet. Sci. Lett.* **2015**, *430*, 284–295. [CrossRef]

33. Sunagawa, I. Growth and morphology of diamond crystals under stable and metastable conditions. *J. Cryst. Growth* **1990**, *99*, 1156–1161. [CrossRef]
34. Dasgupta, R.; Chi, H.; Shimizu, N.; Buono, A.S.; Walker, D. Carbon solution and partitioning between metallic and silicate melts in a shallow magma ocean: Implications for the origin and distribution of terrestrial carbon. *Geochim. Cosmochim. Acta* **2013**, *102*, 191–212. [CrossRef]
35. Litasov, K.; Ohtani, E. The solidus of carbonated eclogite in the system $CaO-Al_2O_3-MgO-SiO_2-Na_2O-CO_2$ to 32 GPa and carbonatite liquid in the deep mantle. *Earth Planet. Sci. Lett.* **2010**, *295*, 115–126. [CrossRef]
36. Palyanov, Y.N.; Bataleva, Y.V.; Sokol, A.G.; Borzdov, Y.M.; Kupriyanov, I.N.; Reutsky, V.N.; Sobolev, N.V. Mantle–slab interaction and redox mechanism of diamond formation. *Proc. Natl. Acad. Sci. USA* **2013**, *110*, 20408–20413. [CrossRef] [PubMed]
37. Kamenetsky, V.S.; Maas, R.; Kamenetsky, M.B.; Paton, C.; Phillips, D.; Golovin, A.V.; Gornova, M.A. Chlorine from the mantle: Magmatic halides in the Udachnaya-East kimberlite, Siberia. *Earth Planet. Sci. Lett.* **2009**, *285*, 96–104. [CrossRef]
38. Kaminsky, F. Mineralogy of the lower mantle: A review of 'super-deep' mineral inclusions in diamond. *Earth Sci. Rev.* **2012**, *110*, 127–147. [CrossRef]
39. Zedgenizov, D.A.; Ragozin, A.L.; Shatsky, V.S.; Araujo, D.; Griffin, W.L. Fibrous diamonds from the placers of the northeastern Siberian Platform: Carbonate and silicate crystallization media. *Russ. Geol. Geophys.* **2011**, *52*, 1298–1309. [CrossRef]
40. Sharygin, I.S.; Golovin, A.V.; Korsakov, A.V.; Pokhilenko, N.P. Eitelite in sheared peridotite xenoliths from Udachnaya-East kimberlite pipe (Russia)—A new locality and host rock type. *Eur. J. Mineral.* **2013**, *25*, 825–834. [CrossRef]
41. Sokol, A.G.; Borzdov, Y.M.; Palyanov, Y.N.; Khokhryakov, A.F.; Sobolev, N.V. An experimental demonstration of diamond formation in the dolomite-carbon and dolomite-fluid-carbon systems. *Eur. J. Mineral.* **2001**, *13*, 893–900. [CrossRef]
42. Khokhryakov, A.F.; Palyanov, Y.N.; Kupriyanov, I.N.; Nechaev, D.V. Diamond crystallization in a CO_2-rich alkaline carbonate melt with a nitrogen additive. *J. Cryst. Growth* **2016**, *449*, 119–128. [CrossRef]
43. Harte, B.; Fitzsimons, I.C.W.; Harris, J.W.; Otter, M.L. Carbon isotope ratios and nitrogen abundances in relation to cathodoluminescence characteristics for some diamonds from the Kaapvaal Province, S. Africa. *Mineral. Mag.* **1999**, *63*, 829–856. [CrossRef]
44. Kowalski, P.M.; Jahn, S. Prediction of equilibrium Li isotope fractionation between minerals and aqueous solutions at high P and T: An efficient ab initio approach. *Geochim. Cosmochim. Acta* **2011**, *75*, 6112–6123. [CrossRef]
45. Kowalski, P.M.; Wunder, B.; Jahn, S. Ab initio prediction of equilibrium boron isotope fractionation between minerals and aqueous fluids at high P and T. *Geochim. Cosmochim. Acta* **2013**, *101*, 285–301. [CrossRef]
46. Bigeleisen, J.; Mayer, M.G. Calculation of equilibrium constants for isotopic exchange reactions. *J. Chem. Phys.* **1947**, *15*, 261–267. [CrossRef]
47. Marx, D.; Hutter, J. Ab initio molecular dynamics: Theory and implementation. In *Modern Methods and Algorithms of Quantum Chemistry*; Grotendorst, J., Ed.; NIC: FZ Juelich, Germany, 2000; pp. 301–449.
48. Becke, A.D. Density-functional exchange-energy approximation with correct asymptotic behavior. *Phys. Rev.* **1988**, *38*, 3098–3100. [CrossRef]
49. Liu, Y.; Tossell, A. Ab initio molecular orbital calculations for boron isotope fractionations on boric acids and borates. *Geochim. Cosmochim. Acta* **2005**, *69*, 3995–4006. [CrossRef]
50. Goedecker, S.; Teter, M.; Hutter, J. Separable dual-space Gaussian pseudopotentials. *Phys. Rev. B* **1996**, *54*, 1703–1710. [CrossRef]
51. Straumanis, M.E.; Aka, E.Z. Precision determination of lattice parameter, coefficient of thermal expansion and atomic weight of carbon in diamond. *J. Am. Chem. Soc.* **1951**, *73*, 5643–5646. [CrossRef]

Article

Nanoscale Sensing Using Point Defects in Single-Crystal Diamond: Recent Progress on Nitrogen Vacancy Center-Based Sensors

Ettore Bernardi, Richard Nelz, Selda Sonusen and Elke Neu *

Faculty for Natural Sciences and Technology, Physics Department, Saarland University, 66123 Saarbrücken, Germany; ettore.bernardi@physik.uni-saarland.de (E.B.); richard.nelz@uni-saarland.de (R.N.); selda.sonusen@physik.uni-saarland.de (S.S.)

* Correspondence: elkeneu@physik.uni-saarland.de; Tel.: +49-681-302-2739

Academic Editor: Yuri Palyanov
Received: 8 April 2017; Accepted: 24 April 2017; Published: 28 April 2017

Abstract: Individual, luminescent point defects in solids, so-called color centers, are atomic-sized quantum systems enabling sensing and imaging with nanoscale spatial resolution. In this overview, we introduce nanoscale sensing based on individual nitrogen vacancy (NV) centers in diamond. We discuss two central challenges of the field: first, the creation of highly-coherent, shallow NV centers less than 10 nm below the surface of a single-crystal diamond; second, the fabrication of tip-like photonic nanostructures that enable efficient fluorescence collection and can be used for scanning probe imaging based on color centers with nanoscale resolution.

Keywords: diamond; color center; magnetic sensing; scanning probes; nanostructures

---

1. Introduction

Nanotechnology has led to many significant technological and scientific advances in recent years. For instance, two-dimensional or nanoscale materials such as carbon nanotubes or graphene are investigated for next generation electronics and photonics [1]. Simultaneously, functionalized nanoparticles, e.g., for drug delivery, are promising to enhance various therapies [2]. Moreover, electronic systems like transistors are being miniaturized and controlled down to the single electron level [3].

Simultaneously to developing nanotechnology, a need for sensing techniques that work on the nanoscale has been arising to investigate nanoscale materials and to foster their further development. Quantities of interest are magnetic fields, often created as a result of electrical currents [4], electric fields [5], temperatures [6], pressure or crystal strain [7], as well as the presence of individual fluorescent markers, e.g., molecules [8]. For sensing with nanoscale spatial resolution, in general, the sensor needs to fulfill several demanding prerequisites:

- The sensor's size or active area has to be small compared to the structure under investigation. If this is not the case, spatial averaging over the detector area may mask information from the sample's nanostructure. Consequently, sensors approaching atomic dimensions (<1 nm) are desirable for nanoscale sensing.
- The sensor's geometry must allow for close proximity in-between the investigated object and the sensor. In most cases, controlled proximity to the sample is ensured by manufacturing the sensor in a tip-like geometry and approaching it to the sample via a scanning probe mechanism. This mechanism often keeps the force between the sample and the tip constant (for pioneering work see, e.g., [9–11]). Alternatively, the sensor can consist of a nanoparticle that is, e.g., inserted into a cell for sensing [6].

- The sensor needs to provide sufficient sensitivity to capture the weak signals that arise from nanoscopic or atomic objects. To illustrate this demanding point, the magnetic field of a single electron spin even at a distance of 50 nm amounts to only $\approx$9 nT [12]. The field of magnetic dipoles decays with the distance r from the dipole like r^{-3} [13]. For the near-field energy transfer between two point-like dipoles, which is a valuable imaging resource, as well, even a r^{-6} decay has to be considered [14]. Thus, bringing the sensor and the sample in close proximity is not only mandatory for high resolution, but also enables detecting weak signals from nanoscale objects.

Using individual, optically-active point defects in solids as sensors allows simultaneously fulfilling the prerequisites listed above. Moreover, such defects are versatile sensors for several quantities, including magnetic and electric fields and temperature.

Point defects alter the host crystal's periodic lattice only in one or a few neighboring lattice sites. Mostly, impurity atoms enter the crystal lattice and can form complexes with vacancies. Electrons, or more precisely the electronic wave-functions, localize at the defect within a few lattice constants and thus on atomic scales; typically within less than 1 nm [15]. This manifests also in the existence of such defects $\approx$1 nm below crystal surfaces [16]. At such distances, crystal surfaces strongly influence the defects and may cause instability [17]. Leading contenders for sensing are optically-active point defects in the wide-bandgap semiconductors diamond [13,18,19] and silicon carbide [20].

This review focuses on sensors based on diamond and especially one of the most prominent point defects in diamond, the nitrogen vacancy (NV) color center. This defect consists of a nitrogen atom replacing a carbon atom and a neighboring lattice vacancy [21]. Figure 1 summarizes its basic properties. NV centers, in general, provide long-term photo-stable fluorescence. Their bright emission with high luminescence efficiency [22] lies in the red and near-infrared spectral range and spans about 100 nm. NV color centers form emitting electric dipoles [23]. Their emission is bright enough to allow for a straightforward detection of individual, isolated centers in a confocal fluorescence microscope. Thus, NV centers have been investigated as solid-state sources of single photons [24]. Single photons in turn are a valuable resource for nanoscale sensing, e.g., in scanning near-field optical microscopy as a nanoscopic light source [25]. Alternatively, the dipole of the NV center can interact with other dipoles and transfer energy via optical near fields (Förster resonance energy transfer (FRET) [8,26]). Via this process, NV centers can reveal the presence of other dipoles.

In addition to their optical properties, NV centers provide highly-coherent, optically-readable electronic spins (see Figure 1; first observation in [32]). The NV centers' long spin coherence times (T_2^* or T_2) show that coherent superposition states of the 0 and ± 1 ground state spin levels retain their phase for a long time even at room temperature. Here, NV centers profit from the fact that the diamond lattice naturally features a low magnetic noise. Thus, it protects these superposition states from decoherence: the most abundant carbon isotope ^{12}C does not have a nuclear spin. Additionally, the concentration of paramagnetic ^{13}C isotopes (1.1%) can be reduced by isotopically pure diamond synthesis [33]. In such an isotopically-engineered diamond, T_2 can be as high as 1.8 ms. The NV center's T_2 is a valuable sensing resource: its decrease, for example, directly reveals the presence of magnetic molecules on the diamond surface [34]. The use of spin coherence as a sensing resource opens up novel sensing schemes with potentially enhanced sensitivity compared to classical techniques. These novel approaches are typically summarized under the term of quantum sensing (for a recent review, see, e.g., [35]). However, sensing approaches that rely on the direct measurement of spin resonances (optically-detected magnetic resonance (ODMR); for an explanation, see Figure 1) and their shift in magnetic fields (Figure 1) also profit from highly-coherent NV centers: the NV's sensitivity η to static magnetic fields is given by [13]:

$$\eta \approx \frac{h}{g\mu_B} \frac{\Delta\nu}{\sqrt{I_0}C} \, ,\tag{1}$$

where $\Delta\nu$ is the ODMR linewidth and I_0 is the detected photon rate from the NV center. C is the fluorescence contrast in ODMR. The latter is an intrinsic property of NV centers and can hardly be modified as it is determined by the internal dynamics. $\Delta\nu$ is fundamentally limited by the inverse of the coherence time T_2^* and thus connects the sensitivity to the coherence properties.

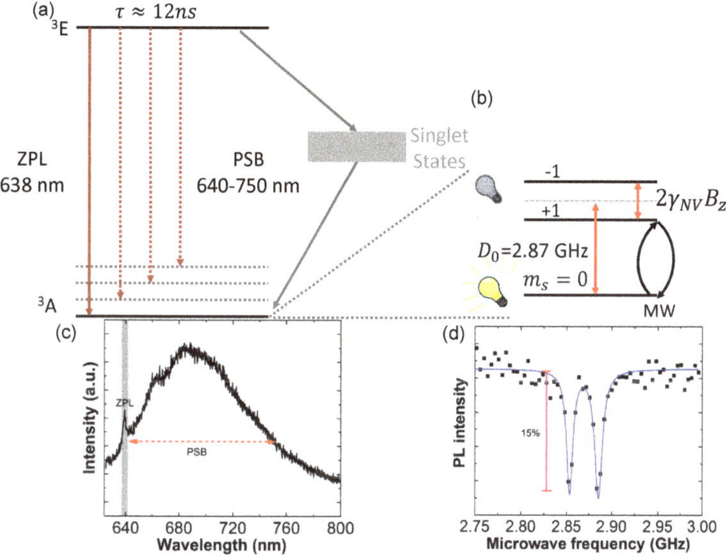

Figure 1. Illustration of the basic properties of nitrogen vacancy (NV) centers in diamond. (**a**) Strongly-radiative, electric dipole transitions between the NV excited (^3E) and ground (^3A) states create a photo-luminescence in the red and near-infrared spectral region shown in (**c**). For a detailed investigation of the NV level structure and dynamics, see, e.g., [27]. The excited state lifetime is $\tau \approx 12$–13 ns in bulk diamond [28]. The purely electronic transition at 638 nm (zero-phonon-line (ZPL)) is marked in the spectrum in (**c**) and features a width of about 1 nm at room temperature. Transitions to vibrationally-excited states create additional, broad phonon sidebands (PSB) as indicated in the schematics and in the spectrum. (**b**) Detailed description of the ground state spin levels. The triplet state has three spin sub-levels: $m_s = 0$ and $m_s = \pm 1$. If no external field is applied, the levels with the projection spin quantum number 0 and ± 1 are split by 2.87 GHz due to spin-spin interactions. In the presence of a magnetic field, the Zeeman effect splits the +1 and −1 states by $2\gamma_{NV}B_z$ as schematically shown in (**b**) and discernible from the measurement in (**d**). Note that only the magnetic field B_z projected onto the NV's high symmetry axis (connecting line between vacancy and nitrogen, <111> direction) leads to a splitting [13]. If the NV center spin is in one of the ± 1 states, the probability is enhanced that the center undergoes an inter-system relaxation to the singlet levels (see (**a**)). As these levels have a lifetime that is more than one order of magnitude longer than for the triplet levels [29], the NV's photo-luminescence is reduced in the ± 1 states enabling the optical read-out of the NV spin state (optically-detected magnetic resonance (ODMR)). For sensing applications, transitions between the 0 state and +1 or −1 state are typically driven using circularly-polarized microwaves (MW) [23,30]. Using green laser light (532 nm), the NV center is initialized to its $m_s = 0$ state via optical pumping within roughly 1 μs [31].

Note that Equation (1) refers to non-resonant spin read-out, typically using green laser light to excite the NV center. This read-out scheme is feasible at room temperature and thus advantageous for sensing. At cryogenic temperature, resonantly-addressing spin-selective transitions within the ZPL enables single-shot read-out of the electronic spin state of NV centers [36].

The carbon lattice of diamond itself provides only a weak source of decoherence. However, paramagnetic impurities (e.g., nitrogen in its substitutional form) and spins on the diamond surface can significantly reduce the coherence time of NV centers [37]. Reduced coherence times render NV centers less sensitive for magnetic fields, as discernible from Equation (1). Consequently, it is vital to control this loss of coherence to use NV centers as highly-sensitive sensors for fields outside the diamond crystal.

In essence, realizing an optimal NV sensor narrows to three main aspects:

- It is mandatory to reliably create stable NV centers with a controlled density buried less than $\approx$10 nm below diamond surfaces (shallow NV centers). These NV centers need to retain spin coherence for optimal sensitivity.
- ODMR and optical sensing schemes demand efficient collection of fluorescence light and high photon rates from single centers. This in turn demands the incorporation of color centers into nanophotonic structures.
- To realize a sensor that probes the sample surface and to realize controlled positioning of the NV sensor requires realizing a tip-like sensor and scanning probe sensing schemes.

This review is structured according to these main aspects. In Section 2, we summarize recent progress in creating and optimizing shallow NV centers. In Section 3, we turn to the photonic nanostructures for sensing and their fabrication. In Section 4, we illustrate scanning probe-based approaches, as well as recent advances in NV-based sensing.

2. Shallow NV Centers for Sensing

2.1. Creation Methods and Creation Yield

NVs with a controlled distance to diamond surfaces have been created using two approaches, namely ion implantation and δ-doping. We discuss both in the following:

- Ion implantation: This approach relies on the commercial availability of chemical vapor deposition (CVD) diamond with low nitrogen (N) content (<5 ppb substitutional N) and an almost negligible density of in situ-created, native NV centers ([38], element six, electronic grade diamond). This diamond is irradiated with N ions at energies typically between 4 and 8 keV. These implantation energies correspond to mean implantation depths between $\approx$7 nm and $\approx$12 nm, as calculated by Monte Carlo simulations [39]. Coherence times T_2 of NVs created deep (>50 nm) inside this material by N implantation reach $T_2 \approx 200$ μs comparable to in situ-created NV centers [40]. Such T_2 times indicate the high purity of the material, whereas in diamonds with 100 ppm N, T_2 generally reduces to about 1 μs [13]. After the implantation, high temperature annealing repairs crystal damage, mobilizes vacancies via diffusion and forms NV complexes via N impurities capturing vacancies. In contrast, impurities like, for example, N are not expected to diffuse at these temperatures [16]. Two approaches are reported: annealing in vacuum at pressures below 5×10^{-7} mbar [16,40,41], where a high vacuum is needed to avoid etching of the surface [42]. Alternatively, annealing in forming gas (4% H_2 in Ar) is used (e.g., [43,44]). The conversion efficiency from implanted N to NV (mostly NV$^-$) is called yield. It amounts to typically only 1% for an implantation energy of 5 keV (depth $\approx$ 8 nm) [41,45]. For micrometer-deep implanted NVs ($E = 18$ MeV), the yield increases to 45% [45]. For shallow NVs, vacancies are partially captured by the diamond surface [42]. Furthermore, higher energy implantation increases the number of vacancies produced [45]. Very recent studies, however, indicate that increasing the number of vacancies by co-implanting, e.g., carbon, does not increase the NV yield for shallow implantation [46].
- The δ-doping method for creating shallow NVs consists of the following steps; see Figure 2:
  - An N-doped, several nm-thick layer, the δ-doped layer [47,48], is created in situ by the controlled introduction of N_2 gas during slow, plasma-enhanced CVD growth of single-crystal diamond (growth rate $\approx$ 0.1 nm/min; Figure 2a). Changing the N_2 flow tunes the resulting N densities. In order to decrease the magnetic noise, diamond growth can be performed using

isotopically-purified $^{12}CH_4$ as the carbon source [47,49,50]. Very recently, overgrowth of a nitrogen-terminated diamond surface has been employed for δ-doping [51].

- Vacancies are created ex situ by implanting helium ions [52,53], carbon ions [49] or irradiating with electrons [47,50]. A subsequent annealing at high temperature causes vacancy diffusion and creates NV centers (Figure 2c). Varying the thickness of an undoped diamond layer, overgrown before the implantation step, allows controlling the depth of the NV layer (Figure 2b). The vacancy profile depends on the species used: for electron irradiation, it is flat and extends throughout the grown film and substrate [49]. In contrast, implanted carbon ions are localized in the lattice. A shallow layer ($\approx$5 nm) of implanted ^{12}C ions has been used as the source of vacancies for a deeper ($\approx$50 nm) δ-doped layer [49], taking advantage of annealing-induced diffusion. With this method, it is possible to increase the NV density in the δ-doped layer by increasing the ^{12}C dose without activating NV centers in the substrate. An alternative method, applied to low energy helium implantation, consists of growing a thicker N-doped layer ($d = 18$ nm) and adding a last etching step to create NV centers near the surface [52].

NV yields reached values of 15% [52] and 50% [50], however, remaining quite low (1.9%) in [49]. As discussed below, often, δ-doping enhances the coherence times of shallow NVs compared to N implantation. Moreover, it potentially localizes NV centers in a more defined depth.

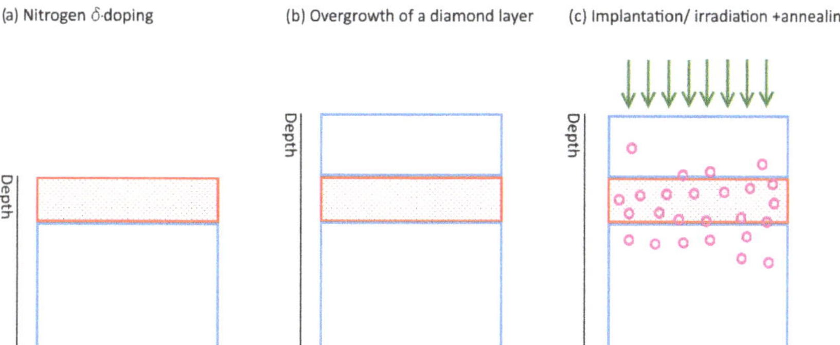

Figure 2. Schematic representation of the δ-doping method. (**a**) Creation of an N-doped, several nm-thick layer, the δ-doped layer (red, spotted rectangle). (**b**) Overgrowth with an undoped, epitaxial diamond layer (blue rectangle). (**c**) Implantation/irradiation and annealing. Pink circles indicate lattice vacancies close to the δ-doped layer.

2.2. Photochromism, Quantum Efficiencies and Coherence Time

NV centers exist in two luminescent charge states: negative (NV$^-$) and neutral (NV^0) [54]; only NV$^-$ exhibits ODMR. To create an NV$^-$, an electron has to be captured by the NV center. The NV$^-$ population never exceeds 75–80% (excitation wavelength 450–610 nm) [55–57], due to photo-induced ionization [57]. In the band from 450–575 nm, NV$^-$ and NV^0 absorb light. Consequently, the loop of NV$^-$ excitation-ionization and NV^0 excitation-recombination is closed, and the NV cycles between both charge states. The NV$^-$ steady state population is maximized for optical excitation in the band 510–540 nm [55]. Besides this photochromism, the NV charge state is directly influenced by the Fermi level position in diamond. For shallow NVs, the Fermi level position closely relates to the electron affinity of the surface (see Section 2.3).

An important characteristic for light-emitting systems is their quantum efficiency (QE). It is defined as $QE = \frac{k_r}{(k_r+k_{nr})}$, where k_r, k_{nr} are the radiative and non-radiative decay rate of the NV center, respectively. A higher QE, in general, leads to brighter emission from the centers. For shallow

NVs, QE can be as high as 70% (82%) for a depth of 4.1 nm (8 nm), respectively [22]. The work in [22] estimates QE > 96% in bulk diamond. This result implies decreased QE for shallow NVs, which was attributed to non-radiative decays induced by surface strain. Generally, careful control of strain and crystalline quality seems necessary to obtain high QE: QE in the range of only 25–60% was found in ion-damaged diamond (H-implantation, fluence $F = 1 \times 10^{15}$ cm^{-2} corresponding to an estimated induced vacancy density $d = 60$ ppm, [58]). NV centers in 25-nm nanodiamonds even may show QE < 20% [59].

Generally, shallowly-implanted NV centers show reduced T_2 together with broadened ODMR lines [16,41,60,61]. Table 1 illustrates this reduction in comparison to deep, native centers. Degraded coherence properties are related to noise created by the proximity to the surface. The work in [60] presents a spectroscopic analysis of the noise (N dose 10^8 cm^{-2}). High-frequency noise arises due to surface-modified phonons and low frequency noise due to the surface electronic spin bath. For ion doses of 10^{11} cm^{-2}, noise produced by implanted defects close to NV centers adds to surface-related noise. Indeed, N bombardment introduces N impurities and vacancy complexes that act as paramagnetic centers, degrading the coherence time of NV centers [62,63].

Table 1. Typical value of T_2 for shallowly-implanted and native, bulk NV from [16]. Note that for centers at a depth of around 50 nm, the coherence time of deep centers is restored [40].

	Photocounts	T_2 (µs)	Linewidth (MHz)
Very Shallow NV$^-$ (2.1 nm)	5×10^4	12.2 ± 0.6	2
Shallow NV$^-$ (7.7 nm)	4×10^5	40.4 ± 0.8	1.2
Native NV$^-$ (6 µm)	4×10^5	128 ± 10	1

2.3. Methods to Improve Stability and Photoluminescence

The work in [44] shows that the NV$^-$ population is decreased at depths up to 200 nm compared to bulk. This effect is attributed to the presence of an electronic depletion layer at the etched diamond surface. In this depletion region, N donors cannot donate an electron to the NV center because they are ionized. Stabilizing the NV$^-$ state close to the surface involves controlling the Fermi level in diamond [64–67]. The Fermi level, and consequently the NV$^-$ population, can be controlled via chemical functionalization of the surface [64,65,68,69] (see Figure 3a,b) by applying an electrolyte gate voltage [70] (see Figure 3c) using in-plane gate nanostructures [71] or via doping the diamond [67].

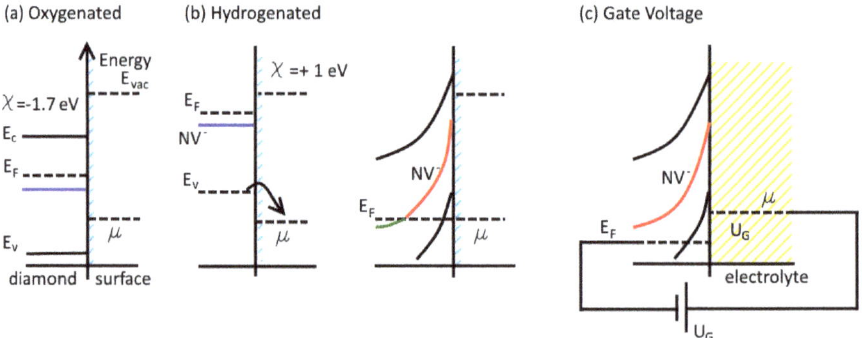

Figure 3. Energy band schematic of diamond energy levels for: (**a**) oxygenated surface; (**b**) hydrogenated surface. (**c**) Applying a gate voltage in the presence of an electrolyte.

First, we consider the decrease in the NV$^-$ population of shallow NV centers close to a hydrogen (H)-terminated surface compared to an oxygenated surface [64,65]. For H-terminated diamond, the conduction and valence bands shift upwards due to the negative electron affinity of the surface. The energy of the valence band maximum E_V is higher than the chemical potential μ of the electronic states created by adsorbed water. Electrons migrate to the adsorbate layer. In equilibrium, the diamond bands bend upwards; a 2D hole gas is created; and the NV^- charge state is depleted [65]. Thus, H-termination is not useful to stabilize NV^-.

In contrast, fluorine (F)-terminated surfaces will lead to a downward bending of the diamond bands due to the large electron affinity of F [72]. In this case, the NV^- population increases compared to H-terminated diamond [68] and also to oxygenated diamond [69]. However, it should be noted that so far, coherence measurements of shallow NV centers close to F-terminated surfaces are missing.

For H-terminated diamond, electrolytic gate electrodes can directly control the Fermi level [70] (see Figure 3c). An increase in the NV^- ensemble population was observed for high implantation dose and positive gate voltage. Unfortunately, no single NV charge switching was observed, due to the high gate voltage needed for charge-state conversion at low implantation dose. On the other hand, this result was achieved in [71] by applying an electric field using in-plane nanostructures. The gate-voltage was applied to a structure formed by H-terminated areas and an O-terminated line, resulting in an offset between the Fermi level in the H-terminated areas on both sides of the line. In this way, switching the charge state of a single NV center from NV^0 to NV^- was established.

Deterministic electrical control of the charge state of a single NV center has been achieved using a p-i-n diode [66]. The NV center is positioned in the intrinsic region of the diode, and a current of holes is induced from the p-region. However, this technique converts NV centers to NV^0. Using in plane Al-Schottky diodes, based on H-terminated diamond, charge state switching of single NV centers (NV^+ to NV^0 to NV^-) has been obtained [73]. Implementing diode structures in scanning probe sensing seems, however, highly challenging.

Another way to change the Fermi level of diamond is by boron or phosphorus (P) doping [67,74]. In particular, P impurities donate an electron to NV^0, due to the fact that the P activation energy is low ($E_A = 0.57$ eV) compared to NV acceptor states ($E_{NV} = 2.58$ eV). In this way, a five-fold increase in luminescence and a pure NV^- state was observed for a single NV [67] and a P doping $p = 5 \times 10^{16}$ cm^{-3}. It should be noted that this single NV had a short coherence time $T_2 = (19.77 \pm 0.27)$ µs, and the depth of the NV center in this case was not reported.

2.4. Methods to Improve T_2

In this section, we discuss how N δ-doping [47,51] and ion implantation [52,75,76] or electron irradiation [50] can lead to improved spin coherence. Figure 4 summarizes recently obtained T_2 values for different methods and works. As discussed above, implanted defects are sources of noise. Implanted defects can be of two types:

- Vacancy complexes [63];
- Ion impurities and crystal defects [77].

The formation of vacancy complexes during annealing is inhibited by Coulomb repulsion, if their charge state is changed from neutral to positive [63]. Vacancy charging is accomplished in a p-i junction by implanting N in the junction's space charge layer. A two-fold increase in yield and ten-fold increase in T_2 for very shallow NVs has been reported ($T_2 = 30$ µs, NV depth d = 3 nm).

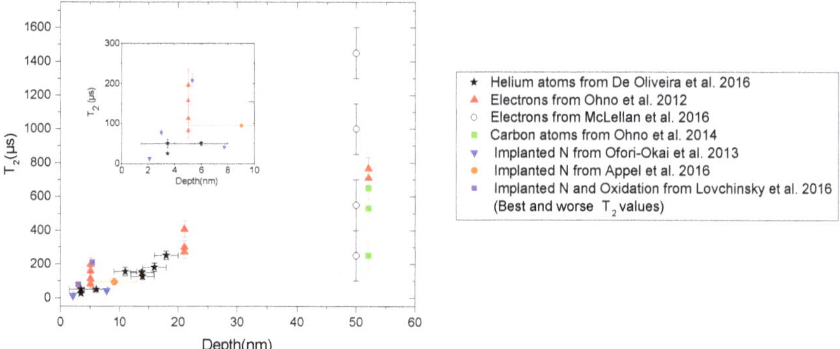

Figure 4. Coherence time T_2 for NV centers created via δ-doping methods and implantation with different inert species. For comparison, T_2 of centers created via N implantation are given. The inset highlights T_2 for an NV center less than 10 nm below the surface, which are especially important for sensing applications. Note the spread of T_2 in the different works.

Noise from implantation-induced impurities and crystal defects can be reduced using δ-doping and ex situ creation of vacancies (see Section 2.1). To be used in scanning probe sensing, NV centers should have a depth below $\approx$10 nm. As mentioned before, this can be achieved by tuning the dimension of the cap layer [47] or adding a last etching step [52]. In this depth range, the maximum reached T_2 is 200 µs obtained when irradiating an isotopically-purified sample with electrons [47]. In this case, however, the NV density was too low for applications requiring the fabrication of single-crystal scanning probes, as discussed below.

We remark that NV areal densities around $D_{NV} = 3 \times 10^9$ cm^{-2} are typically necessary for single crystal scanning probes: this corresponds to one NV center in a circular area with a diameter of 200 nm and thus to a single NV in typical nanostructures used for sensing (see Section 3). In [47], the δ-doped layer has an N concentration of $C_N = 3 \times 10^{16}$ cm^{-3}, considering a layer thickness of 2 nm; this results in an areal density of N of 6×10^9 cm^{-2}, so in order to have an acceptable NV areal density, one should have a very high yield, around 50%. A similar concentration of N ($C_N = 0.8 \pm 0.6 \times 10^{16}$ cm^{-3}) is reported in [49], leading to a low NV area density ($D_{NV} = 10^6$ cm^{-2}). A slightly too low NV density is reported in [50] $D_{NV} = 1.4 \times 10^9$ cm^{-2} and in [52] $D_{NV} = 1 \times 10^8$ cm^{-2}. It is worth noticing that an N volume concentration of $C_N = 1.8 \times 10^{20}$ cm^{-3} is reported in [51]; this would result in an NV density of $D_{NV} = 4 \times 10^{12}$ cm^{-2}, considering a thickness layer of 2 nm and a yield of 10%. In contrast, the N implantation method described in [63] seems more promising from this point of view: tuning the fluence of implanted N from 10^{10} cm^{-2}–10^{12} cm^{-2} (E = 5 keV) allows accessing the optimal densities of shallow NV centers.

We underline that till now, δ-doping has not been applied in the manufacturing of diamond probes for nanosensing. This is probably due to the fact that δ-doping involves the realization of sophisticated CVD methods. In contrast, creating single NVs by ion implantation can be achieved using commercial diamond material and implantation facilities.

Finally, we remark that surface oxidation has been used recently to improve T_2 [78]. Wet oxidative chemistry and sample annealing at 465 °C in a dry oxygen atmosphere increased T_2 by an order of magnitude (max. $T_2 = 208$ µs for a single NV, depth (5.3 ± 0.1) nm, $T_2 > 100$ µs observed for 6 NVs). Recent work [79], however, was not able to reproduce these results for shallower NVs: in [79], a similar oxidation procedure resulted in $T_2 \leq 6$ µs for calculated depths of (2.6 ± 1.1) nm.

3. Nanostructures for Photonics and Scanning Probe Operation

NV centers close to the surface of bulk, single-crystal diamond are, on the one hand, advantageous as they reside in high-quality, high-purity, potentially low-stress diamond material mostly synthesized by the CVD method [80]. On the other hand, these NV centers do not fulfill all prerequisites for nanoscale sensing: first, positioning an NV created in a macroscopic diamond crystal within nanometer distance from a sample is hardly feasible. In principle, attaching the 'sample', e.g., the substance under investigation, to a scanning probe tip while keeping the NV center stationary can circumvent this issue. However, this limits the technique to microscopic samples, e.g., specific molecules or ions (e.g., [81]) for which attachment to a nano-sized tip is feasible.

Additionally, even for an optimal dipole orientation, a typical air microscope objective (NA 0.8 [82]) collects only about 5% of the fluorescence from an NV center in bulk diamond: as diamond exhibits a high refractive index of 2.4 for visible light [18], only light incident at angles <24.6° (partially) leaves the diamond as illustrated in Figure 5a. Other light rays are lost due to total internal reflection. Additionally, the dipolar emission from color centers is typically directed towards the optically-dense medium. Thus, in our case, emission is directed towards the bulk diamond, additionally hindering fluorescence detection [83,84]. Consequently, coupling to nanophotonic structures is mandatory to enhance the photon rates from individual color center sensors in single-crystal diamond. Alternatively, attaching nanodiamonds (NDs) containing NV centers to a scanning probe tip provides a scannable NV platform (e.g., [26,85]). However, NDs show non-ideal material properties, typically due to excess nitrogen or crystal strain resulting from milling of the material [86]. Thus, incorporated color centers may suffer from short coherence time, strong inhomogeneous line spreads and reduced stability [87]. In this review, we thus focus on single-crystal diamond-based sensing techniques.

Scanning probe sensing demands nanophotonic structures with several properties:

- Broadband operation: Section 1 and Figure 1c illustrate the 100 nm-broad room temperature emission spectrum of NV centers. Room temperature spin read-out under non-resonant excitation is most efficient if the integral emission is detected. Thus, the photonic structure should allow for efficient fluorescence detection from the complete NV emission spectrum.
- Tip-like geometry and suitability for NV centers close to surfaces: a tip-like geometry ensures close proximity of the scannable NV center to the sample even in the presence of alignment uncertainties. The photonic structure needs to be functional for NV centers very close to diamond surfaces.
- Attachment to the feedback system: the nanostructure has to be attached to a force sensor, i.e., a tuning fork or a cantilever. A miniaturized structure is needed, thus avoiding strong shifting of the resonance frequency due to the additional mass or severe damping of the oscillating system.

Considering the first criterion, resonant photonic structures, like, e.g., photonic crystal cavities [88], which selectively enhance a specific transition within the emission spectrum, are not fully suitable for sensing applications under ambient conditions. For quantum information, in contrast, these systems are promising as they potentially form a coherent interface between spins and photons in quantum networks [89]. In our case, waveguide-like structures that channel the NV's broad emission into certain spatial modes and direct the light towards the collection optics (nanopillars; see Figure 5b) seem more suitable mainly due to their broadband operation. Certain waveguide-based structures, i.e., dielectric antennas [90], have been found to enable potentially near unity collection efficiency for NV fluorescence [91]. However, they consist of planar multilayer structures and are thus not optimized for scanning probe sensing. In contrast, they might be very suitable for approaches where a diamond chip is used for sensing (wide field approaches, lower spatial resolution, e.g., [92]). Taking into account the first two criteria, also solid immersion lenses, where color centers are buried inside diamond half spheres, are not suitable [93]. In this review, we focus on the nanostructure types suitable for scanning probe sensing. For more complete recent reviews of diamond nanophotonics in the context of quantum information, see [94–97].

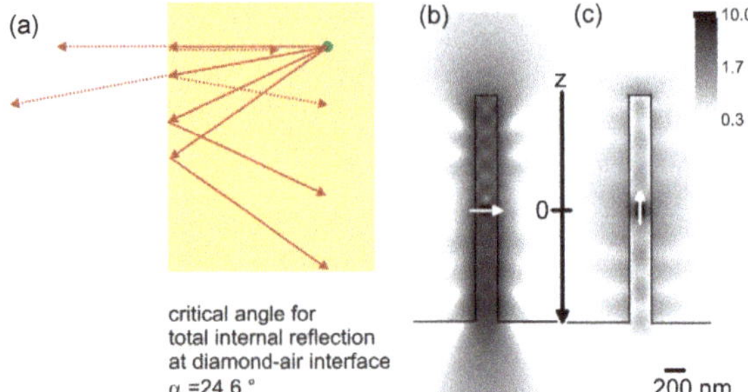

Figure 5. (a) Illustration of total internal reflection: light incident under low angles cannot leave the diamond; for normal incidence, roughly 17% of the incident power is reflected (Fresnel reflection). (b,c) These images show the magnitude of the electric field (simulated for a light wavelength of $\lambda = 700$ nm) inside diamond nanopillars. Note that darker areas correspond to higher fields, as indicated in the color bar. The field is being channeled in the direction of the pillar axis, witnessing wave-guiding via the photonic modes of the pillars. Reprinted from [82] under CC-BY3.0. The arrow within the structures indicates the dipole orientation. For the optimal dipole orientation, where the dipole axis is perpendicular to the pillar axis, collection efficiencies up to 40% can be reached (collection into NA0.8 from the air side).

In essence, photonic nanostructures suitable for scanning probe sensing consist of a roughly tip-shaped nanophotonic structure on a thin (typically <1 µm) diamond mounting structure, as shown in Figure 6a. First, operational sensors have been presented in 2012 [61]. The work in [41,98] presents more recent nanofabrication approaches. In a further very recent approach, mounting structures holding an array of pillars are used. Consequently, several NV centers are usable, and vector magnetometry might be feasible [99]. However, this approach introduces the risk that a pillar that is not in direct contact with the sample is used and, thus, an unwanted stand-off distance strongly reduces the resolution.

We start by addressing how large-area, thin, single-crystal diamond membranes can be formed. High-purity, CVD diamond with <5 ppb substitutional nitrogen and an intrinsically very low density of NV centers is commercially available (see Section 2.1). Starting from this material, thin, single-crystal diamond membranes can be formed by thinning down diamond plates obtained by polishing (typically several tens of µm thick). To thin the plates down, reactive-ion etching with different reactive gases in the plasma has been used among them chlorine and oxygen-based recipes [41,61], fully-chlorine-based chemistry [101], oxygen/fluorine-based chemistry [98,102] and fully-fluorine-based chemistry [7]. For all of these approaches except in [98], argon is added to the plasma to introduce physical etching (sputter etching) to the process. Alternatively, membranes can be created by damaging a buried layer of diamond via ion implantation. When this layer is graphitized upon annealing, it can be chemically etched, and membranes are lifted-off. However, due to strong ion damage, this material is not directly usable and has to be overgrown with a pristine layer of CVD diamond, adding technological complexity to the process [103]. We note that it is not possible to obtain thin, low-stress, single-crystal membranes directly via growth on a non-diamond substrate that might allow for wet-chemical etching of the substrate: for all diamond heteroepitaxy, the starting phase of the growth is highly defective, and forming high-quality, thin membranes includes removing this initial growth areas [104]. Reliable, large-area fabricating of single-crystal membranes is essential for high-yield fabrication of scanning probe devices and potential future up-scaling of the technology.

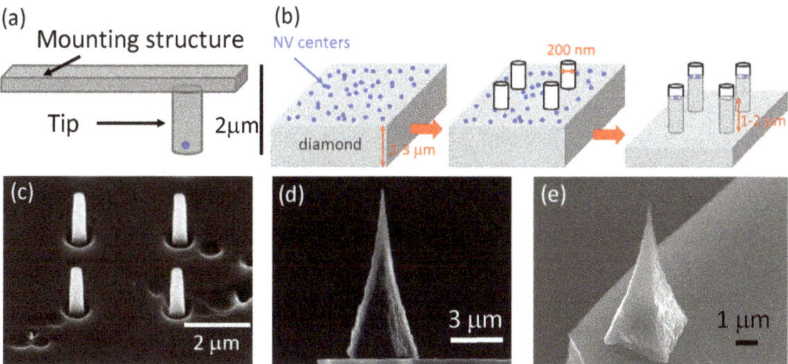

Figure 6. (**a**) Schematics of an all diamond scanning probe nanostructure, consisting of a cylindrical diamond tip on a thin mounting structure. (**b**) Illustration of the fabrication process of diamond pillars for sensing: shallowly-buried NV centers are created (blue dots: NV centers; grey: diamond). An etch mask (white) is created (typically via electron beam lithograph in Hydrogen silsesquioxane resist (here Dow Corning, Fox16)). The pillars are etched in a highly-anisotropic etch. NV centers not protected by the etch mask are removed with the surrounding diamond. If the optimal density is chosen, roughly one out of three pillars contains a single NV center [41]. (**c**) Electron microscopy image of diamond nanopillars etched into CVD diamond. (**d,e**) Diamond pyramids formed during a specialized CVD growth process and mounted to silicon cantilevers for scanning probe applications [100].

The nanostructure forming the tip is either almost cylindrical (termed nanopillar [41,82,84,105,106]), pyramidal [100] or features a truncated cone shape with a taper angle [84,107]. The work in [107] demonstrated that tapering the pillar enhances fluorescence rates from single NVs. Lithographically-defined etch masks are used to create the pillars, as illustrated in Figure 6b. These masks undergo faceting and erosion [108], thus limiting the pillar length and influencing its shape. For an electron microscopy image of diamond nanopillars, see Figure 6c. Mask erosion is especially critical, as a high-density plasma (mostly oxygen and argon) is needed to enable highly-anisotropic etching of diamond. For the coupling to the nanostructure's photonic modes, not only the placement of the NV center is essential, but also the orientation of the NV's electric dipoles. As indicated in Figure 5b, placing the NV's dipoles perpendicularly to the pillar axis is the most advantageous configuration. This situation can be reached in <111>-oriented diamond. Here, NV centers created during CVD growth preferentially align in the growth direction, with their dipoles in the plane perpendicular to that direction [109–111]. However, CVD growth in the <111> crystal direction is challenging [112] and still leads to a material with lower quality than the standard <100> growth. In <100> diamond, NV centers align along all four equivalent <111> directions and have an oblique angle of 54.7° with the pillar axis. As an alternative, CVD growth in the <113> direction has been investigated [113]. In this material, 73% of NVs form an angle of 29.5° with the potential pillar axis and thus bring their dipoles closer to the ideal orientation. The work in [82] demonstrates pillars fabricated into <111>-oriented material. However, so far, optimized crystal orientations have not been used for scanning probe sensing.

To manufacture nanophotonic structures, two fundamentally different approaches exist: first, the structures can be sculpted from bulk diamond material (top-down approach) as discussed above [41,82,84,105,108]. In contrast, nanostructures might also be created directly during (CVD) synthesis of diamond (bottom up approach [100,114,115]). Pyramidal nanostructures formed via bottom-up approaches (see Figure 6d,e) also have advantageous photonic properties [100]. However, they currently show high levels of color center incorporation during growth. This results from growth on non-diamond materials or incorporation of impurities from masks used in the growth. Consequently, these structures are not usable for sensing with single color centers so far. In top-down approaches, plasma-based etching forms the nano-structures. The work in [116] shows that a plasma can damage the diamond surface and reduce NV coherence. From this point of view, bottom-up approaches are appealing as they offer the opportunity to avoid plasma etching of diamond.

As discussed in Section 2, shallow NV creation is a crucial step towards functional sensing devices. So far, NVs in nanopillars have been created by ion implantation [41,61,106,107] followed by structuring the nanopillars via lithography and etching (see Figure 6b). Alternatively, NV centers created during growth, but at non-defined depth, were used [82,105]. Thus, the latter pillars are not suitable for scanning probe sensing. A promising alternative approach is spatially localized ion implantation into high-purity nanostructures. Three different approaches have been presented: the first uses focused nitrogen ion beams for localized implantation [117]. In the second approach, a pierced AFM tip is used as an implantation aperture [118]. In the third approach, single ion traps are planned to be used as deterministic sources of implanted ions [119]. Also in the context of δ-doping, three-dimensional localization of NV centers has been reported: [49] uses a mask with nanometric apertures for single NV creation in a δ-doped layer. For the smallest aperture of (114 ± 21) nm, single NVs were created in 18% of the apertures. In [50], a transmission electron microscope is used for irradiation, achieving sub-micron accuracy in lateral positioning.

We point out that in parallel to the efforts in nanophotonics, recently electrical read-out of the NV spin state has been proposed alternatively to the challenging optical read-out: NV centers are more likely to undergo multi-photon photo-ionization as long as they are not being shelved to the singlet states. Thus, an NV center in the $m_s = 0$ spin state is more likely to be ionized and leads to a higher photo-current. Demonstrations include continuous [120], as well as pulsed detection [121]. Whereas the read-out contrast can approach 20%, demonstrations of single NV read-out are still pending.

4. Recent Advances in NV Sensing

In this section, we review selected recent advances in sensing with single NV centers. In Section 4.1, we summarize recent steps towards scanning near-field optical microscopy (SNOM) with a single NV center. In Section 4.2, we present recent results using the magnetic sensing capabilities of NV centers (atomic size sensor, high sensitivity, low-invasiveness) to study physical phenomena at low temperature and/or few atoms level.

4.1. Near-Field Microscopy with NV Centers

In recent years, the highly-photostable emission of individual NV centers has triggered several efforts to realize near-field sensing based on NV centers [8,26,122,123]. In general, near-field-based imaging, where a sample is illuminated by the near-field of a light source, allows a higher resolution than far field-based techniques and is able to beat the Abbé limit of resolution.

Single NV centers were used as nano-sized light-source that is brought within <100 nm of the sample in a SNOM setup [25,123,124]. A nanodiamond (ND) attached to the end of a SNOM tip achieved a resolution of around 50 nm when imaging metallic nanostructures, thus beating the Abbé limit of resolution (see Figure 7). The resolution was limited by the vertical NV-to-sample distance, which is constrained by the ND size and by the fact that the ND was not put in contact with the sample to avoid too strong friction forces at the tip apex.

Förster resonance energy transfer (FRET) was first described as non-radiative energy transfer between a pair of single molecules (dipoles) in close proximity [14,125]. To enable FRET, the emission band of the donor molecule has to overlap with the absorption band of the acceptor molecule, allowing energy transfer from the donor to the acceptor molecule. The transfer efficiency reaches 50% for a characteristic distance, the so-called Förster distance R_0, and drops with the inverse sixth power of the distance. NV centers in a FRET pair with a fluorescent dye have shown $R_0 = 5.6$ nm [126]. Thus, this effect promises truly nanoscale resolution in near-field-based imaging.

In recent experiments, an ND placed close to the apex of a commercial AFM tip was scanned close to a surface covered with graphene flakes [26]. The NV center and the graphene flake form a FRET pair. A Förster distance of 15 nm was found. Sekatskii and co-workers [122] reported on unsuccessful experiments on FRET transfer between a scanning NV and a dye molecule, despite previous demonstration of FRET to dye molecules covalently bonded to NDs [8]. This finding might relate to a varying quantum efficiency (QE) for NV centers in NDs [59], the need for accurate control of the ND surface (graphite layers), as well as to large stand-off distances when attaching an ND to a scanning probe tip. These issues may be addressable in the future using scanning probe devices sculpted out of single-crystal diamond [41,61]. So far, no experiments on near-field sensing based on single-crystal scanning probes have been reported.

Finally, we mention that super-resolution images of NV centers in bulk diamond were recorded using stimulated emission depletion microscopy [127], ground state depletion microscopy [128], stochastic optical reconstruction microscopy [129] and by sampling second- and third- order photon correlation function [130]. These techniques beat the Abbé resolution limit without the need for near-field imaging.

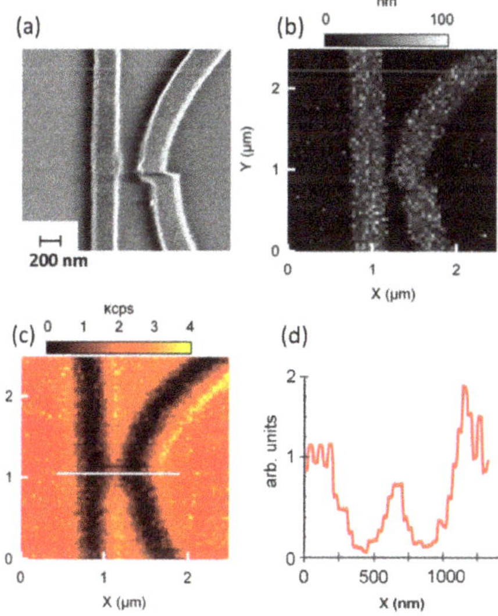

Figure 7. (**a**) Scanning electron micrograph of chromium structures patterned on a fused silica cover slip. (**b**) Numerically-flattened topography of the same region. (**c**) Fluorescence scanning near-field optical microscopy (SNOM) image recorded using the single photon tip. (**d**) cross-cut of the optical intensity along the white line in (c). Reprinted from [124], copyright the Optical Society of America (2009).

4.2. Recent Applications of Single NV Sensing

The implementation of scanning NV-based microscopy at cryogenic temperatures forms a recent milestone: in a preliminary realization, cryogenic operation was achieved with a single NV center in bulk material and a scanning magnetic tip [131]. More recently, versatile approaches with NV centers in single crystal scanning probes have been presented [99,132]. This technology significantly broadens the range of samples for which NV magnetometry can be applied: solid materials present unique phenomena that only occur at cryogenic temperatures such as superconductivity. An ideal superconductor is supposed to expel all magnetic fields (Meissner effect). However, real superconductors present points at which magnetic fields penetrate and magnetic vortices form. NV nano-magnetometry was applied for a high-resolution, low-invasive imaging of superconducting vortices (6 K, [99], 4.2 K [132]). To achieve cryogenic operation, the NV microscope is enclosed in a liquid ^{4}He cryostat with optical access. Particular care was taken to avoid heating by microwave (MW) and laser excitation. Superconducting vortices were imaged with a resolution below 100 nm with magnetic field sensitivity of 30 μT·Hz$^{1/2}$ [99] and 11.9 μT·Hz$^{1/2}$ [132], allowing in [132] to verify vortex models beyond the monopole approximation.

Very recently, NV microscopy imaged nanoscale ferrimagnetic domains in antiferromagnetic random access memories [133]. We underline that part of these measurements were performed in zero field cooling (ZFC), taking advantage of low-invasive sensing using NV centers.

As discussed above, NV microscopy is often employed to sense static magnetic fields with nanoscale spatial resolution. However, NV centers are also sensitive to magnetic fields oscillating at GHz frequencies: single-crystal NV scanning probes imaged magnetic fields generated by MW currents with nanoscale resolution and sensitivity of a few nA·Hz$^{1/2}$ [30]. The basic idea is to tune an NV spin resonance to the investigated MW frequency using a static magnetic field. The induced coherent spin oscillation frequency (Rabi frequency) then maps the magnetic field amplitude.

Single NV centers also form atomic-sized probes for nuclear magnetic resonance (NMR) signals [134–136]. The NV spin state is sensitive to the stochastic transverse magnetization at the Larmor frequency of the investigated ensemble of nuclei. Using suitable pulse sequences, it is possible to filter NMR signals and recover NMR spectra. Using NV probes, NMR investigations of small ensembles of nuclei are feasible. In contrast, conventional NMR spectroscopy is limited to macroscopic, thermally-polarized ensembles in high fields. Recent work presents NMR studies of a single protein [78] and of atomically-thin hexagonal boron nitride (h-BN) layers [137]. ^{2}H and ^{13}C NMR spectra of ubiquitin proteins were measured [78], paving the way for experimental studies of biological systems at the single-molecule level. Conventional nuclear quadrupole resonance (NQR) [138] spectroscopy yields important information on chemical properties of macroscopic samples. However, it suffers from poor sensitivity due to low thermal polarization. This issue can be overcome using single NV centers that are sensitive to stochastic nuclear magnetization. The stochastic polarization is proportional to $\sqrt{n}$, where n is the number of nuclei, but does not depend on the applied field. Single NV NQR was applied to study thin h-BN flakes, verifying a correlation between the number of atomic layers and a shift in NQR spectra [137].

NMR imaging has already been partially combined with scanning NV microscopy, in the sense that samples on a tip were scanned above a stationary NV. This configuration has been used for magnetic resonance imaging of ^{1}H nuclei with a spatial resolution of 12 nm [139]. ^{19}F nuclei were imaged in a calibration grating [140]; widths of (29 ± 2) nm were resolved [140]. To the best of our knowledge, to date, there are no reports on the use of single-crystal diamond scanning probes for NMR imaging. However, this fact does not represent a general limitation. It most probably relates to the fact that because of challenging nanofabrication procedures, only a few research groups have access to single-crystal scanning probes so far.

5. Conclusions

We have summarized recent achievements and the basics of NV center-based sensing. The fabrication of suitable photonic nanostructures that allow scanning probe sensing and high photon rates from single centers, as well as the creation of NV color centers close to the diamond surface are still challenging subjects of research. To date, many recent advances like the creation of NV centers via δ-doping methods, optimized surface treatments and doping of diamond, as well as diamond crystals with optimized orientation have not yet been used in the challenging fabrication of single-crystal diamond scanning probes. Thus, enhanced sensitivities and resolution in NV-based imaging are feasible in the future, given that the recent advances in material sciences are fully transferred to the sensor fabrication. As discussed in this review, many approaches to optimal, shallow NV centers are investigated. The results considering creation yield and NV properties are not always fully consistent, illustrating the complexity of the field. Moreover, the field could tremendously profit from simplified procedures to upscale the fabrication of optimized sensing nanostructures.

Acknowledgments: The authors acknowledge funding via a NanoMatFutur grant of the German Ministry of Education and Research. Elke Neu acknowledges funding via a PostDoc fellowship of the Daimler and Benz foundation.

Author Contributions: Ettore Bernardi and Elke Neu wrote the manuscript. Selda Sonusen and Richard Nelz performed the experiments and contributed experimental data (NV spectroscopy, ODMR, nanowires). All authors discussed and commented on the manuscript.

Conflicts of Interest: The authors declare no conflict of interest.

References

1. Dresselhaus, M.S. A revolution of nanoscale dimensions. *Nat. Rev. Mater.* **2016**, *1*, 15017.
2. Wilhelm, S.; Tavares, A.J.; Dai, Q.; Ohta, S.; Audet, J.; Dvorak, H.F.; Chan, W.C. Analysis of nanoparticle delivery to tumours. *Nat. Rev. Mater.* **2016**, *1*, 16014.
3. Wagner, T.; Strasberg, P.; Bayer, J.C.; Rugeramigabo, E.P.; Brandes, T.; Haug, R.J. Strong suppression of shot noise in a feedback-controlled single-electron transistor. *Nat. Nanotechnol.* **2017**, *12*, 218–222.
4. Nowodzinski, A.; Chipaux, M.; Toraille, L.; Jacques, V.; Roch, J.F.; Debuisschert, T. Nitrogen-Vacancy centers in diamond for current imaging at the redistributive layer level of Integrated Circuits. *Microelectron. Reliab.* **2015**, *55*, 1549–1553.
5. Dolde, F.; Doherty, M.W.; Michl, J.; Jakobi, I.; Naydenov, B.; Pezzagna, S.; Meijer, J.; Neumann, P.; Jelezko, F.; Manson, N.B.; et al. Nanoscale Detection of a Single Fundamental Charge in Ambient Conditions Using the NV^- Center in Diamond. *Phys. Rev. Lett.* **2014**, *112*, 097603.
6. Kucsko, G.; Maurer, P.; Yao, N.; Kubo, M.; Noh, H.; Lo, P.; Park, H.; Lukin, M. Nanometre-scale thermometry in a living cell. *Nature* **2013**, *500*, 54–58.
7. Ali Momenzadeh, S.; de Oliveira, F.F.; Neumann, P.; Bhaktavatsala Rao, D.D.; Denisenko, A.; Amjadi, M.; Chu, Z.; Yang, S.; Manson, N.B.; Doherty, M.W.; et al. Thin Circular Diamond Membrane with Embedded Nitrogen-Vacancy Centers for Hybrid Spin-Mechanical Quantum Systems. *Phys. Rev. Appl.* **2016**, *6*, 024026.
8. Tisler, J.; Reuter, R.; Lämmle, A.; Jelezko, F.; Balasubramanian, G.; Hemmer, P.R.; Reinhard, F.; Wrachtrup, J. Highly Efficient FRET from a Single Nitrogen-Vacancy Center in Nanodiamonds to a Single Organic Molecule. *ACS Nano* **2011**, *5*, 7893–7898.
9. Binnig, G.; Quate, C.F.; Gerber, C. Atomic Force Microscope. *Phys. Rev. Lett.* **1986**, *56*, 930–933.
10. Martin, Y.; Wickramasinghe, H.K. Magnetic imaging by force microscopy with 1000 Å resolution. *Appl. Phys. Lett.* **1987**, *50*, 1455–1457.
11. Ohnesorge, F.; Binnig, G. True Atomic Resolution by Atomic Force Microscopy Through Repulsive and Attractive Forces. *Science* **1993**, *260*, 1451–1456.
12. Grinolds, M.; Hong, S.; Maletinsky, P.; Luan, L.; Lukin, M.; Walsworth, R.; Yacoby, A. Nanoscale magnetic imaging of a single electron spin under ambient conditions. *Nat. Phys.* **2013**, *9*, 2145.
13. Rondin, L.; Tetienne, J.P.; Hingant, T.; Roch, J.F.; Maletinsky, P.; Jacques, V. Magnetometry with nitrogen-vacancy defects in diamond. *Rep. Prog. Phys.* **2014**, *77*, 056503.
14. Förster, T. Zwischenmolekulare Energiewanderung und Fluoreszenz. *Ann. Phys.* **1948**, *437*, 55–75.
15. Acosta, V.; Hemmer, P. Nitrogen-vacancy centers: Physics and applications. *MRS Bull.* **2013**, *38*, 127–130.
16. Ofori-Okai, B.K.; Pezzagna, S.; Chang, K.; Loretz, M.; Schirhagl, R.; Tao, Y.; Moores, B.A.; Groot-Berning, K.; Meijer, J.; Degen, C.L. Spin properties of very shallow nitrogen vacancy defects in diamond. *Phys. Rev. B* **2012**, *86*, 081406.
17. Bradac, C.; Gaebel, T.; Pakes, C.I.; Say, J.M.; Zvyagin, A.V.; Rabeau, J.R. Effect of the Nanodiamond Host on a Nitrogen-Vacancy Color-Centre Emission State. *Small* **2013**, *9*, 132–139.
18. Zaitsev, A. *Optical Properties of Diamond: A Data Handbook*; Springer: Berlin/Heidelberg, Germany, 2001.
19. Aharonovich, I.; Castelletto, S.; Simpson, D.; Su, C.; Greentree, A.; Prawer, S. Diamond-based single-photon emitters. *Rep. Prog. Phys.* **2011**, *74*, 076501.
20. Kraus, H.; Soltamov, V.; Fuchs, F.; Simin, D.; Sperlich, A.; Baranov, P.; Astakhov, G.; Dyakonov, V. Magnetic field and temperature sensing with atomic-scale spin defects in silicon carbide. *Sci. Rep.* **2014**, *4*, 5303.
21. Davies, G.; Hamer, M. Optical studies of the 1.945 eV vibronic band in diamond. *Philos. Trans. Roy. Soc. A* **1976**, *348*, 285.

22. Radko, I.P.; Boll, M.; Israelsen, N.M.; Raatz, N.; Meijer, J.; Jelezko, F.; Andersen, U.L.; Huck, A. Determining the internal quantum efficiency of shallow-implanted nitrogen-vacancy defects in bulk diamond. *Opt. Express* **2016**, *24*, 27715–27725.

23. Alegre, T.P.M.; Santori, C.; Medeiros-Ribeiro, G.; Beausoleil, R.G. Polarization-selective excitation of nitrogen vacancy centers in diamond. *Phys. Rev. B* **2007**, *76*, 165205.

24. Kurtsiefer, C.; Mayer, S.; Zarda, P.; Weinfurter, H. Stable Solid-State Source of Single Photons. *Phys. Rev. Lett.* **2000**, *85*, 290–293.

25. Kühn, S.; Hettich, C.; Schmitt, C.; Poizat, J.P.; Sandoghdar, V. Diamond colour centres as a nanoscopic light source for scanning near-field optical microscopy. *J. Microsc.* **2001**, *202*, 2–6.

26. Tisler, J.; Oeckinghaus, T.; Stöhr, R.J.; Kolesov, R.; Reuter, R.; Reinhard, F.; Wrachtrup, J. Single defect center scanning near-field optical microscopy on graphene. *Nano Lett.* **2013**, *13*, 3152–3156.

27. Manson, N.B.; Harrison, J.P.; Sellars, M.J. Nitrogen-vacancy center in diamond: Model of the electronic structure and associated dynamics. *Phys. Rev. B* **2006**, *74*, 104303.

28. Collins, A.; Thomaz, M.; Jorge, M. Luminescence decay time of the 1.945 eV centre in type Ib diamond. *J. Phys. C* **1983**, *16*, 2177–2181.

29. Robledo, L.; Bernien, H.; van der Sar, T.; Hanson, R. Spin dynamics in the optical cycle of single nitrogen-vacancy centres in diamond. *New J. Phys.* **2011**, *13*, 025013.

30. Appel, P.; Ganzhorn, M.; Neu, E.; Maletinsky, P. Nanoscale microwave imaging with a single electron spin in diamond. *New J. Phys.* **2015**, *17*, 112001.

31. Dréau, A.; Lesik, M.; Rondin, L.; Spinicelli, P.; Arcizet, O.; Roch, J.F.; Jacques, V. Avoiding power broadening in optically detected magnetic resonance of single NV defects for enhanced dc magnetic field sensitivity. *Phys. Rev. B* **2011**, *84*, 195204.

32. Gruber, A.; Dräbenstedt, A.; Tietz, C.; Fleury, L.; Wrachtrup, J.; von Borczyskowski, C. Scanning confocal optical microscopy and magnetic resonance on single defect centers. *Science* **1997**, *276*, 2012–2014.

33. Balasubramanian, G.; Neumann, P.; Twitchen, D.; Markham, M.; Kolesov, R.; Mizuochi, N.; Isoya, J.; Achard, J.; Beck, J.; Tissler, J.; et al. Ultralong spin coherence time in isotopically engineered diamond. *Nat. Mater.* **2009**, *8*, 383–387.

34. Ermakova, A.; Pramanik, G.; Cai, J.M.; Algara-Siller, G.; Kaiser, U.; Weil, T.; Tzeng, Y.K.; Chang, H.C.; McGuinness, L.P.; Plenio, M.B.; et al. Detection of a Few Metallo-Protein Molecules Using Color Centers in Nanodiamonds. *Nano Lett.* **2013**, *13*, 3305–3309.

35. Degen, C.; Reinhard, F.; Cappellaro, P. Quantum sensing. *arXiv* **2016**, arXiv:1611.02427.

36. Robledo, L.; Childress, L.; Bernien, H.; Hensen, B.; Alkemade, P.F.; Hanson, R. High-fidelity projective read-out of a solid-state spin quantum register. *Nature* **2011**, *477*, 574–578.

37. Luan, L.; Grinolds, M.S.; Hong, S.; Maletinsky, P.; Walsworth, R.L.; Yacoby, A. Decoherence imaging of spin ensembles using a scanning single-electron spin in diamond. *Sci. Rep.* **2015**, *5*, 8119.

38. Ryan, C.A.; Hodges, J.S.; Cory, D.G. Robust Decoupling Techniques to Extend Quantum Coherence in Diamond. *Phys. Rev. Lett.* **2010**, *105*, 200402.

39. Ziegler, J.F.; Ziegler, M.D.; Biersack, J.P. SRIM—The stopping and range of ions in matter (2010). *Nucl. Instrum. Methods Phys. Res. Sect. B Beam Interact. Mater. At.* **2010**, *268*, 1818–1823.

40. Wang, J.; Zhang, W.; Zhang, J.; You, J.; Li, Y.; Guo, G.; Feng, F.; Song, X.; Lou, L.; Zhua, W.; Wang, G. Coherence times of precise depth controlled NV centers in diamond. *Nanoscale* **2016**, *8*, 5780–5785.

41. Appel, P.; Neu, E.; Ganzhorn, M.; Barfuss, A.; Batzer, M.; Gratz, M.; Tschöpe, A.; Maletinsky, P. Fabrication of all diamond scanning probes for nanoscale magnetometry. *Rev. Sci. Instrum.* **2016**, *87*, 063703.

42. Antonov, D.; Häußermann, T.; Aird, A.; Roth, J.; Trebin, H.R.; Müller, C.; McGuinness, L.; Jelezko, F.; Yamamoto, T.; Isoya, J.; et al. Statistical investigations on nitrogen-vacancy center creation. *Appl. Phys. Lett.* **2014**, *104*, 012105.

43. Orwa, J.O.; Santori, C.; Fu, K.M.C.; Gibson, B.; Simpson, D.; Aharonovich, I.; Stacey, A.; Cimmino, A.; Balog, P.; Markham, M.; et al. Engineering of nitrogen-vacancy color centers in high purity diamond by ion implantation and annealing. *J. Appl. Phys.* **2011**, *109*, 083530.

44. Santori, C.; Barclay, P.E.; Fu, K.M.C.; Beausoleil, R.G. Vertical distribution of nitrogen-vacancy centers in diamond formed by ion implantation and annealing. *Phys. Rev. B* **2009**, *79*, 125313.

45. Pezzagna, S.; Naydenov, B.; Jelezko, F.; Wrachtrup, J.; Meijer, J. Creation efficiency of nitrogen-vacancy centres in diamond. *New J. Phys.* **2010**, *12*, 065017.

46. Fávaro de Oliveira, F.; Momenzadeh, S.A.; Antonov, D.; Fedder, H.; Denisenko, A.; Wrachtrup, J. On the efficiency of combined ion implantation for the creation of near-surface nitrogen-vacancy centers in diamond. *Phys. Status Solidi A* **2016**, *213*, 2044–2050.

47. Ohno, K.; Heremans, F.J.; Bassett, L.C.; Myers, B.A.; Toyli, D.M.; Jayich, A.C.B.; Palmstrom, C.J.; Awschalom, D.D. Engineering shallow spins in diamond with nitrogen delta-doping. *Appl. Phys. Lett.* **2012**, *101*, 082413.

48. Osterkamp, C.; Lang, J.; Scharpf, J.; Müller, C.; McGuinness, L.P.; Diemant, T.; Behm, R.J.; Naydenov, B.; Jelezko, F. Stabilizing shallow color centers in diamond created by nitrogen delta-doping using SF6 plasma treatment. *Appl. Phys. Lett.* **2015**, *106*, 113109.

49. Ohno, K.; Joseph Heremans, F.; de las Casas, C.F.; Myers, B.A.; Alemán, B.J.; Bleszynski Jayich, A.C.; Awschalom, D.D. Three-dimensional localization of spins in diamond using ^{12}C implantation. *Appl. Phys. Lett.* **2014**, *105*, 052406.

50. McLellan, C.A.; Myers, B.A.; Kraemer, S.; Ohno, K.; Awschalom, D.D.; Bleszynski Jayich, A.C. Patterned formation of highly coherent nitrogen-vacancy centers using a focused electron irradiation technique. *Nano Lett.* **2016**, *16*, 2450–2454.

51. Chandran, M.; Michaelson, S.; Saguy, C.; Hoffman, A. Fabrication of a nanometer thick nitrogen delta doped layer at the sub-surface region of (100) diamond. *Appl. Phys. Lett.* **2016**, *109*, 221602.

52. Favaro de Oliveira, F.; Momenzadeh, S.A.; Antonov, D.; Scharpf, J.; Osterkamp, C.; Naydenov, B.; Jelezko, F.; Denisenko, A.; Wrachtrup, J. Toward Optimized Surface δ-Profiles of Nitrogen-Vacancy Centers Activated by Helium Irradiation in Diamond. *Nano Lett.* **2016**, *16*, 2228–2233.

53. Kleinsasser, E.E.; Stanfield, M.M.; Banks, J.K.Q.; Zhu, Z.; Li, W.D.; Acosta, V.M.; Watanabe, H.; Itoh, K.M.; Fu, K.M.C. High density nitrogen-vacancy sensing surface created via He+ ion implantation of C-12 diamond. *Appl. Phys. Lett.* **2016**, *108*, 202401.

54. Mita, Y. Change of absorption spectra in type-Ib diamond with heavy neutron irradiation. *Phys. Rev. B* **1996**, *53*, 11360–11364.

55. Aslam, N.; Waldherr, G.; Neumann, P.; Jelezko, F.; Wrachtrup, J. Photo-induced ionization dynamics of the nitrogen vacancy defect in diamond investigated by single-shot charge state detection. *New J. Phys.* **2013**, *15*, 013064.

56. Chen, X.D.; Sun, F.W.; Zou, C.L.; Cui, J.M.; Zhou, L.M.; Guo, G.C. Vector magnetic field sensing by a single nitrogen vacancy center in diamond. *EPL* **2013**, *101*, 67003.

57. Beha, K.; Batalov, A.; Manson, N.B.; Bratschitsch, R.; Leitenstorfer, A. Optimum Photoluminescence Excitation and Recharging Cycle of Single Nitrogen-Vacancy Centers in Ultrapure Diamond. *Phys. Rev. Lett.* **2012**, *109*, 097404.

58. Gatto Monticone, D.; Quercioli, F.; Mercatelli, R.; Soria, S.; Borini, S.; Poli, T.; Vannoni, M.; Vittone, E.; Olivero, P. Systematic study of defect-related quenching of NV luminescence in diamond with time-correlated single-photon counting spectroscopy. *Phys. Rev. B* **2013**, *88*, 155201.

59. Mohtashami, A.; Koenderink, A.F. Suitability of nanodiamond nitrogen—Vacancy centers for spontaneous emission control experiments. *New J. Phys.* **2013**, *15*, 043017.

60. Romach, Y.; Müller, C.; Unden, T.; Rogers, L.J.; Isoda, T.; Itoh, K.M.; Markham, M.; Stacey, A.; Meijer, J.; Pezzagna, S.; et al. Spectroscopy of Surface-Induced Noise Using Shallow Spins in Diamond. *Phys. Rev. Lett.* **2015**, *114*, 017601.

61. Maletinsky, P.; Hong, S.; Grinolds, M.; Hausmann, B.; Lukin, M.; Walsworth, R.; Loncar, M.; Yacoby, A. A robust scanning diamond sensor for nanoscale imaging with single nitrogen-vacancy centres. *Nat. Nanotechnol.* **2012**, *7*, 320–324.

62. Yamamoto, T.; Umeda, T.; Watanabe, K.; Onoda, S.; Markham, M.L.; Twitchen, D.J.; Naydenov, B.; McGuinness, L.P.; Teraji, T.; Koizumi, S.; Dolde, F.; et al. Extending spin coherence times of diamond qubits by high-temperature annealing. *Phys. Rev. B* **2013**, *88*, 075206.

63. de Oliveira, F.F.; Antonov, D.; Wang, Y.; Neumann, P.; Momenzadeh, S.A.; Häußermann, T.; Pasquarelli, A.; Denisenko, A.; Wrachtrup, J. Tailoring spin defects in diamond. *arXiv* **2017**, arXiv:1701.07055.

64. Fu, K.M.C.; Santori, C.; Barclay, P.E.; Beausoleil, R.G. Conversion of neutral nitrogen-vacancy centers to negatively charged nitrogen-vacancy centers through selective oxidation. *Appl. Phys. Lett.* **2010**, *96*, 121907.

65. Hauf, M.V.; Grotz, B.; Naydenov, B.; Dankerl, M.; Pezzagna, S.; Meijer, J.; Jelezko, F.; Wrachtrup, J.; Stutzmann, M.; Reinhard, F.; et al. Chemical control of the charge state of nitrogen-vacancy centers in diamond. *Phys. Rev. B* **2011**, *83*, 081304(R).

66. Doi, Y.; Makino, T.; Kato, H.; Takeuchi, D.; Ogura, M.; Okushi, H.; Morishita, H.; Tashima, T.; Miwa, S.; Yamasaki, S.; et al. Deterministic Electrical Charge-State Initialization of Single Nitrogen-Vacancy Center in Diamond. *Phys. Rev. X* **2014**, *4*, 011057.

67. Doi, Y.; Fukui, T.; Kato, H.; Makino, T.; Yamasaki, S.; Tashima, T.; Morishita, H.; Miwa, S.; Jelezko, F.; Suzuki, Y.; et al. Pure negatively charged state of the NV center in *n*-type diamond. *Phys. Rev. B* **2016**, *93*, 081203.

68. Shanley, T.W.; Martin, A.A.; Aharonovich, I.; Toth, M. Localized chemical switching of the charge state of nitrogen-vacancy luminescence centers in diamond. *Appl. Phys. Lett.* **2014**, *105*, 063103.

69. Cui, S.; Hu, E.L. Increased negatively charged nitrogen-vacancy centers in fluorinated diamond. *Appl. Phys. Lett.* **2013**, *103*, 051603.

70. Grotz, B.; Hauf, M.V.; Dankerl, M.; Naydenov, B.; Pezzagna, S.; Meijer, J.; Jelezko, F.; Wrachtrup, J.; Stutzmann, M.; Reinhard, F.; et al. Charge state manipulation of qubits in diamond. *Nat. Commun.* **2012**, *3*, 729.

71. Hauf, M.V.; Simon, P.; Aslam, N.; Pfender, M.; Neumann, P.; Pezzagna, S.; Meijer, J.; Wrachtrup, J.; Stutzmann, M.; Reinhard, F.; et al. Addressing single nitrogen-vacancy centers in diamond with transparent in-plane gate structures. *Nano Lett.* **2014**, *14*, 2359–2364.

72. Rietwyk, K.J.; Wong, S.L.; Cao, L.; O'Donnell, K.M.; Ley, L.; Wee, A.T.S.; Pakes, C.I. Work function and electron affinity of the fluorine-terminated (100) diamond surface. *Appl. Phys. Lett.* **2013**, *102*, 091604.

73. Schreyvogel, C.; Polyakov, V.; Wunderlich, R.; Meijer, J.; Nebel, C.E. Active charge state control of single NV centres in diamond by in-plane Al-Schottky junctions. *Sci. Rep.* **2015**, *5*, 12160.

74. Groot-Berning, K.; Raatz, N.; Dobrinets, I.; Lesik, M.; Spinicelli, P.; Tallaire, A.; Achard, J.; Jacques, V.; Roch, J.F.; Zaitsev, A.M.; et al. Passive charge state control of nitrogen-vacancy centres in diamond using phosphorous and boron doping. *Phys. Status Solidi A* **2014**, *211*, 2268–2273.

75. Huang, Z.; Li, W.D.; Santori, C.; Acosta, V.M.; Faraon, A.; Ishikawa, T.; Wu, W.; Winston, D.; Williams, R.S.; Beausoleil, R.G. Diamond nitrogen-vacancy centers created by scanning focused helium ion beam and annealing. *Appl. Phys. Lett.* **2013**, *103*, 081906.

76. Naydenov, B.; Richter, V.; Beck, J.; Steiner, M.; Neumann, P.; Balasubramanian, G.; Achard, J.; Jelezko, F.; Wrachtrup, J.; Kalish, R. Enhanced generation of single optically active spins in diamond by ion implantation. *Appl. Phys. Lett.* **2010**, *96*, 163108.

77. Wang, P.; Ju, C.; Shi, F.; Du, J. Optimizing ultrasensitive single electron magnetometer based on nitrogen-vacancy center in diamond. *Chin. Sci. Bull.* **2013**, *58*, 2920–2923.

78. Lovchinsky, I.; Sushkov, A.; Urbach, E.; de Leon, N.; Choi, S.; De Greve, K.; Evans, R.; Gertner, R.; Bersin, E.; Müller, C.; et al. Nuclear magnetic resonance detection and spectroscopy of single proteins using quantum logic. *Science* **2016**, *351*, 836–841.

79. Yamano, H.; Kawai, S.; Kato, K.; Kageura, T.; Inaba, M.; Okada, T.; Higashimata, I.; Haruyama, M.; Tanii, T.; Yamada, K.; et al. Charge state stabilization of shallow nitrogen vacancy centers in diamond by oxygen surface modification. *Jpn. J. Appl. Phys.* **2017**, *56*, 04CK08.

80. Balmer, R.S.; Brandon, J.R.; Clewes, S.L.; Dhillon, H.K.; Dodson, J.M.; Friel, I.; Inglis, P.N.; Madgwick, T.D.; Markham, M.L.; Mollart, T.P.; et al. Chemical vapour deposition synthetic diamond: Materials, technology and applications. *J. Phys. Condens. Matter* **2009**, *21*, 364221.

81. Pelliccione, M.; Myers, B.A.; Pascal, L.M.A.; Das, A.; Bleszynski Jayich, A.C. Two-Dimensional Nanoscale Imaging of Gadolinium Spins via Scanning Probe Relaxometry with a Single Spin in Diamond. *Phys. Rev. Appl.* **2014**, *2*, 054014.

82. Neu, E.; Appel, P.; Ganzhorn, M.; Miguel-Sanchez, J.; Lesik, M.; Mille, V.; Jacques, V.; Tallaire, A.; Achard, J.; Maletinsky, P. Photonic nano-structures on (111)-oriented diamond. *Appl. Phys. Lett.* **2014**, *104*, 153108.

83. Lukosz, W.; Kunz, R. Light-emission By Magnetic And Electric Dipoles Close To A Plane Interface. 1. Total Radiated Power. *J. Opt. Soc. Am.* **1977**, *67*, 1607–1615.

84. Hausmann, B.J.; Khan, M.; Zhang, Y.; Babinec, T.M.; Martinick, K.; McCutcheon, M.; Hemmer, P.R.; Loncar, M. Fabrication of diamond nanowires for quantum information processing applications. *Diam. Relat. Mater.* **2010**, *19*, 621–629.

85. Tetienne, J.P.; Hingant, T.; Martinez, L.; Rohart, S.; Thiaville, A.; Diez, L.H.; Garcia, K.; Adam, J.P.; Kim, J.V.; Roch, J.F.; et al. The nature of domain walls in ultrathin ferromagnets revealed by scanning nanomagnetometry. *Nat. Commun.* **2015**, *6*, 6733.

86. Santori, C.; Barclay, P.E.; Fu, K.M.C.; Beausoleil, R.G.; Spillane, S.; Fisch, M. Nanophotonics for quantum optics using nitrogen-vacancy centers in diamond. *Nanotechnology* **2010**, *21*, 274008.

87. Bradac, C.; Gaebel, T.; Naidoo, N.; Sellars, M.J.; Twamley, J.; Brown, L.J.; Barnard, A.S.; Plakhotnik, T.; Zvyagin, A.V.; Rabeau, J.R. Observation and control of blinking nitrogen-vacancy centres in discrete nanodiamonds. *Nat. Nanotechnol.* **2010**, *5*, 345–349.

88. Faraon, A.; Santori, C.; Huang, Z.; Acosta, V.M.; Beausoleil, R.G. Coupling of Nitrogen-Vacancy Centers to Photonic Crystal Cavities in Monocrystalline Diamond. *Phys. Rev. Lett.* **2012**, *109*, 033604.

89. Li, L.; Schröder, T.; Chen, E.H.; Walsh, M.; Bayn, I.; Goldstein, J.; Gaathon, O.; Trusheim, M.E.; Lu, M.; Mower, J.; et al. Coherent spin control of a nanocavity-enhanced qubit in diamond. *Nat. Commun.* **2015**, *6*, 6173.

90. Lee, K.; Chen, X.; Eghlidi, H.; Kukura, P.; Lettow, R.; Renn, A.; Sandoghdar, V.; Götzinger, S. A planar dielectric antenna for directional single-photon emission and near-unity collection efficiency. *Nat. Photonics* **2011**, *5*, 166–169.

91. Riedel, D.; Rohner, D.; Ganzhorn, M.; Kaldewey, T.; Appel, P.; Neu, E.; Warburton, R. J.; Maletinsky, P. Low-Loss Broadband Antenna for Efficient Photon Collection from a Coherent Spin in Diamond. *Phys. Rev. Appl.* **2014**, *2*, 064011.

92. Le Sage, D.; Arai, K.; Glenn, D.; DeVience, S.; Pham, L.; Rahn-Lee, L.; Lukin, M.; Yacoby, A.; Komeili, A.; Walsworth, R. Optical magnetic imaging of living cells. *Nature* **2013**, *496*, 486–489.

93. Marseglia, L.; Hadden, J.P.; Stanley-Clarke, A.C.; Harrison, J.P.; Patton, B.; Ho, Y.L.D.; Naydenov, B.; Jelezko, F.; Meijer, J.; Dolan, P.R.; et al. Nanofabricated solid immersion lenses registered to single emitters in diamond. *Appl. Phys. Lett.* **2011**, *98*, 133107.

94. Hausmann, B.J.M.; Choy, J.T.; Babinec, T.M.; Shields, B.J.; Bulu, I.; Lukin, M.D.; Loncar, M. Diamond nanophotonics and applications in quantum science and technology. *Phys. Status Solidi A* **2012**, *209*, 1619–1630.

95. Beha, K.; Fedder, H.; Wolfer, M.; Becker, M.C.; Siyushev, P.; Jamali, M.; Batalov, A.; Hinz, C.; Hees, J.; Kirste, L.; et al. Diamond nanophotonics. *Beilstein J. Nanotechnol.* **2012**, *3*, 895–908.

96. Aharonovich, I.; Neu, E. Diamond Nanophotonics. *Adv. Opt. Mater.* **2014**, *2*, 911–928.

97. Schroder, T.; Mouradian, S.L.; Zheng, J.; Trusheim, M.E.; Walsh, M.; Chen, E.H.; Li, L.; Bayn, I.; Englund, D. Quantum nanophotonics in diamond. *J. Opt. Soc. Am. B Opt. Phys.* **2016**, *33*, B65–B83.

98. Kleinlein, J.; Borzenko, T.; Münzhuber, F.; Brehm, J.; Kiessling, T.; Molenkamp, L. NV-center diamond cantilevers: Extending the range of available fabrication methods. *Microelectron. Eng.* **2016**, *159*, 70–74.

99. Pelliccione, M.; Jenkins, A.; Ovartchaiyapong, P.; Reetz, C.; Emmanouilidou, E.; Ni, N.; Jayich, A.C.B. Scanned probe imaging of nanoscale magnetism at cryogenic temperatures with a single-spin quantum sensor. *Nat. Nanotechnol.* **2016**, *11*, 700–705.

100. Nelz, R.; Fuchs, P.; Opaluch, O.; Sonusen, S.; Savenko, N.; Podgursky, V.; Neu, E. Color center fluorescence and spin manipulation in single crystal, pyramidal diamond tips. *Appl. Phys. Lett.* **2016**, *109*, 193105.

101. Tao, Y.; Boss, J.; Moores, B.; Degen, C. Single-crystal diamond nanomechanical resonators with quality factors exceeding one million. *Nat. Commun.* **2014**, *5*, 3638.

102. Jung, T.; Kreiner, L.; Pauly, C.; Mücklich, F.; Edmonds, A.M.; Markham, M.; Becher, C. Reproducible fabrication and characterization of diamond membranes for photonic crystal cavities. *Phys. Status Solidi A* **2016**, *213*, 3254–3264.

103. Piracha, A.H.; Ganesan, K.; Lau, D.W.M.; Stacey, A.; McGuinness, L.P.; Tomljenovic-Hanic, S.; Prawer, S. Scalable fabrication of high-quality, ultra-thin single crystal diamond membrane windows. *Nanoscale* **2016**, *8*, 6860–6865.

104. Riedrich-Möller, J.; Kipfstuhl, L.; Hepp, C.; Neu, E.; Pauly, C.; Mücklich, F.; Baur, A.; Wandt, M.; Wolff, S.; Fischer, M.; et al. One- and two-dimensional photonic crystal microcavities in single crystal diamond. *Nat. Nanotechnol.* **2012**, *7*, 69.

105. Babinec, T.; Hausmann, B.; Khan, M.; Zhang, Y.; Maze, J.; Hemmer, P.; Loncar, M. A diamond nanowire single-photon source. *Nat. Nanotechnol.* **2010**, *5*, 195–199.

106. Hausmann, B.; Babinec, T.; Choy, J.; Hodges, J.; Hong, S.; Bulu, I.; Yacoby, A.; Lukin, M.; Lončar, M. Single-color centers implanted in diamond nanostructures. *New J. Phys.* **2011**, *13*, 045004.

107. Momenzadeh, S.A.; Stöhr, R.J.; de Oliveira, F.F.; Brunner, A.; Denisenko, A.; Yang, S.; Reinhard, F.; Wrachtrup, J. Nanoengineered Diamond Waveguide as a Robust Bright Platform for Nanomagnetometry Using Shallow Nitrogen Vacancy Centers. *Nano Lett.* **2015**, *15*, 165–169.

108. Jiang, Q.; Li, W.; Tang, C.; Chang, Y.; Hao, T.; Pan, X.; Ye, H.; Li, J.; Gu, C. Large scale fabrication of nitrogen vacancy-embedded diamond nanostructures for single-photon source applications. *Chin. Phys. B* **2016**, *25*, 118105.

109. Lesik, M.; Tetienne, J.P.; Tallaire, A.; Achard, J.; Mille, V.; Gicquel, A.; Roch, J.F.; Jacques, V. Perfect preferential orientation of nitrogen-vacancy defects in a synthetic diamond sample. *Appl. Phys. Lett.* **2014**, *104*, 113107.

110. Michl, J.; Teraji, T.; Zaiser, S.; Jakobi, I.; Waldherr, G.; Dolde, F.; Neumann, P.; Doherty, M.W.; Manson, N.B.; Isoya, J.; et al. Perfect alignment and preferential orientation of nitrogen-vacancy centers during chemical vapor deposition diamond growth on (111) surfaces. *Appl. Phys. Lett.* **2014**, *104*, 102407.

111. Fukui, T.; Doi, Y.; Miyazaki, T.; Miyamoto, Y.; Kato, H.; Matsumoto, T.; Makino, T.; Yamasaki, S.; Morimoto, R.; Tokuda, N.; et al. Perfect selective alignment of nitrogen-vacancy centers in diamond. *Appl. Phys Express* **2014**, *7*, 055201.

112. Tallaire, A.; Achard, J.; Boussadi, A.; Brinza, O.; Gicquel, A.; Kupriyanov, I.; Palyanov, Y.; Sakr, G.; Barjon, J. High quality thick CVD diamond films homoepitaxially grown on (111)-oriented substrates. *Diam. Relat. Mater.* **2014**, *41*, 34–40.

113. Lesik, M.; Plays, T.; Tallaire, A.; Achard, J.; Brinza, O.; William, L.; Chipaux, M.; Toraille, L.; Debuisschert, T.; Gicquel, A.; et al. Preferential orientation of NV defects in CVD diamond films grown on (113)-oriented substrates. *Diam. Relat. Mater.* **2015**, *56*, 47–53.

114. Aharonovich, I.; Lee, J.C.; Magyar, A.P.; Bracher, D.O.; Hu, E.L. Bottom-up engineering of diamond micro- and nano-structures. *Laser Photonics Rev.* **2013**, *7*, L61–L65.

115. Zhang, X.; Hu, E.L. Templated growth of diamond optical resonators via plasma-enhanced chemical vapor deposition. *Appl. Phys. Lett.* **2016**, *109*, 081101.

116. de Oliveira, F.F.; Momenzadeh, S.A.; Wang, Y.; Konuma, M.; Markham, M.; Edmonds, A.M.; Denisenko, A.; Wrachtrup, J. Effect of low-damage inductively coupled plasma on shallow nitrogen-vacancy centers in diamond. *Appl. Phys. Lett.* **2015**, *107*, 073107.

117. Lesik, M.; Spinicelli, P.; Pezzagna, S.; Happel, P.; Jacques, V.; Salord, O.; Rasser, B.; Delobbe, A.; Sudraud, P.; Tallaire, A.; et al. Maskless and targeted creation of arrays of colour centres in diamond using focused ion beam technology. *Phys. Status Solidi A* **2013**, *210*, 2055–2059.

118. Meijer, J.; Pezzagna, S.; Vogel, T.; Burchard, B.; Bukow, H.; Rangelow, I.; Sarov, Y.; Wiggers, H.; Plümel, I.; Jelezko, F.; et al. Towards the implanting of ions and positioning of nanoparticles with nm spatial resolution. *Appl. Phys. A Mater. Sci. Process.* **2008**, *91*, 567–571.

119. Jacob, G.; Groot-Berning, K.; Wolf, S.; Ulm, S.; Couturier, L.; Dawkins, S.T.; Poschinger, U.G.; Schmidt-Kaler, F.; Singer, K. Transmission microscopy with nanometer resolution using a deterministic single ion source. *Phys. Rev. Lett.* **2016**, *117*, 043001.

120. Bourgeois, E.; Jarmola, A.; Siyushev, P.; Gulka, M.; Hruby, J.; Jelezko, F.; Budker, D.; Nesladek, M. Photoelectric detection of electron spin resonance of nitrogen-vacancy centres in diamond. *Nat. Commun.* **2015**, *6*, 8577.

121. Hrubesch, F.M.; Braunbeck, G.; Stutzmann, M.; Reinhard, F.; Brandt, M.S. Efficient Electrical Spin Readout of NV$^-$ Centers in Diamond. *Phys. Rev. Lett.* **2017**, *118*, 037601.

122. Sekatskii, S.; Dukenbayev, K.; Mensi, M.; Mikhaylov, A.; Rostova, E.; Smirnov, A.; Suriyamurthy, N.; Dietler, G. Single molecule fluorescence resonance energy transfer scanning near-field optical microscopy: Potentials and challenges. *Faraday Discuss.* **2015**, *184*, 51–69.

123. Drezet, A.; Sonnefraud, Y.; Cuche, A.; Mollet, O.; Berthel, M.; Huant, S. Near-field microscopy with a scanning nitrogen-vacancy color center in a diamond nanocrystal: A brief review. *Micron* **2015**, *70*, 55–63.

124. Cuche, A.; Drezet, A.; Sonnefraud, Y.; Faklaris, O.; Treussart, F.; Roch, J.F.; Huant, S. Near-field optical microscopy with a nanodiamond-based single-photon tip. *Opt. Express* **2009**, *17*, 19969–19980.

125. Dexter, D.L. A Theory of Sensitized Luminescence in Solids. *J. Chem. Phys.* **1953**, *21*, 836–850.

126. Mohan, N.; Tzeng, Y.; Yang, L.; Chen, Y.; Hui, Y.; Fang, C.; Chang, H. Sub-20-nm Fluorescent Nanodiamonds as Photostable Biolabels and Fluorescence Resonance Energy Transfer Donors. *Adv. Mater.* **2010**, *22*, 843–847.

127. Rittweger, E.; Han, K.Y.; Irvine, S.E.; Eggeling, C.; Hell, S.W. STED microscopy reveals crystal colour centres with nanometric resolution. *Nat. Photonics* **2009**, *3*, 144–147.

128. Han, K.Y.; Kim, S.K.; Eggeling, C.; Hell, S.W. Metastable Dark States Enable Ground State Depletion Microscopy of Nitrogen Vacancy Centers in Diamond with Diffraction-Unlimited Resolution. *Nano Lett.* **2010**, *10*, 3199–3203.

129. Pfender, M.; Aslam, N.; Waldherr, G.; Neumann, P.; Wrachtrup, J. Single-spin stochastic optical reconstruction microscopy. *Proc. Natl. Acad. Sci. USA* **2014**, *111*, 14669–14674.

130. Monticone, D.G.; Katamadze, K.; Traina, P.; Moreva, E.; Forneris, J.; Ruo-Berchera, I.; Olivero, P.; Degiovanni, I.; Brida, G.; Genovese, M. Beating the Abbe diffraction limit in confocal microscopy via nonclassical photon statistics. *Phys. Rev. Lett.* **2014**, *113*, 143602.

131. Schaefer-Nolte, E.; Reinhard, F.; Ternes, M.; Wrachtrup, J.; Kern, K. A diamond-based scanning probe spin sensor operating at low temperature in ultra-high vacuum. *Rev. Sci. Instrum.* **2014**, *85*, 013701.

132. Thiel, L.; Rohner, D.; Ganzhorn, M.; Appel, P.; Neu, E.; Müller, B.; Kleiner, R.; Koelle, D.; Maletinsky, P. Quantitative nanoscale vortex imaging using a cryogenic quantum magnetometer. *Nat. Nanotechnol.* **2016**, *11*, 677–681.

133. Kosub, T.; Kopte, M.; Hühne, R.; Appel, P.; Shields, B.; Maletinsky, P.; Hübner, R.; Liedke, M.O.; Fassbender, J.; Schmidt, O.G.; et al. Purely antiferromagnetic magnetoelectric random access memory. *Nat. Commun.* **2017**, *8*, 13985.

134. Wrachtrup, J.; Finkler, A. Single spin magnetic resonance. *J. Magn. Reson.* **2016**, *269*, 225–236.

135. Mamin, H.J.; Kim, M.; Sherwood, M.H.; Rettner, C.T.; Ohno, K.; Awschalom, D.D.; Rugar, D. Nanoscale Nuclear Magnetic Resonance with a Nitrogen-Vacancy Spin Sensor. *Science* **2013**, *339*, 557–560.

136. Staudacher, T.; Shi, F.; Pezzagna, S.; Meijer, J.; Du, J.; Meriles, C.A.; Reinhard, F.; Wrachtrup, J. Nuclear Magnetic Resonance Spectroscopy on a $(5\,\text{nm})^3$ Sample Volume. *Science* **2013**, *339*, 561–563.

137. Lovchinsky, I.; Sanchez-Yamagishi, J.; Urbach, E.; Choi, S.; Fang, S.; Andersen, T.; Watanabe, K.; Taniguchi, T.; Bylinskii, A.; Kaxiras, E.; et al. Magnetic resonance spectroscopy of an atomically thin material using a single-spin qubit. *Science* **2017**, *355*, 503–507.

138. Das, T.P.; Hahn, E.L. *Nuclear Quadrupole Resonance Spectroscopy*; Academic Press Publishers: New York, NY, USA, 1958.

139. Rugar, D.; Mamin, H.J.; Sherwood, M.H.; Kim, M.; Rettner, C.T.; Ohno, K.; Awschalom, D.D. Proton magnetic resonance imaging using a nitrogen-vacancy spin sensor. *Nat. Nanotechnol.* **2015**, *10*, 120–124.

140. Häberle, T.; Schmid-Lorch, D.; Reinhard, F.; Wrachtrup, J. Nanoscale nuclear magnetic imaging with chemical contrast. *Nat. Nanotechnol.* **2015**, *10*, 125–128.

Review

Incorporation of Large Impurity Atoms into the Diamond Crystal Lattice: EPR of Split-Vacancy Defects in Diamond

Vladimir Nadolinny [1,*], Andrey Komarovskikh [1,2] and Yuri Palyanov [2,3]

[1] Nikolaev Institute of Inorganic Chemistry, Siberian Branch of Russian Academy of Sciences, Lavrentiev ave. 3, 630090 Novosibirsk, Russia; komarovskikh@niic.nsc.ru

[2] Sobolev Institute of Geology and Mineralogy, Siberian Branch of Russian Academy of Sciences, Koptyug ave. 3, 630090 Novosibirsk, Russia; palyanov@igm.nsc.ru

[3] Department of Geology and Geophysics, Novosibirsk State University, 630090 Novosibirsk, Russia

* Correspondence: spectr@niic.nsc.ru; Tel.: +7-383-330-9515

Academic Editor: Helmut Cölfen

Received: 27 June 2017; Accepted: 28 July 2017; Published: 31 July 2017

Abstract: Diamond is a unique mineral widely used in diverse fields due to its remarkable properties. The development of synthesis technology made it possible to create diamond-based semiconductor devices. In addition, doped diamond can be used as single photon emitters in various luminescence applications. Different properties are the result of the presence of impurities or intrinsic defects in diamond. Thus, the investigation of the defect formation process is of particular interest. Although hydrogen, nitrogen, and boron have been known to form different point defects, the possibility for large impurity atoms to incorporate into the diamond crystal structure has been questioned for a long time. In the current paper, the paramagnetic nickel split-vacancy defect in diamond is described, and the further investigation of nickel-, cobalt-, titanium-, phosphorus-, silicon-, and germanium-related defects is discussed.

Keywords: diamond; high-pressure high-temperature synthesis; split-vacancy structure; electron paramagnetic resonance; photoluminescence

---

1. Introduction

Diamond has always attracted the interest of scientists due to its extraordinary properties, such as high hardness and thermal conductivity. The crystal structure of diamond was first established by Bragg and Bragg in 1913 using the X-ray diffraction method [1]. The diamond crystal structure was attributed to the cubic crystal family and the hexoctahedral point group. Robertson et al. carried out an optical study of a large amount of different natural diamonds. Based on this study, they divided diamonds into two types: diamonds of different groups have different absorption spectra in infrared (IR) and ultraviolet (UV) ranges [2]. Mass spectrometry measurements of Kaiser and Bond [3] showed a large amount of a nitrogen impurity in diamond crystals of type I. The optical investigation allowed the authors to correlate the nitrogen content in diamond with unique absorption in IR and UV regions. Also, the authors tried to establish the structural forms of nitrogen impurity defects in diamond. A lattice expansion was detected in diamond with increasing nitrogen concentration, and it was concluded that nitrogen presumably formed substitutional defects. The maximum concentration of the nitrogen impurity in type I diamond detected by mass spectrometry was ~10^{20} cm^{-3}. After the comparison [4] of the IR and UV absorption spectra with the electron paramagnetic resonance (EPR) spectra, type I diamonds were also subdivided into two groups. Diamonds classified as type Ib have a high concentration of single substitutional nitrogen defects (P1 defects). The P1 defect is paramagnetic, and demonstrates a specific hyperfine structure (HFS) related to the ^{14}N nucleus (nuclear spin $I = 1$, natural abundance 99.6%) in the EPR spectra [5]. The main features of the single substitutional nitrogen defect in the IR absorption spectra are the peaks at 1130 and 1344 cm^{-1} [6,7]. Later, the peaks at 1332 cm^{-1}, 1046 cm^{-1}, and 950 cm^{-1} in the IR absorption spectra were attributed to the positively charged substitutional nitrogen defect N$^+$ [8]. Diamonds of type Ia contain nitrogen mainly in the A form. The A defect (neutral nearest-neighbor pair of nitrogen atoms) is diamagnetic, and gives rise to IR absorption at 1282 cm^{-1} [9]. Type II diamonds have insufficient nitrogen to be detected. Type IIa diamonds are non-conducting. Type IIb diamonds containing boron as a substitutional impurity are p-type semiconductors [10]. In the late 90s, progress in diamond synthesis technology allowed Evans and Qi [11] to carry out experiments on high-pressure high-temperature (HPHT) annealing at 1800–2700 K. It was found that A centers were formed by the aggregation of single nitrogen atoms. This solid-state reaction obeyed the second-order kinetics, with an activation energy of 5 eV. As the annealing temperature increases, the concentration of A centers decreases, and the characteristic spectrum related to nitrogen B centers appears in the IR absorption region. It was assumed that the B centers formed as a result of the aggregation of A centers. The second-order kinetics are also applicable to the formation of B centers. Later, the first-principles studies [12,13] confirmed that the B center consisted of four substitutional nitrogen atoms around a vacancy. After high-temperature annealing, apart from the B centers, the paramagnetic P2 center is observed. This center has electron spin $S = 1/2$ and the HFS of three equivalent nitrogen atoms. An analysis of HFS from ^{14}N and ^{13}C atoms made it possible to propose that the P2 center consisted of three nitrogen atoms near the vacancy [14,15].

Several other impurities were unambiguously established to incorporate into the diamond crystal lattice. First of all, boron was known to form substitutional defects that give rise to p-type diamonds [10]. Hydrogen yields different types of defects that could be detected in the IR absorption spectra [16]. Attempts were made to incorporate other impurity atoms into the diamond lattice using the HPHT technique or ion implantation; however, the spectral characterization was rather poor. Thus, in the late 90s, it was assumed that only impurity atoms with small atomic radii formed point defects in diamonds.

2. Nickel-Containing Centers in Diamonds

The first observation of a nickel paramagnetic center in diamond was reported by Samilovich et al. in 1971 [17]. The isotropic line with a *g*-factor of 2.032 was observed, and for the samples grown in the ^{61}Ni enriched system, the HFS of one ^{61}Ni atom was observed (W8 center). Based on the observations, the EPR signal could be attributed to point defects or nickel inclusions in diamond. Only 20 years later, Isoya et al. succeeded in growing diamond crystals of appropriate quality, and the EPR spectrum of the W8 center was analyzed [18]. The HFS of ^{61}Ni and four ^{13}C atoms were detected: $A(\text{Ni}) = 0.65$ mT, $A_{||}(^{13}\text{C}) = 1.339$ mT, $A_{\perp}(^{13}\text{C}) = 0.340$ mT, and the ^{13}C hyperfine tensor was axially symmetric about the [111] axis. The $S = 3/2$ spin state was determined by Fourier-transform nutational EPR spectroscopy. The obtained data proved that the EPR signal was from the negatively charged substitutional nickel defect.

Further works on the diamond synthesis technology were focused on the synthesis in Ni, Co, and Fe solvent–catalyst systems, rather than the synthesis in an N-C system. The use of metal melts significantly reduces the diamond nucleation temperature. At the same time, the HPHT annealing procedure of diamonds was elaborated in detail. Using these new techniques, diamond crystals were synthesized in the Fe-Ni-C and Ni-C systems in Novosibirsk, Russia [19–22]. The P1 and W8 centers were observed in the EPR spectra of diamond crystals synthesized at a temperature of 1700 K and a pressure of 5.5 GPa. Then, the obtained diamond samples were annealed at different temperatures for 3–5 h. After annealing the diamond crystals at 1800 K, the W8 content decreased, and a new spectrum of the NE4 center (the abbreviation NE means nickel exhibition) with an anisotropic *g*-tensor ($g_1 = 2.0227$, $g_2 = g_3 = 2.0988$) appeared. The analysis of the angular dependence of the NE4 spectrum showed that g_1 corresponded to the <111> direction. Two possible models were proposed for this defect:

a substitutional impurity atom with a distortion along <111>, or an impurity atom at a split-vacancy site. Large values of the g-tensor (with respect to the free electron g-factor g_e = 2.0023) with large anisotropy are typical of a d ion with more than a half-filled d shell. Thus, the NE4 paramagnetic center was proposed to have a nickel atom in its structure. It was supposed that because of the large atomic radius, there was great strain around the nickel atom at the substitutional site. On annealing, the nickel atom pushed one of the nearest carbon atoms out to an interstitial position, and moved to a vacancy, forming a nickel split-vacancy defect (in some contributions this site is known as a double semi-vacancy).

Subsequent annealing of as-grown diamonds at 2100 K leads to the disappearance of the NE4 spectrum, while a new NE1 spectrum appears. The new spectrum is characterized by spin S = 1/2, the anisotropic g-tensor, and the HFS of two equivalent nitrogen atoms (see EPR parameters in Table 1). The g_2 direction (14° from <110> in the (1–10) plane) corresponds to the C_1–C_2 direction in the split-vacancy defect (Figure 1). More information about the structure of the NE1 defect was obtained when diamond crystals were synthesized in the Ni-C system with a source material containing 5% of ^{13}C. In the EPR spectra, the HFS of four equivalent ^{13}C atoms was found for the NE1 center. The spectrum simulation showed that the observed ratio between the lines $I(^{13}C)/I(^{12}C)$ corresponded to 4.6% ^{13}C enrichment. The HFS tensor for ^{13}C has an axial symmetry, with A_1 parallel to the <111> direction. In accordance with the obtained g-tensor and HFS of ^{14}N and ^{13}C atoms, the following model was proposed for the NE1 center: the nickel atom at a split-vacancy site, with two nitrogen atoms at diametrically opposite ligand positions. The unpaired electron occupies the d($3z^2$-r^2) orbital oriented along the N_1–N_2 direction.

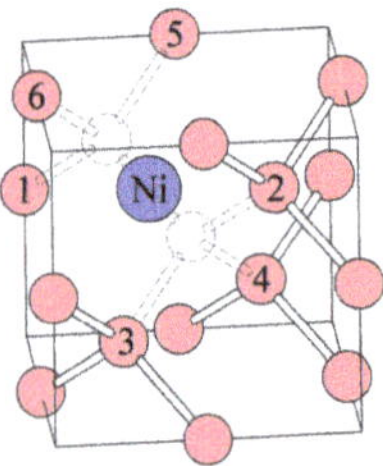

Figure 1. The diamond unit cell with a nickel atom at a split-vacancy site.

After annealing at 2100 K, the spectrum of the NE5 center appeared along with the spectrum of the NE1 center. The new EPR spectrum can be described with S = 1/2, the anisotropic g-tensor (values larger than g_e), and the HFS of two equivalent nitrogen atoms. The equivalence of nitrogen atoms in the structure of the NE5 center was established by analyzing the angular dependence of the spectrum.

An increase in the annealing temperature up to 2200 K results in the formation of two other paramagnetic centers: NE2 and NE3. For these centers, the HFS of three nitrogen atoms was distinguished. The EPR parameters of these nickel–nitrogen defects are listed in Table 1. NE2 is similar to NE1 in the g-tensor (but has lower symmetry (C_1)), and the HFS tensor, for two of the nitrogen atoms. The lower symmetry of the NE2 center is associated with the additional nitrogen atom.

Finally, after annealing the diamond crystals at 2300 K, two other nitrogen–nickel paramagnetic NE8 and NE9 centers were found. The complicated EPR spectrum of NE8 was analyzed and described with the anisotropic g-factor and the HFS from four equivalent nitrogen atoms (Table 1). The ^{14}N HFS parameters indicate that NE8 is based on a nickel split-vacancy unit with four N atoms in the plane almost normal to the g_2 direction.

The second nitrogen-nickel NE9 center has the C_{3v} symmetry, S = 1/2, and resembles the P2 paramagnetic center. The NE9 spectrum was fitted with g_1 = 2.1670, g_2 = g_3 = 2.0910, and the HFS of three equivalent nitrogen atoms $A(^{14}N)_1$ = 1.22 mT, $A(^{14}N)_2$ = 0.76 mT, $A(^{14}N)_3$ = 0.64 mT.

The discovered NE9 center was proposed to have the structure of three nitrogen atoms around the nickel split-vacancy unit.

Table 1. Parameters of the spin Hamiltonian for different nickel-containing centers with $S = 1/2$ in diamond [19–22]. The directions are for a C_{2h} symmetry site with (1–10) as its plane of reflection symmetry. g_1 corresponds to (1–10), and α is the angle between g_2 and <110>. NE2 has the C_1 symmetry, and g_1 is 20° from (1–10). β is the angle between A_1 and <111>, A_3 corresponds to (1–10). γ is the angle between A_1 and (1-1-1) or (-11-1) that rotates towards (1–10), A_2 is perpendicular to the (110) plane. For the NE2 center, the hyperfine structure (HFS) from ^{13}C was determined unambiguously only for the (100) orientation of the crystal with respect to the magnetic field.

| Center | g-Value | Constant A (mT) | | | | | |
| | | Atom Number | | | | | |
		1	2	3	4	5	6
NE1	$g_1 = 2.1282$ $g_2 = 2.0070$ $g_3 = 2.0908$ $\alpha = 14°$	$A^N_1 = 2.09$ $A^N_2 = 1.43$ $A^N_3 = 1.45$ $\beta = 5°$	$A^N_1 = 2.09$ $A^N_2 = 1.43$ $A^N_3 = 1.45$ $\beta = 5°$	$A^C_1 = 1.74$ $A^C_2 = 1.12$ $A^C_3 = 1.12$ $\gamma = 0°$	$A^C_1 = 1.74$ $A^C_2 = 1.12$ $A^C_3 = 1.12$ $\gamma = 0°$	$A^C_1 = 1.74$ $A^C_2 = 1.12$ $A^C_3 = 1.12$ $\gamma = 0°$	$A^C_1 = 1.74$ $A^C_2 = 1.12$ $A^C_3 = 1.12$ $\gamma = 0°$
NE2	$g_1 = 2.1301$ $g_2 = 2.0100$ $g_3 = 2.0931$ $\alpha = 14°$	$A^N_1 = 2.10$ $A^N_2 = 1.42$ $A^N_3 = 1.41$ $\beta = 3°$	$A^N_1 = 1.87$ $A^N_2 = 1.18$ $A^N_3 = 1.25$ $\beta = 0°$	$A^N_1 = 0.18$ $A^N_2 = 0.35$ $A^N_3 = 0.25$ $\beta = 0°$	$A^C = 1.12$	$A^C = 1.12$	$A^C = 1.12$
NE3	$g_1 = 2.0729$ $g_2 = 2.0100$ $g_3 = 2.0476$ $\alpha = 14°$	$A^N_1 = 1.60$ $A^N_2 = 1.24$ $A^N_3 = 1.15$ $\beta = 4°$		$A^N_1 = 0.66$ $A^N_2 = 0.50$ $A^N_3 = 0.50$ $\gamma = 0°$	$A^N_1 = 0.66$ $A^N_2 = 0.50$ $A^N_3 = 0.50$ $\gamma = 0°$		
NE5	$g_1 = 2.0329$ $g_2 = 2.0898$ $g_3 = 2.0476$ $\alpha = 27.5°$			$A^N_1 = 1.22$ $A^N_2 = 0.98$ $A^N_3 = 0.89$ $\gamma = 3.8°$	$A^N_1 = 1.22$ $A^N_2 = 0.98$ $A^N_3 = 0.89$ $\gamma = 3.8°$	$A^C_1 = 4.14$ $A^C_2 = 2.58$ $A^C_3 = 2.49$ $\gamma = 0°$	$A^C_1 = 4.14$ $A^C_2 = 2.58$ $A^C_3 = 2.49$ $\gamma = 0°$
NE8	$g_1 = 2.0439$ $g_2 = 2.1722$ $g_3 = 2.0476$ $\alpha = 27.5°$			$A^N_1 = 1.14$ $A^N_2 = 0.78$ $A^N_3 = 0.75$ $\gamma = 9.8°$	$A^N_1 = 1.14$ $A^N_2 = 0.78$ $A^N_3 = 0.75$ $\gamma = 9.8°$	$A^N_1 = 1.14$ $A^N_2 = 0.78$ $A^N_3 = 0.75$ $\gamma = 9.8°$	$A^N_1 = 1.14$ $A^N_2 = 0.78$ $A^N_3 = 0.75$ $\gamma = 9.8°$

3. Specific Features of the Formation of Nitrogen–Nickel Defects

As discussed previously, the transformation of the W8 center into the NE4 center is caused by strain due to a larger Ni-C bond length (~2 Å) with respect to the C-C bond length in diamond (1.54 Å). On annealing, strain around the substitutional nickel atom is relaxed by a transition of one carbon neighbor to an interstitial position; the nickel atom moves towards the created vacancy and forms a nickel split-vacancy defect. In the paper [23], it was shown that the main reason for nitrogen diffusion towards the nickel-related defects was the Coulomb interaction. The substitutional nitrogen defect donates an electron to the nickel defect. Also, there is thermal activation of nitrogen diffusion upon annealing. Based on the obtained experimental data, the formation mechanism of nitrogen–nickel defects was proposed. The evolution of the nitrogen–nickel defects consists of several stages:

1. $Ni_s^- \rightarrow NE4 + C_i$
2. $NE4 + 2N \rightarrow NE1$ $NE4 + 2N \rightarrow NE5$
3. $NE1 + N \rightarrow NE2$ $NE5 + N \rightarrow NE3$
4. $NE5 + N \rightarrow NE9$ $NE2 + N \rightarrow NE8$

All the mentioned nitrogen–nickel defects have electron spin $S = 1/2$. The electronic state of nickel–nitrogen centers was described by Ludwig–Woodbury formalism (Figure 2), which was designed for transition metal ions in silicon [24]. Considering the W8 center, nickel should donate four of its 10 electrons to form bonds with four neighboring atoms. The remaining electrons, in accordance with Hund's rule, should fill the lower orbital doublet, and then the higher orbital triplet. The W8 center is negatively charged because the substitutional nitrogen atom donates an electron. Photoillumination experiments confirmed that nitrogen centers acted as a bulk charge compensator for W8 centers [25].

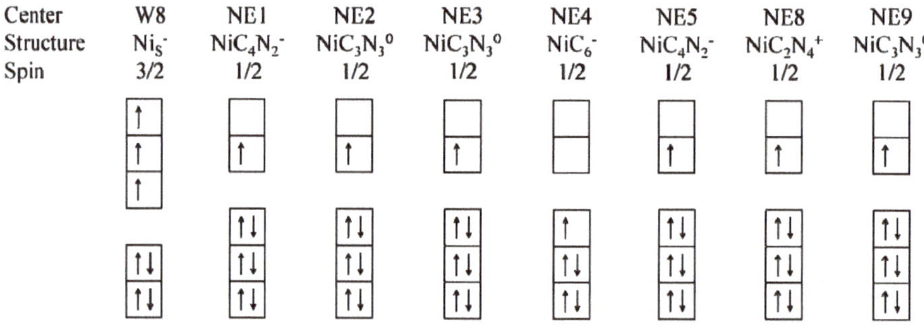

Center	W8	NE1	NE2	NE3	NE4	NE5	NE8	NE9
Structure	Ni_S^-	$NiC_4N_2^-$	$NiC_3N_3^0$	$NiC_3N_3^0$	NiC_6^-	$NiC_4N_2^-$	$NiC_2N_4^+$	$NiC_3N_3^0$
Spin	3/2	1/2	1/2	1/2	1/2	1/2	1/2	1/2

Figure 2. Electronic state of nickel-containing centers in diamond.

For nickel split-vacancy defects, nickel should donate six of its 10 electrons to form bonds with six neighboring atoms. The remaining electrons should fill the lower orbital triplet, and then the higher orbital doublet. For the NE5 and NE1 centers, the two nitrogen atoms replace two carbon atoms, and bring two extra electrons into the system. Since the NE1 and NE5 centers have electron spin $S = 1/2$, these centers are supposed to be negatively charged. Substitutional nitrogen defects were proposed to be electron donors. It should be noted that the Coulomb interaction between the negatively charged NE1 and NE5 defects and positively charged substitutional nitrogen defects leads to the formation of more complex nitrogen–nickel centers.

Another scientific group synthesized diamonds in Fe-Ni-C and Ni-C systems [26–29]. The diamonds crystals have large cubic faces. After annealing the samples at 1900–2300 K for four hours, the NE1, NE2, and NE3 paramagnetic centers were observed. The other presumably nickel-containing paramagnetic centers AB1–AB5 were also detected. For the AB1–AB5 centers, the HFS of nitrogen atoms was not distinguished (Table 2).

Table 2. Spin-Hamiltonian parameters of the nitrogen-free nickel defects in diamond [26–29].

Center	Spin	g-Value	Zero-Field Splitting (T)
AB1	$S = 1/2$	$g_{||} = 2.0024 \, || \, (111), g_\perp = 2.0920 \perp (111)$	-
AB2	$S = 1/2$	$g_{||} = 2.0072 \, || \, (111), g_\perp = 2.0672 \perp (111)$	-
AB3	$S = 1/2$	$g_1 = 2.1105 \, || \, (100), g_2 = 2.0663 \, || \, (011), g_3 = 2.0181 \, || \, (0–11)$	-
AB4	$S = 1/2$	$g_1 = 2.0220 \, || \, (100), g_2 = 2.0094 \, || \, (011), g_3 = 2.0084 \, || \, (0–11)$	-
AB5	$S = 1$	$g_{||} = 2.037 \, || \, (111), g_\perp = 2.022 \perp (111)$	$D = \pm 1.132$
NOL1 (NIRIM5)	$S = 1$	$g_{||} = 2.0235 \, || \, (111), g_\perp = 2.002 \perp (111)$	$D = -6.10$

The structure of the AB1–AB5 centers was not established. However, the authors assumed these defects to have a complex structure and to be formed as a result of aggregation of the known nickel defects (substitutional nickel atom or nickel split-vacancy defect), with some intrinsic defects (vacancies or interstitial carbon atoms). It should be noted that the formation of a <100>-split interstitial defect (R2) is energetically favorable. The R2 center has a very high mobility, and can aggregate with nickel-containing defects. Strain around the nickel atom in a tetrahedral position could be relaxed if two R2 defects were captured. It is a very tentative model of the AB3 paramagnetic center.

A defect with the same symmetry can arise if the substitutional nickel atom captures four R2 centers. This model can be proposed for the AB4 center; a smaller distortion around nickel (with respect to the AB3 center) corresponds to lower g-tensor values. The transformation of substitutional nickel defects into split-vacancy nickel defects can be a source of interstitial carbon atoms.

Since the AB1, AB2, and AB5 centers have trigonal symmetry, a possible model for these centers is the aggregation of a substitutional nickel atom with another impurity atom (for example, nitrogen) as the next neighbor. For example, the NOL1 paramagnetic center (the abbreviation NOL stands for "Novosibirsk, Oxford, London") has trigonal symmetry about the <111> axis, and the structure of the Ni-B pair at the neighboring carbon sites [29]. Alternatively, the <111> split interstitial structure can occur near the substitutional nickel atom; in this case, the strain around nickel is relaxed. This type of defects should have a trigonal symmetry; according to the Ludwig–Woodbury formalism, the neutral charge state of such a center corresponds to spin $S = 1$.

The microscopic properties of nickel-related centers in diamond were investigated with density functional theory (DFT) calculations. The obtained results on the symmetry and electronic state for the $Ni_s{}^-$ center are fully consistent with the experimental data. Moreover, the calculated spin-Hamiltonian parameters ($A(Ni) = 0.64$ mT, $A(^{13}C)_{||} = 1.46$ mT, $A(^{13}C)_\perp = 0.29$ mT) are in good agreement with the experimental ones ($A(Ni) = 0.65$ mT, $A(^{13}C)_{||} = 1.339$ mT, $A(^{13}C)_\perp = 0.340$ mT) [30]. Considering the NE4 and NE8 defects with a nickel atom at a split-vacancy site, the calculated results are consistent with the experimental data in terms of spin and symmetry [31].

4. Titanium-Containing Centers in Diamond

The structure of N3 and OK1 paramagnetic centers has been discussed for 40 years. Both centers have electron spin $S = 1/2$, and the HFS of one nitrogen atom (Table 3) [32,33]. In the photoluminescence spectra, the OK1 paramagnetic centers correspond to the S1 system (zero-phonon lines at 503.4 nm and 510.7 nm); the N3 center has a zero-phonon line at 440.3 nm. Small HFS constants suggest a low spin density on the nitrogen atom for both centers. Initially, nitrogen-vacancy complexes were proposed as possible models for these defects [34]. These models were ruled out later, since the nitrogen-vacancy centers were shown to have other EPR parameters. Newton and Baker suggested that the OK1 and N3 paramagnetic centers contained an oxygen atom within their structure [35]. A chemical analysis of the crystals containing these centers showed the presence of an oxygen impurity. However, the oxygen HFS could not be detected in the EPR spectra, since ^{17}O ($I = 5/2$, natural abundance 0.038%) is the only stable isotope with a magnetic nucleus. The finding of Novosibirsk geologists of the eclogite sample, with diamonds containing N3 and OK1 centers, stimulated new efforts in the investigation of N3 and OK1 centers [32,33]. The statistics shows that the concentration of OK1 and N3 centers inversely depends on the nitrogen concentration. This suggests that some impurity plays a role of a nitrogen getter, and is present in the structure of OK1 and N3 defects. A chemical analysis of the eclogite and its inclusion in diamond crystals shows a high titanium concentration. After that, diamond crystals were synthesized in a Fe-Ni-C system, with additional titanium as a nitrogen getter. The N3/440.3 nm center was observed in the photoluminescence spectra of the samples [36]. The analysis of the phonon structure of the N3/440.3 nm center showed that the energy of the quasi-local vibration (53 meV) corresponded to the titanium atom mass [33]. Annealing of natural diamond crystals at 2500 K for two hours showed that the N3 paramagnetic center transformed into OK1 paramagnetic centers [37].

Table 3. Electron paramagnetic resonance (EPR) parameters for titanium-containing centers with $S = 1/2$ in diamond. $g_1(A_1)$ corresponds to (1–10), α is the angle between $g_2(A_2)$ and (110). γ is the angle between A_1 and (1-1-1) or (-11-1) that rotates towards (1–10), A_2 is perpendicular to the (110) plane [32,33].

Center	g-Value	$A(^{14}N)$ (mT)	$A(^{13}C)$ (mT)	$A(^{47}Ti, \, ^{49}Ti)$ (mT)
N3	$g_1 = 2.0022, g_2 = 2.0025$ $g_3 = 2.0020$ $\alpha = 32°$	$A_1 = 0.11, A_2 = 0.15$ $A_3 = 0.11$ $\alpha = 26°$		$A_1 = 0.28, A_2 = 0.40$ $A_3 = 0.28$ $\alpha = 26°$
OK1	$g_1 = 2.0031, g_2 = 2.0019$ $g_3 = 2.0025$ $\alpha = 40°$	$A_1 = 0.55, A_2 = 0.77$ $A_3 = 0.54$ $\alpha = 20°$	$A_{||} = 4.38, A_\perp = 2.62$ $\gamma = 0$ Posit. 3, 4	$A = 0.06$

A detailed analysis of the N3 EPR spectrum allowed the authors to assume that the N3 center had the structure of a Ti–N pair at the neighboring carbon sites (Figure 3) [32,33]. The Ti–C bond lengths (~2 Å) are considerably longer than the C–C bond in diamond (1.54 Å), which results in internal strain around the defect. On high temperature annealing, the titanium atom pushes one of the nearest carbon atoms out to the interstitial position (as well as the nickel atom). The formed OK1 center consists of a titanium atom at a split-vacancy site, with one nitrogen atom in the first coordination sphere (Figure 3).

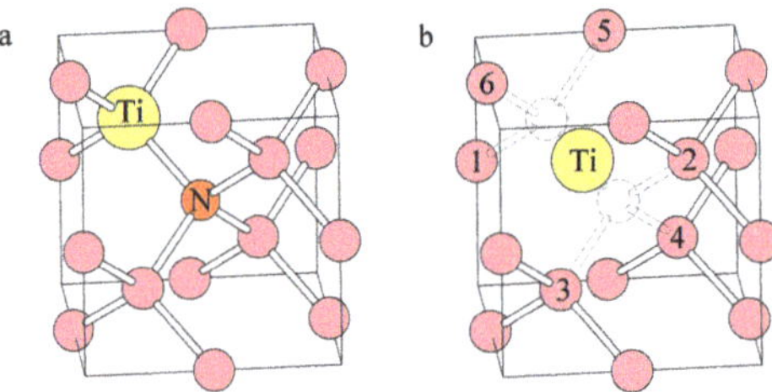

Figure 3. (**a**) Proposed model for the N3 center. (**b**) Proposed model for the OK1 center.

5. Cobalt-Containing Centers in Diamond

The cobalt-containing solvent–catalyst system is also considered to be promising for diamond growth. There were no new cobalt-containing paramagnetic centers in HPHT diamond crystals synthesized in the Co–C system at 1900 K and 6 GPa. However, after annealing at 2200 K for 18 h, three new paramagnetic centers appeared: O4, NLO2, NWO1 (the abbreviation NWO stands for "Novosibirsk, Warwick, Oxford"). These centers have spin $S = 1/2$ and the HFS characteristic of one ^{59}Co atom (cobalt has only one stable isotope). The spin-Hamiltonian parameters of the cobalt-containing spectra were determined (Table 4) [22].

The absence of paramagnetic cobalt-containing centers in as-grown diamonds can be explained using the Ludwig–Woodbury theory [24]. As expected, in as-grown diamond, substitutional cobalt defects formed. The cobalt atom donates four of its nine electrons to make bonds with four neighboring carbon atoms. The remaining electrons should fill the lower orbital doublet. The diamagnetic state corresponds to the positively charged state of this defect. The large atomic radius of cobalt affects distortions near the substitutional cobalt defect. On annealing at 2200 K, the produced strain is relaxed, and the cobalt atom pushes one of the nearest carbon atom to an interstitial position, and moves to the split-vacancy site. Following the well-tested Ludwig–Woodbury formalism, the cobalt split-vacancy defect would have spin $S = 1/2$ if the defect had the «−2» charge state. Another possibility is the presence of two nitrogen atoms in the structure of the defect, which bring two extra electrons. It allows us to imagine tree types of defects with $S = 1/2$, and cobalt positioned at a split-vacancy site. A careful analysis of ^{59}Co HFS showed that only ~10% of the spin density was located on the cobalt atom. The absence of ^{14}N HFS suggests that the spin density is localized, presumably on carbon atoms; thus, nitrogen atoms (if present) should be in the second coordination sphere.

Table 4. EPR parameters for the cobalt-containing centers with $S = 1/2$ in diamond [22]. g_2, A_2 corresponds to (011), and α is the angle between g_1, A_1 and (100).

Center	g-Value	$A(^{59}$Co) (mT)
O4	$g_1 = 2.3463$, $g_2 = 1.8438$, $g_3 = 1.7045$ $\alpha = 29°$	$A_1 = 8.86$, $A_2 = 6.43$, $A_3 = 5.82$ $\alpha = 29°$
NLO2	$g_1 = 2.3277$, $g_2 = 1.7982$, $g_3 = 1.7149$ $\alpha = 28°$	$A_1 = 8.24$, $A_2 = 6.57$, $A_3 = 5.76$ $\alpha = 28°$
NWO1	$g_1 = 2.3463$, $g_3 = 1.9458$ $\alpha = 29°$	$A_1 = 8.86$, $A_3 = 6.69$ $\alpha = 29°$

6. Phosphorus-Containing Centers in Diamond

Doped diamond is a promising material for high-power and high-temperature semiconductor devices due to its remarkable properties, such as a wide band gap, a high breakdown field, and a high thermal conductivity. The phosphorus doping of diamond is of interest, since this type of diamond exhibits n-type conductivity, and can be used in different semiconductor applications. The first work devoted to the investigation of synthetic diamonds grown with phosphorus additions was published in 1991 [38]. The MA1 center was found with $S = 1/2$ and the HFS from one phosphorus atom. The phosphorus HFS constants were small, and there were no data on which atom the spin density was localized. Other attempts were made to incorporate phosphorus via ion implantation (with subsequent annealing), or during the growth. Isoya et al. measured a new EPR spectrum of NIRIM8 (NIRIM denotes the National Institute for Research in Inorganic Materials, Japan) in HPHT diamonds grown from the phosphorus–carbon system. The authors stated that the new spectrum had the HFS from one phosphorus atom and one nitrogen atom; however, no information on the parameters was provided [39]. Finally, in the late 90s, phosphorus doped (111) homoepitaxial diamond layers with n-type conductivity were obtained [40,41].

In the works of Nadolinny et al. [42,43], the HPHT diamond crystals synthesized in the P-C medium in the temperature range 1900–2100 K were studied by EPR spectroscopy. High-pressure annealing of phosphorus-containing diamonds was performed at temperatures of 2100–2600 K. For diamonds grown at 1900 K, the MA1 center was detected in the EPR spectra (see EPR parameters in Table 5). It was established that the MA1 center consisted of a substitutional phosphorus atom, and it was suggested that the spin density was localized on the C_1 carbon atom that was far enough from the phosphorus atom. The s/p hybridization parameter for the C_1 carbon atom was found to be similar to that of the undistorted lattice. Annealing at a higher temperature (the high-pressure annealing was performed at temperatures of 2100 and 2300 K for two hours) leads to the aggregation of P1 centers (substitutional nitrogen defects) with MA1 centers (substitutional phosphorus defects), and different phosphorus–nitrogen pairs form (Figure 4). On the first step, the nitrogen–phosphorus pairs separated by two carbon atoms are formed (NIRIM8 or NP1 center). The abbreviation NP corresponds to nitrogen–phosphorus defects. With an increase in the annealing temperature, the nitrogen and phosphorus atoms move closer, and form nitrogen-phosphorus pairs separated by one carbon atom (NP2 centers) and close nitrogen–phosphorus pairs (NP3 centers). On the one hand, this behavior can be explained by the Coulomb interaction between the phosphorus and nitrogen atoms. In the theoretical work of Anderson et al. [44], it was shown that substitutional nitrogen could act as an electron acceptor for substitutional phosphorus. In addition, the possibility for nitrogen and phosphorus to form a close pair was discussed in this paper [44]. On the other hand, the increased annealing temperature allows the nitrogen atom diffusing towards the phosphorus atom to overcome the energy barrier. The energy barrier arises as a result of strain around the phosphorus atom having a large atomic radius.

It is interesting to note that the NP1–NP3 centers have electron spin $S = 1/2$, which means that these centers must be positively charged. The substitutional nitrogen defects can play the role of the acceptor for these nitrogen–phosphorus defects. In the work of Jones et al. [45], the energy state of negatively-charged substitutional nitrogen was calculated, and shown to be close to the energy state of neutral substitutional nitrogen.

Annealing of diamond crystals at 2600 K for 10 minutes leads to the disappearance of the NP1–NP3 centers, and the detection of new paramagnetic centers (NP4, NP5, and NP6) (Table 5) [46]. The NP4, NP5, and NP6 centers have the HFS of only one phosphorus atom. It is noteworthy that the EPR spectrum of the NP5 and NP6 centers appears after X-ray irradiation of the diamond crystals. Annealing at 800 K results in the reversible disappearance of the NP5 and NP6 spectra.

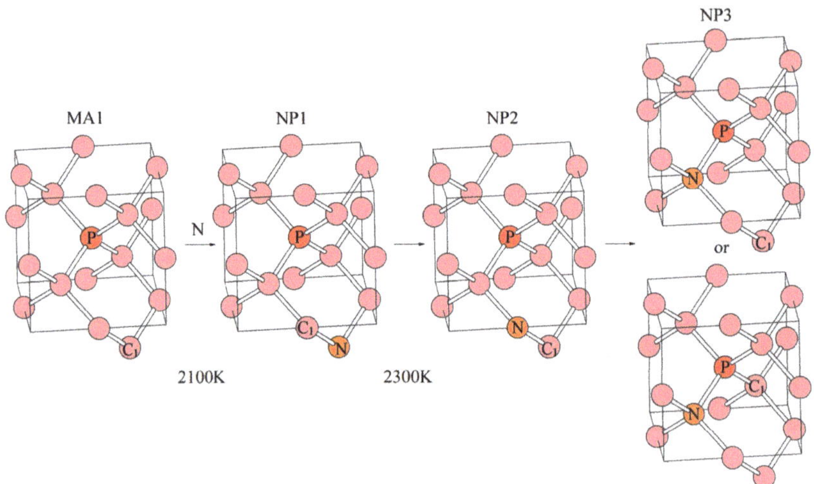

Figure 4. Supposed structures of MA1, NP1 (NIRIM8), NP2, and NP3 centers in synthetic microdiamonds grown in the C–P system.

It was supposed that when a large phosphorus atom occupied the substitutional site, it generated high strain around a defect. On annealing at 2600 K, the NP1–NP3 centers transform into the NP4–NP6 centers. The analysis of the ^{31}P HFS for the NP4–NP6 centers revealed a low spin density on the phosphorus atom. The obtained result is consistent with the data obtained in the theoretical work of Jones et al. [47]; it was shown that there was a low spin density on the P atom for the phosphorus split-vacancy defect. One can assume that the NP4–NP6 centers differ in distances between the nitrogen atom and the split-vacancy fragment. For the NP4 center consisting of a phosphorus split-vacancy defect, the unpaired electron is localized on a dangling bond of the carbon atom. In the case of NP5 and NP6 centers, the presence of a nitrogen atom in the second or third coordination spheres results in electron transfer from the nitrogen atom to the dangling bond of the carbon atom, making these centers diamagnetic. However, an electron can be trapped as a result of X-ray irradiation, which makes these two centers paramagnetic.

Two new phosphorus paramagnetic centers were detected in HPHT diamonds synthesized in the phosphorus–carbon system with aluminum introduced into the system as a getter (Table 5) [48]. Also, a decrease in the temperature of the spectral measurement leads to the appearance of a new EPR spectrum in the form of a Dyson line associated with conductivity electrons. Thus, it can be concluded that nitrogen prevents the formation of n-type conductivity in diamond.

It should be noted that the tetrahedral environment of a phosphorus atom transforms into an octahedral environment at a very high annealing temperature. The high formation temperature of the NP4–NP6 defects can be explained by the fact that impurity nitrogen can slightly relax strain around the substitutional phosphorus defect. The same tendency can be noted for the titanium-containing centers; the OK1 center (in which a titanium atom is located at a split-vacancy site) is formed at a temperature above 2300 K.

Table 5. EPR parameters of the phosphorus-containing centers with $S = 1/2$ in diamond [46,48].

Center	g-Value	Constant A (mT)
MA1	$g = 2.0025$	$A(^{31}\mathrm{P})_{\|\|} = 2.32, A(^{31}\mathrm{P})_{\perp} = 1.96$ $A(^{13}\mathrm{C}_1)_{\|\|} = 18.13, A(^{13}\mathrm{C}_1)_{\perp} = 13.92$
NIRIM8 (NP1)	$g_1 = 2.00243, g_2 = 2.0028, g_3 = 2.0026$	$A(^{31}\mathrm{P})_1 = 2.08, A(^{31}\mathrm{P})_2 = 2.02, A(^{31}\mathrm{P})_3 = 2.18$ $A(^{14}\mathrm{N})_1 = 4.08, A(^{14}\mathrm{N})_2 = 3.10, A(^{14}\mathrm{N})_3 = 3.00$
NP2	$g = 2.0025$	$A(^{31}\mathrm{P})_{\|\|} = 2.34, A(^{31}\mathrm{P})_{\perp} = 2.09$ $A(^{14}\mathrm{N})_{\|\|} = 6.42, A(^{14}\mathrm{N})_{\perp} = 3.09$
NP3	$g = 2.0025$	$A(^{31}\mathrm{P})_{\|\|} = 17.48, A(^{31}\mathrm{P})_{\perp} = 18.23$ $A(^{14}\mathrm{N})_{\|\|} = 0.10, A(^{14}\mathrm{N})_{\perp} = 0.33$
NP4	$g_1 = 2.0009, g_2 = 2.0012, g_3 = 2.00047$	$A(^{31}\mathrm{P})_1 = 5.456, A(^{31}\mathrm{P})_2 = 3.838, A(^{31}\mathrm{P})_3 = 3.80$
NP5	$g_{\|\|} = 2.00087, g_{\perp} = 2.0009$	$A(^{31}\mathrm{P})_{\|\|} = 6.522, A(^{31}\mathrm{P})_{\perp} = 1.024$
NP6	$g_{\|\|} = 2.00085, g_{\perp} = 2.00083$	$A(^{31}\mathrm{P})_1 = 7.585, A(^{31}\mathrm{P})_2 = 2.942, A(^{31}\mathrm{P})_3 = 2.328$
NP8	$g_{\|\|} = 2.0048, g_{\perp} = 2.0016$	$A(^{31}\mathrm{P})_{\|\|} = 5.6, A(^{31}\mathrm{P})_{\perp} = 3.2;$
NP9	$g_{\|\|} = 2.0030, g_{\perp} = 2.0038$	$A(^{31}\mathrm{P}_1)_{\|\|} = 13.6, A(^{31}\mathrm{P}_1)_{\perp} = 8.8$ $A(^{31}\mathrm{P}_2)_{\|\|} = 1.4, A(^{31}\mathrm{P}_2)_{\perp} = 2.2$

7. Silicon-Vacancy Defect in Diamond

During the last 15 years, optically active centers in diamond have been extensively studied. Nitrogen-vacancy, silicon-vacancy, and germanium-vacancy centers are the most known color centers. These defects are thought to be promising as single photon emitters in different quantum applications. The NV^0 and NV^- centers have zero-phonon lines at 575 and 637 nm. The NV^- center is paramagnetic, and has the EPR spectrum fitted with $S = 1$, g = 2.0028, $D = 103$ mT, $A(^{14}\mathrm{N}) = 0.08$ mT, $A(^{13}\mathrm{C})_{\|\|} = 7.31$ mT, $A(^{13}\mathrm{C})_{\perp} = 4.40$ mT [49].

The NV^- center shows a significant phonon sideband in photoluminescence spectra, and this defect has a long excited state lifetime (~10 ns [50]). Thus, other optically active centers were sought. In 1980, Vavilov et al. detected a 737 nm optical system (attributed later to the SiV^- center) for diamond films grown on a silicon substrate [51]. Silicon ion implantation resulted in the appearance of the 737 nm system [52]. In subsequent papers [53,54], it was shown that electron irradiation of silicon-containing diamonds followed by annealing (at a temperature of 1000 K) resulted in an increase in the 737 nm system. At a temperature of 1000 K, the vacancy becomes mobile in diamond. It was concluded that the center responsible for the 737 nm luminescence was a vacancy trapped in silicon. The hypothesis about the negative charge state of the defect was suggested by Collins et al. [52]. They showed that the 737 nm system was more intense in silicon-implanted diamonds that had a high concentration of substitutional nitrogen, which usually acted as an electron donor.

Different possible charges and spin states of the silicon-vacancy defect made it interesting to study silicon doped diamonds by EPR. The neutral silicon-vacancy defect was shown to be paramagnetic. The corresponding spectrum (KUL1) has an axial symmetry around <111> and was fitted with the following spin-Hamiltonian parameters: $S = 1, g_{\|\|} = 2.0040, g_{\perp} = 2.0035, D = 35.8$ mT, $E = 0$ [55–57]. In a later work [58], the HFS of one $^{29}\mathrm{Si}$ atom ($A(^{29}\mathrm{Si})_{\|\|} = 2.73$ mT, $A(^{29}\mathrm{Si})_{\perp} = 2.82$ mT) and the HFS of six equivalent $^{13}\mathrm{C}$ atoms ($A(^{13}\mathrm{C})_{\|\|} = 2.36$ mT, $A(^{13}\mathrm{C})_{\perp} = 1.08$ mT) were detected. It was shown that approximately 75% of the spin density was localized on these six carbon atoms. Based on the obtained data, this paramagnetic center was concluded to be a neutral silicon split-vacancy defect. A paramagnetic SiV^0 center was also studied at the DFT level; the structure and electronic properties of the defect were analyzed [59]. The calculated EPR parameters ($A(^{29}\mathrm{Si})_{\|\|} = 2.79$ mT, $A(^{29}\mathrm{Si})_{\perp} = 2.93$ mT, $A(^{13}\mathrm{C})_{\|\|} = 1.82$ mT, $A(^{13}\mathrm{C})_{\perp} = 0.43$ mT) are in good agreement with those determined experimentally.

The negatively charged silicon-vacancy center (responsible for the 737 nm optical system) was also expected to be paramagnetic. It was reported [56,57] that the KUL8 spectrum ($S = 1/2, g_1 = 2.00368, g_2 = 2.00336, g_3 = 2.00336$) corresponded to the SiV^- center. However, the KUL8 spectrum was not detected in a series of experiments on charge transfer between SiV^- and SiV^0, which was induced by UV photoexcitation or heating [60].

Along with the peak at 737 nm, a weak peak at 720 nm is often observed in the photoluminescence spectra of diamond crystals synthesized in the silicon-containing medium. It was supposed that the peak at 720 nm could be attributed to an unknown silicon-related center. Since boron is usually present in diamond as a trace impurity, the possibility of the formation of a silicon-boron defect was considered. The HPHT diamond crystals were synthesized in the Mg–C system, with the addition of various amounts of Si and B [61]. In the photoluminescence spectra of the samples, the intense 720 nm system was detected. At the same time, a new EPR spectrum fitted with $S = 1/2$ and an anisotropic g-factor was observed: $g_1 = 2.0033$, $g_2 = 2.0004$, $g_3 = 2.0024$. The g_3 value corresponds to (01-1). The angle between the principal direction of g_1 and (011) is $50°$. Thus, the principal direction of g_2 is close to the (1-1-1) axis. It can be noted that the g-tensor values and principle axes are similar to those of the N3 paramagnetic center, which consisted of titanium and nitrogen atoms at the adjacent carbon sites. Thus, the 720 nm system was tentatively ascribed to the center containing silicon and boron atoms at the adjacent carbon sites (Figure 5). Experiments on HPHT annealing of the diamond crystals showed that the new center was stable up to 2200 K.

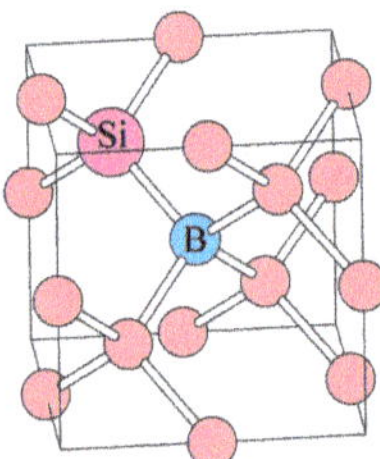

Figure 5. Proposed model for the Si-B center.

8. Germanium-Vacancy Defect in Diamond

The EPR study of diamond crystals synthesized in the Mg-Ge-C system revealed the spectrum of an unknown paramagnetic center. The analysis of the angular dependence of the EPR spectrum showed that it was characterized by an axial symmetry around <111> and had the following spin-Hamiltonian parameters: $S = 1$, $g_{||} = 2.0025$, $g_\perp = 2.0027$, $D = 80.3$ mT, $E = 0$ [61]. As shown previously, several models can be proposed for a defect with a trigonal symmetry around <111> For the newly observed center, the germanium split-vacancy model is more preferred because of the large atomic radius of germanium. The new paramagnetic center has the same symmetry and spin state as the SiV^0 center [58]; thus, the new spectrum was ascribed to the neutral germanium split-vacancy defect.

To confirm the assumption that the new paramagnetic center contains a germanium atom in the structure, additional growth experiments were performed using germanium enriched with the ^{73}Ge isotope ($I = 9/2$). For the GeV^0 paramagnetic center, the HFS of one ^{73}Ge atom was detected with an almost isotropic constant $A(^{73}Ge) \approx 1.64$ mT. The hyperfine interaction was almost isotropic for ^{73}Ge as for the ^{29}Si atom in the structure of the SiV^0 center.

The new center has a large zero-field splitting parameter D, which was suggested to be a result of the spin–orbit interaction. As is known, the zero-field splitting has two contributions: a term arising from the spin–spin dipole–dipole interaction, and a term arising from the spin–orbit coupling. Indeed, Ge has a significantly larger spin–orbit coupling constant than Si ($\lambda(Ge) = 940$ cm^{-1}, $\lambda(Si) = 149$ cm^{-1} [62]), which may result in a larger D value.

The structure and electronic properties of the GeV^0 defect were analyzed at the DFT level [63]. The EPR parameters (g-tensor, hyperfine interaction constants, and zero-field splitting (ZFS) parameters) were calculated. It was shown that the spin–orbit coupling interaction actually gave a dominant contribution to the total D parameter. In Table 6, the calculated and experimentally determined spin-Hamiltonian parameters are presented with good agreement.

Table 6. Experimental and calculated g-, D-, and A-tensors for the GeV^0 ($S = 1$) paramagnetic center [63].

Data	g-Value	D (mT)	E (mT)	A (^{73}Ge) (mT)
experiment	$g_{||} = 2.0025, g_{\perp} = 2.0027$	80.3 mT	0	$A = 1.64$ mT
DFT calculation	$g_{||} = 2.0015, g_{\perp} = 2.0012$	70.9 mT	0	$A_{||} = 1.47$ mT, $A_{\perp} = 1.55$ mT

9. Conclusions

The obtained results on the HPHT synthesis have shown that atoms with large atomic radii can be introduced into the diamond crystal lattice. Nickel-, cobalt-, titanium-, phosphorus-, germanium-, and silicon-related point defects were detected in diamond. As a rule, in the diamond crystals synthesized at relatively low temperatures, large impurity atoms occupy substitutional positions. In this case, the bond length between the impurity and carbon atoms is sufficiently longer than the carbon–carbon bond length, which causes distortions and strain around the defect. Among the defects with a large impurity atom at a substitutional position, only substitutional phosphorus (in the neutral charge state) and nickel (in the negative charged state) defects were observed. This made the investigation of defects with heavy impurity atoms rather complicated. On HPHT annealing, the thermal vibration results in relaxing the structure; then, the impurity atom pushes one of the nearest atoms out to the interstitial position. The split-vacancy structure with an impurity atom is formed; this defect has the D$_{3d}$ symmetry. The tetrahedral environment of the impurity atom usually transforms into the octahedral one at a temperature enabling the thermal diffusion of nitrogen. The Coulomb interaction can be a driving force for the nitrogen diffusion towards another impurity atom.

For the defects with two impurity atoms at the adjacent carbon sites (Ti-N, Ni-B, P-N and Si-B), the tetrahedral environment of a large impurity atom transforms into the octahedral environment at higher temperatures. This effect can be explained by partial relaxation of the structure around the Ti, Ni, P or Si atoms.

It should be emphasized that the aggregation of phosphorus atoms with nitrogen atoms is also caused by the Coulomb interaction. The experiment on HPHT annealing of diamonds synthesized in the P-C system confirmed the previous theoretical prediction that the substitutional nitrogen defect could act as an electron donor for the substitutional phosphorus defect. Thus, the presence of a nitrogen impurity prevents the formation of n-type conductivity in diamonds synthesized with the addition of phosphorus.

It was established that the electronic state of the transition metal centers was well described by the Ludwig–Woodbury formalism. In the case of the tetrahedral environment of a transition metal atom, it donates four electrons to form bonds with the neighboring atoms. The remaining electrons fill the lower orbital doublet, and then the higher orbital triplet. In the case of the octahedral environment of a transition metal atom, it donates six electrons to form bonds with carbon atoms. The remaining electrons fill the lower orbital triplet, and then the higher orbital doublet.

Finally, many defects with an impurity atom at a split-vacancy site were noticed to be optically active. Some of these centers (SiV$^-$, GeV$^-$) are considered promising single-photon sources for quantum applications [64,65].

Acknowledgments: This work was supported by the Russian Science Foundation under Grant No. 14-27-00054.

Author Contributions: Vladimir Nadolinny, Andrey Komarovskikh and Yuri Palyanov contributed equally to the writing of the manuscript.

References

1. Bragg, W.H.; Bragg, W.L. The Structure of the Diamond. *Nature* **1913**, *2283*, 557. [CrossRef]
2. Robertson, R.; Fox, J.J.; Martin, A.E. Two Types of Diamond. *Philos. Trans. R. Soc. Lond. A* **1934**, *232*, 463–535. [CrossRef]
3. Kaiser, W.; Bond, W.L. Nitrogen, A Major Impurity in Common Type I Diamond. *Phys. Rev.* **1959**, *115*, 857–863. [CrossRef]
4. Dyer, H.B.; Raal, F.A.; Du Preez, L.; Loubser, J.H.N. Optical absorption features associated with paramagnetic nitrogen in diamond. *Philos. Mag.* **1965**, *11*, 763–774. [CrossRef]
5. Smith, W.V.; Sorokin, P.P.; Gelles, I.L.; Lasher, G.J. Electron-Spin Resonance of Nitrogen Donors in Diamond. *Phys. Rev.* **1959**, *115*, 1546–1552. [CrossRef]
6. Woods, G.S.; van Wyk, J.A.; Collins, A.T. The nitrogen content of type Ib synthetic diamond. *Philos. Mag. Part B* **1990**, *62*, 589–595. [CrossRef]
7. Collins, A.T.; Stanley, M.; Woods, G.S. Nitrogen isotope effects in synthetic diamonds. *J. Phys. D Appl. Phys.* **1987**, *20*, 969–974. [CrossRef]
8. Lawson, S.C.; Fisher, D.; Huntz, D.C.; Newton, M.E. On the existence of positively charged single-substitutional nitrogen in diamond. *J. Phys. Condens. Matter* **1998**, *10*, 6171–6180. [CrossRef]
9. Boyd, S.R.; Kiflawi, I.; Woods, G.S. The relationship between infrared absorption and the A defect concentration in diamond. *Philos. Mag. Part B* **1994**, *69*, 1149–1153. [CrossRef]
10. Collins, A.T.; Williams, A.W.S. The nature of the acceptor centre in semiconducting diamond. *J. Phys. C Solid State Phys.* **1971**, *4*, 1789–1800. [CrossRef]
11. Evans, T.; Qi, Z. The kinetics of aggregation of nitrogen atoms in diamond. *Proc. R. Soc. Lond. A* **1982**, *381*, 159–178. [CrossRef]
12. Mainwood, A. Nitrogen and nitrogen-vacancy complexes and their formation in diamond. *Phys. Rev. B* **1994**, *49*, 7934–7940. [CrossRef]
13. Jones, R.; Briddon, P.R.; Öberg, S. First-principles theory of nitrogen aggregates in diamond. *Philos. Mag. Lett.* **1992**, *66*, 67–74. [CrossRef]
14. Shcherbakova, M.Y.; Nadolinny, V.A.; Sobolev, E.V. The N3 center in natural diamonds, from ESR data. *J. Struct. Chem.* **1978**, *19*, 261–269. [CrossRef]
15. Van Wyk, J.A. Carbon-12 hyperfine interaction of the unique carbon of the P2 (ESR) or N3 (optical) centre in diamond. *J. Phys. C Solid State Phys.* **1982**, *15*, L981–L983. [CrossRef]
16. Woods, G.S.; Collins, A.T. Infrared absorption spectra of hydrogen complexes in type I diamonds. *J. Phys. Chem. Solids* **1983**, *44*, 471–475. [CrossRef]
17. Samoilovich, M.I.; Bezrukov, G.N.; Butuzov, V.P. Electron paramagnetic resonance of nickel in synthetic diamond. *JETP Lett.* **1971**, *14*, 379–381.
18. Isoya, J.; Kanda, H.; Norris, J.R.; Tang, J.; Bowman, M.K. Fourier-transform and continuous-wave EPR studies of nickel in synthetic diamond: Site and spin multiplicity. *Phys. Rev. B* **1990**, *41*, 3905–3913. [CrossRef]
19. Nadolinny, V.A.; Yelisseyev, A.P. New paramagnetic centres containing nickel ions in diamond. *Diam. Relat. Mater.* **1993**, *3*, 17–21. [CrossRef]
20. Nadolinny, V.A.; Yelisseyev, A.P. Structure and creation conditions of complex nitrogen-nickel defects in synthetic diamonds. *Diam. Relat. Mater.* **1994**, *3*, 1196–1200. [CrossRef]
21. Nadolinny, V.A.; Yelisseyev, A.P.; Baker, J.M. A study of 13C hyperfine structure in the EPR of nickel-nitrogen-containing centres in diamond and correlation with their optical properties. *J. Phys. Condens. Matter* **1999**, *11*, 7357–7376. [CrossRef]
22. Nadolinny, V.A.; Baker, J.M.; Yuryeva, O.P. EPR Study of the Peculiarities of Incorporating Transition Metal Ions into the Diamond Structure. *Appl. Magn. Reson.* **2005**, *28*, 365–381. [CrossRef]
23. Nadolinny, V.A.; Yelisseyev, A.P.; Baker, J.M.; Twitchen, D.J.; Newton, M.E.; Feigelson, B.N.; Yuryeva, O.P. Mechanisms of nitrogen aggregation in nickel- and cobalt-containing synthetic diamonds. *Diam. Relat. Mater.* **2000**, *9*, 883–886. [CrossRef]
24. Ludwig, G.W.; Woodbury, H.H. Electron Spin Resonance in Semiconductors. *Solid State Phys.* **1962**, *13*, 223–304. [CrossRef]

25. Nadolinny, V.; Yelisseyev, A.; Yurjeva, O.; Hofstaetter, A.; Meyer, B.; Feigelson, B. Relationship between electronic states of nickel-containing centres and donor nitrogen in synthetic and natural diamonds. *Diam. Relat. Mater.* **1998**, *7*, 1558–1561. [CrossRef]

26. Pereira, R.N.; Neves, A.J.; Gehlhoff, W.; Sobolev, N.A.; Rino, L.; Kanda, H. Annealing study of the formation of nickel-related paramagnetic defects in diamond. *Diam. Relat. Mater.* **2002**, *11*, 623–626. [CrossRef]

27. Neves, A.J.; Pereira, R.; Sobolev, N.A.; Nazare, M.H.; Gehlhoff, W.; Naser, A.; Kanda, H. New paramagnetic defects in synthetic diamonds grown using nickel catalyst. *Physica B* **1999**, *273–274*, 651–654. [CrossRef]

28. Neves, A.J.; Pereira, R.; Sobolev, N.A.; Nazare, M.H.; Gehlhoff, W.; Naser, A.; Kanda, H. New paramagnetic centers in annealed high-pressure synthetic diamond. *Diam. Relat. Mater.* **2000**, *9*, 1057–1060. [CrossRef]

29. Rakhmanova, M.I.; Nadolinny, V.A.; Yuryeva, O.P. Impurity Centers in Synthetic and Natural Diamonds with a System of Electron-Vibronic Lines at 418 nm in Luminescence Spectra. *Phys. Solid State* **2013**, *55*, 127–130. [CrossRef]

30. Larico, R.; Assali, L.V.C.; Machado, W.V.M. Isolated nickel impurities in diamond: A microscopic model for the electrically active centers. *Appl. Phys. Lett.* **2004**, *84*, 720–722. [CrossRef]

31. Larico, R.; Justo, J.F.; Machado, W.V.M.; Assali, L.V.C. Electronic properties and hyperfine fields of nickel-related complexes in diamond. *Phys. Rev. B* **2009**, *79*, 115202. [CrossRef]

32. Nadolinny, V.A.; Yuryeva, O.P.; Shatsky, V.S.; Stepanov, A.S.; Golushko, V.V.; Rakhmanova, M.I.; Kupriyanov, I.N.; Kalinin, A.A.; Palyanov, Y.N.; Zedgenizov, D. New Data on the Nature of the EPR OK1 and N3 Centers in Diamond. *Appl. Magn. Reson.* **2009**, *36*, 97–108. [CrossRef]

33. Nadolinny, V.A.; Yuryeva, O.P.; Rakhmanova, M.I.; Shatsky, V.S.; Palyanov, Y.N.; Kupriyanov, I.N.; Zedgenizov, D.A.; Ragozin, A.L. Distribution of OK1, N3, and NU1 defects in diamonds of different habits. *Eur. J. Mineral.* **2012**, *24*, 645–650. [CrossRef]

34. Scherbakova, M.Y.; Sobolev, E.V.; Nadolinny, V.A. Electron paramagnetic resonance of low-symmetry impurity centers in diamond. *Dokl. Acad. Nauk SSSR* **1972**, *204*, 851–854. (In Russian)

35. Newton, M.E.; Baker, J.M. ^{14}N ENDOR of the OK1 centre in natural type Ib diamond. *J. Phys. Condens. Matter* **1989**, *1*, 10549–10561. [CrossRef]

36. Nadolinny, V.; Yuryeva, O.; Chepurov, A.; Shatsky, V. Titanium Ions in the Diamond Structure: Model and Experimental Evidence. *Appl. Magn. Reson.* **2009**, *36*, 109–113. [CrossRef]

37. Nadolinny, V.; Palyanov, Y.; Yuryeva, O.; Zedgenizov, D.; Rakhmanova, M.; Kalinin, A.; Komarovskikh, A. The influence of HTHP treatment on the OK1 and N3 centers in natural diamonds. *Phys. Status Solidi A* **2015**, *212*, 2474–2479. [CrossRef]

38. Samsonenko, N.D.; Tokii, V.V.; Gorban, S.V. Electron paramagnetic resonance of phosphorus in diamond. *Fiz. Tver. Tela* **1991**, *33*, 2496–2498. (In Russian)

39. Isoya, J.; Kanda, H.; Akaishi, M.; Morita, Y.; Ohshima, T. ESR studies of incorporation of phosphorus into high-pressure synthetic diamond. *Diam. Relat. Mater.* **1997**, *6*, 356–360. [CrossRef]

40. Koizumi, S.; Kamo, M.; Sato, Y.; Ozaki, H.; Inuzuka, T. Growth and characterization of phosphorous doped {111} homoepitaxial diamond thin films. *Appl. Phys. Lett.* **1997**, *71*, 1065–1067. [CrossRef]

41. Hasegawa, M.; Teraji, T.; Koizumi, S. Lattice location of phosphorus in *n*-type homoepitaxial diamond films grown by chemical-vapor deposition. *Appl. Phys. Lett.* **2001**, *79*, 3068–3070. [CrossRef]

42. Nadolinny, V.A.; Palyanov, Y.N.; Kupriyanov, I.N.; Newton, M.J.; Kryukov, E.; Sokol, A.G. A new EPR data on the MA1 and NIRIM8 (NP1) phosphorus-related centers in synthetic 1b diamonds. *Appl. Magn. Reson.* **2012**, *42*, 179–186. [CrossRef]

43. Nadolinny, V.A.; Pal'yanov, Y.N.; Kalinin, A.A.; Kupriyanov, I.N.; Veber, S.L.; Newton, M.J. Transformation of As-Grown Phosphorus-Related Centers in HPHT Treated Synthetic Diamonds. *Appl. Magn. Reson.* **2011**, *41*, 371–382. [CrossRef]

44. Anderson, A.B.; Kostadinov, L.N. P and N compensation in diamond molecular orbital theory. *J. Appl. Phys.* **1997**, *81*, 264–267. [CrossRef]

45. Jones, R.; Goss, J.P.; Briddon, P.R. Acceptor level of nitrogen in diamond and the 270-nm absorption band. *Phys. Rev. B* **2009**, *80*, 033205. [CrossRef]

46. Nadolinny, V.; Komarovskikh, A.; Pal'yanov, Y.; Kupriyanov, I. EPR of new phosphorus-containing centers in synthetic diamonds. *Phys. Status Solidi A* **2013**, *210*, 2078–2082. [CrossRef]

47. Jones, R.; Lowther, J.E.; Goss, J. Limitations to *n*-type doping in diamond: The phosphorus-vacancy complex. *Appl. Phys. Lett.* **1996**, *69*, 2489–2491. [CrossRef]

48. Nadolinny, V.; Komarovskikh, A.; Palyanov, Y.; Sokol, A. EPR of synthetic diamonds heavily doped with phosphorus. *Phys. Status Solidi A* **2015**, *212*, 2568–2571. [CrossRef]
49. Loubser, J.H.N.; van Wyk, J.A. Electron spin resonance in the study of diamond. *Rep. Prog. Phys.* **1978**, *41*, 1201–1248. [CrossRef]
50. Collins, A.T.; Thomaz, M.F.; Jorge, M.I.B. Luminescence decay time of the 1.945 eV centre in type Ib diamond. *J. Phys. C Solid State Phys.* **1983**, *16*, 2177–2181. [CrossRef]
51. Vavilov, V.S.; Gippius, A.A.; Zaitsev, A.M.; Deryagin, B.V.; Spitsyn, B.V.; Aleksenko, A.E. Investigation of the cathodoluminescence of epitaxial diamond films. *Fiz. Tekh. Polyprovodn.* **1980**, *14*, 1811–1813. (In Russian)
52. Collins, A.T.; Kamo, M.; Sato, Y. A spectroscopic study of optical centers in diamond grown by microwave-assisted chemical vapor deposition. *J. Mater. Res.* **1990**, *5*, 2507–2514. [CrossRef]
53. Clark, C.D.; Dickerson, C.B. The 1.681 eV centre in polycrystalline diamond. *Surf. Coat. Technol.* **1991**, *47*, 336–343. [CrossRef]
54. Collins, A.T.; Allers, L.; Wort, C.J.H.; Scarsbrook, G.A. The annealing of radiation damage in De Beers colourless CVD diamond. *Diam. Relat. Mater.* **1994**, *3*, 932–935. [CrossRef]
55. Iakoubovskii, K.; Stesmans, A. Characterization of Defects in as-Grown CVD Diamond Films and HPHT Diamond Powders by Electron Paramagnetic Resonance. *Phys. Status Solidi A* **2001**, *186*, 199–206. [CrossRef]
56. Iakoubovskii, K.; Stesmans, A. Characterization of hydrogen and silicon-related defects in CVD diamond by electron spin resonance. *Phys. Rev. B* **2002**, *66*, 195207. [CrossRef]
57. Iakoubovskii, K.; Stesmans, A.; Suzuki, K.; Kuwabara, J.; Sawabe, A. Characterization of defects in monocrystalline CVD diamond films by electron spin resonance. *Diam. Relat. Mater.* **2003**, *12*, 511–515. [CrossRef]
58. Edmonds, A.M.; Newton, M.E.; Martineau, P.M.; Twitchen, D.J.; Williams, S.D. Electron paramagnetic resonance studies of silicon-related defects in diamond. *Phys. Rev. B* **2008**, *77*, 245205. [CrossRef]
59. Goss, J.P.; Briddon, P.R.; Shaw, M.J. Density functional simulations of silicon-containing point defects in diamond. *Phys. Rev. B* **2007**, *76*, 075204. [CrossRef]
60. D'Haenens-Johansson, U.F.S.; Edmonds, A.M.; Green, B.L.; Newton, M.E.; Davies, G.; Martineau, P.M.; Khan, R.U.A.; Twitchen, D.J. Optical properties of the neutral silicon split-vacancy center in diamond. *Phys. Rev. B* **2011**, *84*, 245208. [CrossRef]
61. Nadolinny, V.; Komarovskikh, A.; Palyanov, Y.; Kupriyanov, I.; Borzdov, Y.; Rakhmanova, M.; Yuryeva, O.; Veber, S. EPR study of Si- and Ge-related defects in HPHT diamonds synthesized from Mg-based solvent-catalysts. *Phys. Status Solidi A* **2016**, *213*, 2623–2628. [CrossRef]
62. Pzhezhetskii, S.Y. *EPR Svobodnykh Radikalov v Radiatsionnoi Khimii*, 1st ed.; Khimiya: Moscow, Russia, 1972; p. 30. (In Russian)
63. Komarovskikh, A.; Dmitriev, A.; Nadolinny, V.; Palyanov, Y. A DFT calculation of EPR parameters of a germanium-vacancy defect in diamond. *Diam. Relat. Mater.* **2017**, *76*, 86–89. [CrossRef]
64. Wang, C.; Kurtsiefer, C.; Weinfurter, H.; Burchard, B. Single photon emission from SiV centres in diamond produced by ion implantation. *J. Phys. B* **2006**, *39*, 37–41. [CrossRef]
65. Iwasaki, T.; Ishibashi, F.; Miyamoto, Y.; Doi, Y.; Kobayashi, S.; Miyazaki, T.; Tahara, K.; Jahnke, K.D.; Rogers, L.J.; Naydenov, B.; et al. Germanium-Vacancy Single Color Centers in Diamond. *Sci. Rep.* **2015**, *5*, 12882. [CrossRef] [PubMed]

Article

HPHT Diamond Crystallization in the Mg-Si-C System: Effect of Mg/Si Composition

Yuri Palyanov [1,2,*], Igor Kupriyanov [1,2], Yuri Borzdov [1,2,*], Denis Nechaev [1] and Yuliya Bataleva [1]

[1] Sobolev Institute of Geology and Mineralogy SB RAS, Koptyug ave. 3, 630090 Novosibirsk, Russia; spectra@igm.nsc.ru (I.K.); nechaev@igm.nsc.ru (D.N.); browny@bk.ru (Y.B.)

[2] Department of Geology and Geophysics, Novosibirsk State University, Novosibirsk 630090, Russia

[*] Correspondence: palyanov@igm.nsc.ru (Y.P.); borzdov@igm.nsc.ru (Y.B.); Tel.: +7-383-330-7501 (Y.P.)

Academic Editor: Helmut Cölfen
Received: 31 March 2017; Accepted: 23 April 2017; Published: 25 April 2017

Abstract: Crystallization of diamond in the Mg-Si-C system has been studied at 7.5 GPa and 1800 °C with the Mg-Si compositions spanning the range from Mg-C to Si-C end-systems. It is found that as Si content of the system increases from 0 to 2 wt %, the degree of the graphite-to-diamond conversion increases from about 50 to 100% and remains at about this level up to 20 wt % Si. A further increase in Si content of the system leads to a decrease in the graphite-to-diamond conversion degree down to complete termination of diamond synthesis at Si content >50 wt %. Depending on the Si content crystallization of diamond, joint crystallization of diamond and silicon carbide and crystallization of silicon carbide only are found to take place. The cubic growth of diamond, typical of the Mg-C system, transforms to the cube-octahedron upon adding 1 wt % Si and then to the octahedron at a Si content of 2 wt % and higher. The crystallized diamonds are studied by a suite of optical spectroscopy techniques and the major characteristics of their defect-and-impurity structure are revealed. The correlations between the Si content of the Mg-Si-C system and the properties of the produced diamond crystals are established.

Keywords: diamond; high pressure high temperature; crystallization; crystal morphology; defects; characterization

1. Introduction

Synthesis and growth of diamond at high pressure high temperature (HPHT) conditions has evolved into a mature technology that enables the use of diamond in a variety of scientific and industrial applications. The most efficient and commonly used solvent-catalysts for HPHT diamond synthesis are group VIII transition metals, particularly Fe, Co, Ni, and their alloys [1–3]. The processes of synthesis and growth of diamond crystals from these catalysts, which are frequently referred to as conventional, have been well studied, and the major factors controlling nucleation, growth, morphology, and properties of the crystals have been determined [4–6]. In recent decades, various diamond producing systems have been extensively investigated, that has significantly expanded the range of solvent-catalysts for diamond synthesis and provided additional information on the mechanisms of diamond nucleation and growth. The use of new growth systems enables production of diamond crystals with unusual and unique properties. For example, superconducting diamonds were obtained in the B-C system [7]; phosphorus-doped crystals were synthesized in the P-C system [8,9]; Ge-doped diamonds were produced using germanium as the solvent-catalyst [10].

Recently, considerable attention has been focused to the Mg-based solvent-catalysts [11–15]. This interest is related to the following factors: in the Mg-based systems, diamond crystallizes in the kinetically controlled regime with very high growth rates, reaching 8.5 mm/h [11]; the crystal morphology is very specific and largely controlled by the influence of impurities [11–14]; the produced diamond crystals are nitrogen-free type II. It is also important to emphasize that the Mg-based systems enable effective doping of diamond with silicon [12] and germanium [13] impurities, which create in the diamond lattice optically active silicon-vacancy and germanium-vacancy centers. These centers attract heated interest as promising single-photon emitters to be used in the emerging quantum technologies [16–18]. Previous studies determined the effect of temperature on diamond nucleation and growth in the $Mg_{0.8}Si_{0.2}$-C system [12].

In order to get a better understanding of diamond crystallization processes in Mg-based systems, in this work we have studied the effect of the Mg-Si composition ratio on diamond formation in the Mg-Si-C system at HPHT conditions. In this paper, we show that the Mg/Si ratio significantly affects diamond crystallization, degree of the graphite-to-diamond conversion, morphological stability, optical properties of diamond crystals, as well as phase formation in the Mg-Si-C system.

2. Results

2.1. Diamond Crystallization

A series of experiments in the Mg-Si-C system were carried out at a pressure of 7.5 GPa, a temperature of 1800 °C, and a run time of 30 min. These experimental conditions were chosen based on the results of our previous studies on diamond crystallization from the Mg-based catalysts [11,12]. The starting compositions and experimental results are summarized in Table 1. A schematic of the initial sample assembly and general crystallization schemes revealed in the experiments are presented in Figure 1.

Table 1. Experimental conditions and results.

Run N	P, GPa	T, °C	Time, min	Composition, wt %	α Gr → Dm Conversion	Diamond Morphology	SiC
MS-1	7.5	1800	30	Mg	50	{100}, {100}>>{111}	-
MS-2	7.5	1800	30	$Mg_{99}Si_1$	80	{111}≈{100}	-
MS-3	7.5	1800	30	$Mg_{98}Si_2$	100	{111}>{100}	-
MS-4	7.5	1800	30	$Mg_{95}Si_5$	90	{111}>>{100}	-
MS-5	7.5	1800	30	$Mg_{90}Si_{10}$	90	{111}>>{100}	-
MS-6	7.5	1800	30	$Mg_{80}Si_{20}$	100	{111}>>{100}	3C
MS-7	7.5	1800	30	$Mg_{70}Si_{30}$	80	{111}	3C
MS-8	7.5	1800	30	$Mg_{50}Si_{50}$	20	{111}	3C, 4H
MS-9	7.5	1800	30	$Mg_{25}Si_{75}$	0	-	3C
MS-10	7.5	1800	30	Si	0	-	3C

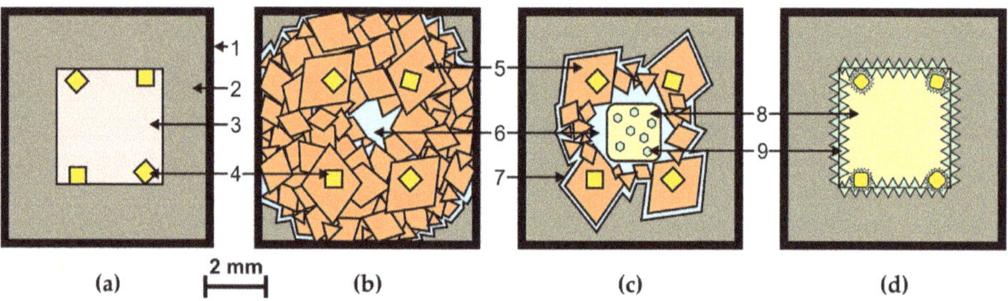

Figure 1. General schemes of crystallization of diamond and silicon carbide in the Mg-Si-C system. (**a**) Initial sample assembly; (**b**) Crystallization of diamond; (**c**) Crystallization of diamond and silicon carbide; (**d**) Crystallization of silicon carbide. 1—molybdenum; 2—graphite capsule; 3—powder mixture of Mg and Si; 4—diamond seed crystals; 5—crystallized diamond; 6—magnesium-rich melt; 7—magnesium-rich melt film; 8—silicon-rich or silicon melt; 9—crystallized silicon carbide.

In the Mg-C system (run MS-1) diamonds of predominantly cubic morphology were formed. The {111} faces were either present as minor faces or completely absent. The degree of the graphite-to-diamond conversion was around 50%, with a significant part of diamonds forming a polycrystalline aggregate of colorless or gray crystals. Addition of 1 wt % Si (run MS-2) resulted in a substantial increase in the degree of graphite-to-diamond conversion (up to 80%) and led to a significant change in the diamond morphology. Although most diamonds were present as cleavage blocks, individual crystals and their fragments demonstrated that the {100} and {111} faces had approximately equal development, and the resulting crystals were cube-octahedrons (Figure 2). Striation along the [110] direction, formed due to the stepped structure of {111} faces, was usually present between the adjacent octahedral faces.

Figure 2. Scanning electron microscopy (SEM) micrographs of a cube-octahedral diamond crystal produced in the $Mg_{99}Si_1$-C system. (**a**) General view; (**b**) Growth features on the {111} faces; (**c**) Pyramidal pits with {111} growth layers on {100} face; (**b**,**c**)—Enlarged views of areas marked with squares in (**a**).

With addition of 2 wt % Si (MS-3), the starting graphite capsule converted to a polycrystalline diamond aggregate consisting of gray and colorless crystals and blocks. The morphology of the produced diamond crystals was determined by the dominant {111} faces, surfaces with crystallographic orientation close to the {110} faces showing pronounced striation along [110], and minor {100} faces at the polyhedron corners (Figure 3). Twinned crystals frequently occurred among the synthesized diamonds (Figure 4). As it can be clearly seen from Figure 4b, the sources of the growth macrolayers on the {111} faces are associated with the twin boundaries.

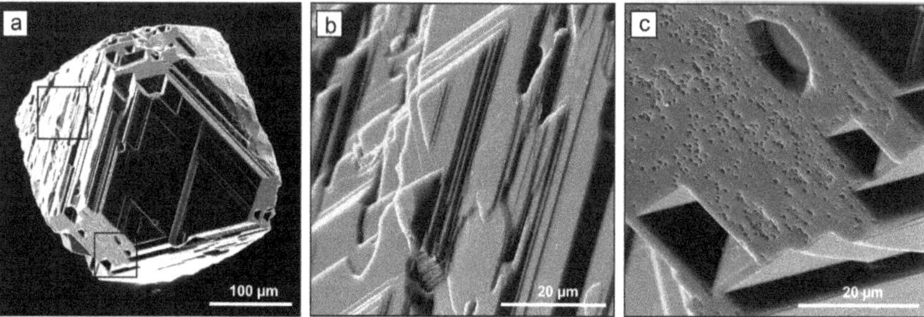

Figure 3. SEM micrographs of a diamond crystal synthesized in the $Mg_{98}Si_2$-C system. (**a**) General view of the crystal showing dominant {111} faces and minor {100} faces; (**b**) Growth layers on the {111} face; (**c**) Growth pits on the {100} face; (**b**,**c**)—Enlarged views of areas marked with squares in (**a**).

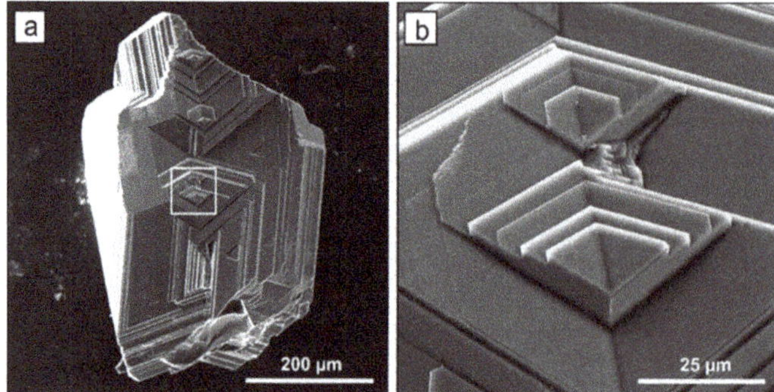

Figure 4. SEM micrographs of a twinned diamond crystal with growth macrolayers on the {111} faces. (**a**) General view; (**b**) Enlarged view of an area marked with square in (**a**).

With the starting compositions $Mg_{95}Si_5$ and $Mg_{90}Si_{10}$, the degree of graphite-to-diamond conversion was estimated at about 90%. Diamonds typically formed polycrystalline aggregates. Individual octahedral crystals with serrate macrolayers propagating from single sources (Figure 5) were rarely observed.

In run MS-6 with 20 wt % Si, along with diamond, tetrahedral silicon carbide crystals were found in the products. All graphite of the starting capsule was converted to diamond. The observed intergrowths of silicon carbide with diamond (Figure 6c) and SiC inclusions in diamond crystals indicate co-crystallization of diamond and silicon carbide. With the $Mg_{70}Si_{30}$ initial composition (run MS-7) the degree of graphite-to-diamond conversion decreased to about 80%. Crystallized diamonds were mainly in the form of blocks consisting of octahedral crystals in approximately parallel orientation (Figure 6a,b). Silicon carbide was present as single crystals, intergrowths, and polycrystalline aggregates in the central part of the capsule. From the Raman scattering measurements, it was found that the silicon carbide crystallized in runs MS-6, and MS-7 corresponded to the cubic 3C-SiC phase (Figure 7).

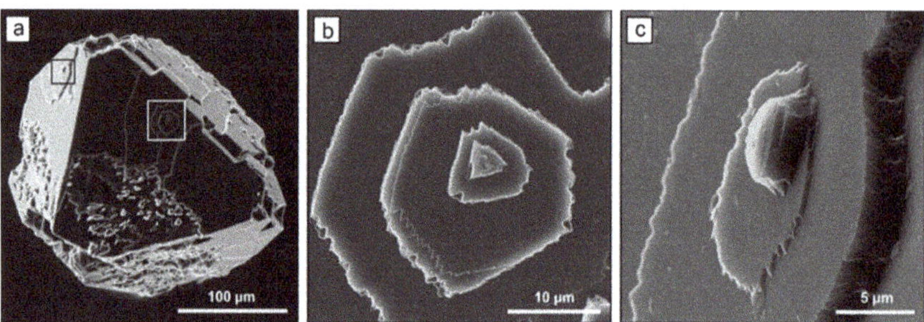

Figure 5. SEM micrographs of an octahedral diamond crystal produced in the $Mg_{95}Si_5$-C. (**a**) General view; (**b,c**) Enlarged views of the {111} faces with serrate macrolayers marked with squares in (**a**).

Figure 6. SEM micrographs of diamond and silicon carbide. (**a**) A diamond block consisting of octahedral crystals in approximately parallel orientation (general view); (**b**) Enlarged view of (**a**); (**c**) Intergrowth of cubic SiC (tetrahedrons) with diamond.

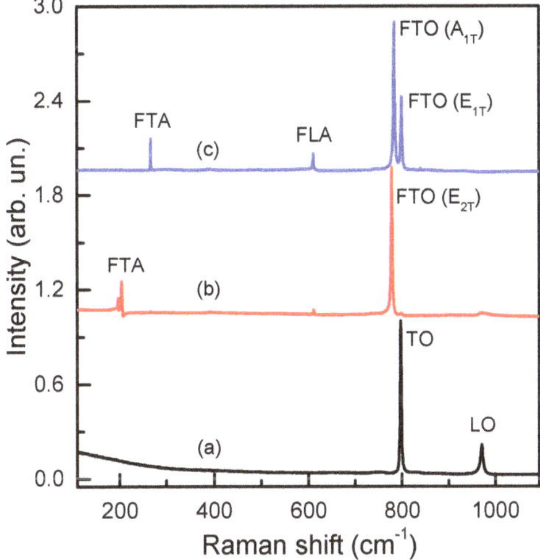

Figure 7. Raman spectra of (**a**) cubic 3C-SiC crystals and (**b,c**) hexagonal 4H-SiC crystals of different morphologies.

In the experiment with the $Mg_{50}Si_{50}$ starting composition (MS-8) it was found that two melts with different properties formed under the experimental conditions. The first melt reacted with graphite to produce diamond with a conversion degree of about 20%. After opening the capsule, the quenching products of this melt were rapidly oxidized in air; i.e., they behaved as in all experiments described above. The second melt was located in the central part of the capsule and the alloy produced upon quenching this melt was stable in air. Furthermore, it was not dissolved in hot nitric acid, but underwent surface oxidization, revealing its eutectoid structure (Figure 8a,b). Numerous crystals of hexagonal silicon carbide were observed inside this quenched alloy. The composition of the first and second quenched alloys was found to be $Mg_{63}Si_{37}$ and $Mg_{35}Si_{65}$, respectively. Silicon carbide formed pyramidal, prismatic, or tabular crystals, depending on the relative development of different crystal faces (Figure 8). Raman spectra recorded for the SiC crystals with different morphologies (Figure 7) demonstrated that in this case silicon carbide corresponded to the hexagonal 4H modification.

In the experiment with the $Mg_{25}Si_{75}$ composition (MS-9), no diamond was established in the run products. Fine silicon carbide crystals, corresponding to the cubic 3C modification, are formed at the melt-graphite ampoule interface. The melt consisted of quenched silicon crystals and phases with the

Mg$_{58}$Si$_{42}$ and Mg$_{32}$Si$_{68}$ compositions which formed eutectoid structures. In the experiment MS-10 conducted in the Si-C system, a quenched Si melt and a polycrystalline silicon carbide aggregate were found at the graphite-silicon interface. Diamond seed crystals were partially corroded and silicon carbide crystal aggregates formed on their surface.

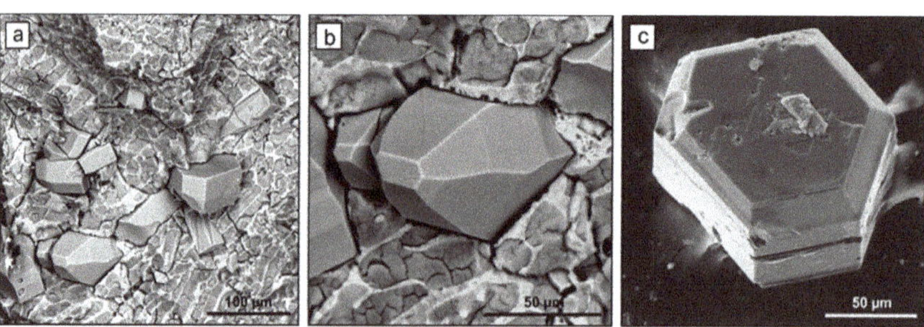

Figure 8. SEM micrographs of hexagonal silicon carbide crystals of different morphologies produced in run MS-8.

2.2. Spectroscopic Characterization

We now consider the results of spectroscopic characterization of the crystallized diamonds. Figure 9 shows typical infrared absorption spectra recorded for the produced diamond crystals. It was found that the crystals synthesized over the entire range of the diamond producing compositions ($\leq$50 wt % Si) showed no detectable features in the IR spectra caused by nitrogen impurities. This characteristic of diamonds grown from the Mg-based catalysts has been demonstrated in previous studies [11,12,19] and attributed to the gettering properties of magnesium. In addition, we can note that silicon, being a carbide-forming element, may also act as the nitrogen getter and contribute to the growth of type II nitrogen-free diamond. The recorded IR spectra frequently exhibited relatively weak absorption peaks due to the uncompensated boron acceptors. The concentration of boron acceptors (N_A) was estimated from the strength of the peak at 2800 cm^{-1} using previously established calibrations [20]. The calculations gave values of N_A from below 0.1 up to 1 ppm. No clear dependence of the boron content in the crystals on the growth system composition was found. For diamonds showing both {111} and {100} growth sectors (runs MS-2, MS-3), no growth sector dependence of boron distribution was found either. It is necessary to note that, in the present study, no boron was deliberately added to the growth system and it was present only as trace impurity in the reagents. Starting from the Mg$_{80}$Si$_{20}$ composition, crystallized diamonds exhibited a distinct feature in the IR spectra comprising a sharp absorption peak at 1338 cm^{-1} (Figure 9). The position of this peak was clearly different from that of the highest energy zone-center phonons of diamond (1332 cm^{-1}). With further increasing Si content of the growth system, the 1338 cm^{-1} peak showed a tendency to increase in strength. The exact nature of this peak has not been clearly established so far. However, there is growing amount of evidence suggesting that the 1338 cm^{-1} localized vibration mode is related to substitutional silicon defects in the diamond lattice and originates due to high local strains surrounding these defects [12,21,22]. Our results give further support to this hypothesis.

Photoluminescence (PL) measurements showed that the PL spectra of diamond crystals synthesized over the entire range of the diamond producing compositions were dominated by the 1.68 eV optical system caused by the negatively charged silicon-vacancy (Si-V) centers (Figure 10). As it was found previously, the 1.68 eV Si-V centers occur as the major PL feature in diamonds synthesized from the pure Mg-C system, which contain silicon only as trace impurities in the starting reagents [11,19]. Comparing PL spectra normalized to the intensity of the diamond Raman scattering peak we found that diamonds produced in the Mg$_{99}$Si$_1$-C and Mg$_{98}$Si$_2$-C systems (runs MS-2 and MS-3) showed, on average, high intensities of emission from the 1.68 eV centers, compared to those synthesized in the pure Mg-C system (MS-1). It is important to note that with these higher PL intensities, the zero-phonon peak of the 1.68 eV vibronic band still demonstrated a clear doublet structure (Figure 10) in the low-temperature (80 K) spectra of diamonds from runs MS-2 and MS-3. This implies that small (1–2 wt %) additives of Si to the Mg-C system improve the luminescent characteristics of the diamonds, without diminishing their internal quality.

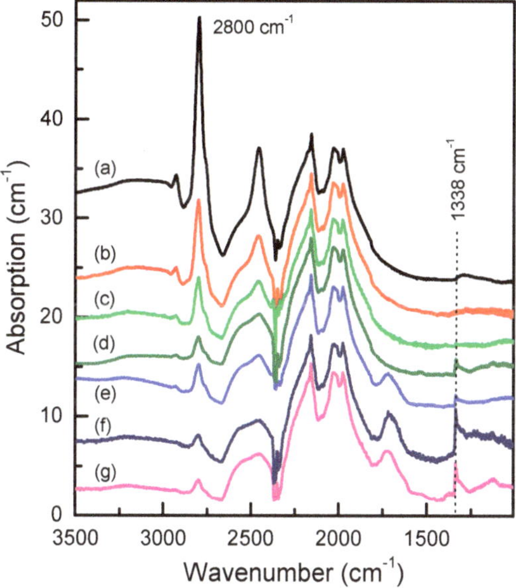

Figure 9. Typical infrared absorption spectra recorded for diamond crystals produced in the Mg-Si-C system with different Si content. (**a,b**) 1; (**c**) 5; (**d,e**) 20; and (**f,g**) 50 wt % Si. The spectra are displaced vertically for clarity.

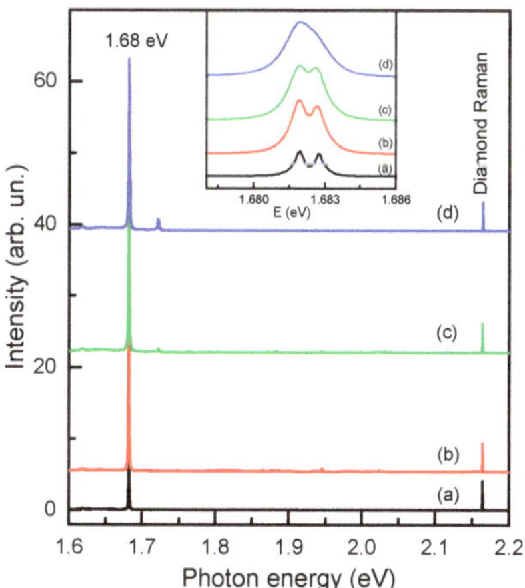

Figure 10. Typical photoluminescence (PL) spectra of diamonds synthesized in the Mg-Si-C system with different Si content. (**a**) 0; (**b**) 1; (**c**) 5 and (**d**) 10 wt % Si. The inset shows the zero-phonon line (ZPL) region of the 1.68 eV PL system. The spectra are measured at 80 K. The spectra are displaced vertically for clarity.

Diamond crystals grown from the compositions with higher Si contents showed changes in the PL spectra (Figure 11). Although the intensity of the 1.68 eV peak did not change significantly, it became broader so that its doublet structure could not be observed. This broadening suggests that the diamonds produced in the Mg-Si-C system with Si content $\geq$10 wt % have a higher level of internal strains. One possible source of these strains could be substitutional silicon defects, which also give rise to the strain-induced vibrational mode at 1338 cm^{-1}. Another distinct feature evident from Figure 11 is that a peak at 1.722 eV, which was very weak for diamonds from the low Si content systems, became prominent in the PL spectra and attained intensities comparable to that of the 1.68 eV peak. The occurrence of the 1.722 eV peak in the spectra of diamonds grown from the Mg-Si-C system was noted previously [12]. Here, we confirm that it appears in the PL spectra of Si-doped HPHT diamonds. The nature of the defect responsible for the 1.722 eV is not clear at the moment and deserves further investigation. Nevertheless, we suppose it is very reasonable to assign this center to a defect related to silicon impurities.

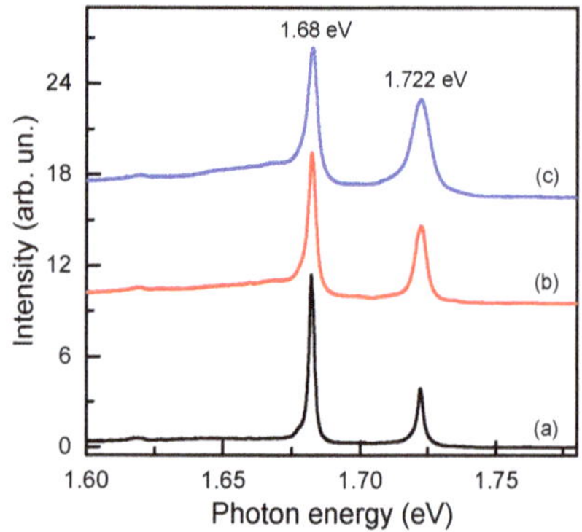

Figure 11. Typical PL spectra of diamonds synthesized in the Mg-Si-C system with different Si content. (**a**) 20, (**b**) 30 and (**c**) 50 wt % Si. The spectra are displaced vertically for clarity.

3. Discussion

The results of this study allow us to assess the effect of the Mg/Si ratio on the crystallization of diamond and accompanying phases in the Mg-Si-C system at fixed P-T-t conditions. Figure 12 shows the degree of the graphite-to-diamond conversion (α) as a function of the silicon content in the Mg-Si-C system. It follows that the addition of even small amounts, about 1 wt %, of silicon results in a substantial increase in the conversion degree. With further increases in Si content in the range of 2 to 20 wt %, α increases to as high as 90−100%. As it was suggested previously [12], the most probable reason for increasing values of α in this range of compositions can be related to increasing carbon solubility in the binary melt of two carbide forming elements, Mg and Si. The observed sharp decrease in α for the Mg$_{50}$Si$_{50}$ starting composition is associated with the formation of two melts with different compositions. The first, magnesium-rich, melt ensures diamond synthesis, while the second, silicon-rich, melt does not exhibit this ability. Because of this catalyst melt partition, the amount of the diamond-forming melt in this experiment is substantially less than in other experiments with lower Si content in the system. In the case of the Mg$_{25}$Si$_{75}$ initial composition, there is a single melt, which is substantially enriched in Si and does not show the catalytic ability for converting graphite to diamond. It should be noted that no carbon or carbon-containing phases were detected in samples of the quenched melt. Furthermore, carbon was not detected as an impurity in all quenched phases. This can be explained by the low carbon solubility in the silicon-rich melt and the formation of silicon carbide at the graphite-melt interface, which blocks further interaction between the graphite and the melt. A similar situation takes place in the Si-C system. It should also be noted that the formation of silicon carbide occurred in all experiments with the Si content from 20 to 100 wt %. SiC was present as the cubic 3C modification in the majority of the experiments, while the hexagonal modification occurred only in the experiment with the $Mg_{50}Si_{50}$ starting composition and crystallized from the melt that did not produce diamond. As is known, preferential crystallization of cubic or hexagonal modifications of silicon carbide is affected by numerous factors, such as temperature, pressure, system composition and impurities [23,24]. The appearance of the 4H-SiC modification in this experiment is possibly related to the system composition, but clarification of this issue lies far beyond the scope of this work.

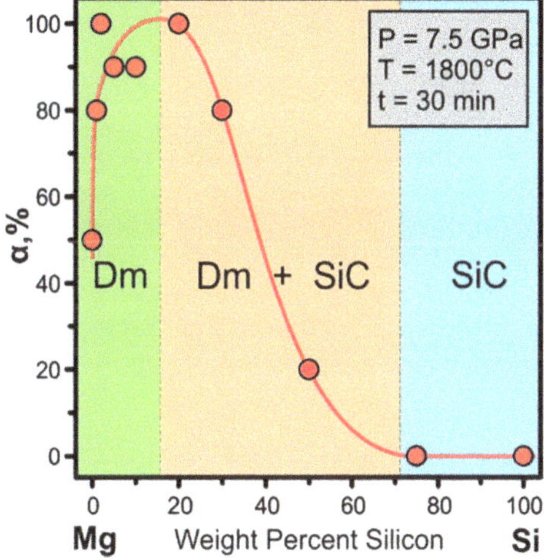

Figure 12. Degree of the graphite-to-diamond transformation as a function of the weight percent silicon in the Mg-Si-C system.

The results of this study also enable us to consider the effect of silicon on the morphology of diamond crystallized in the Mg-Si-C system. As shown previously [11], diamonds with cubic morphology, sometimes having minor {111} faces, form in the Mg-C system at 7 GPa and 1800 °C. The {100} faces are characterized by strong rectangular growth layers, the ends of which are formed exclusively by the {111} faces. Adding 1 wt % silicon leads to a rapid change in the macro- and micromorphology. Cubic morphology transforms to cube-octahedral (Figure 2a,b). Macrolayers on the {100} faces disappear. The surface relief of the {100} faces is determined by plane mirror-smooth {100} areas and numerous pyramidal pits formed exclusively by the {111} microfaces. The growth nature of the pits is confirmed by the distinct growth layers on the bounding {111} microfaces (Figure 2c). On the {111} faces of cube-octahedral crystals, macrolayers parallel to the face edges form. The ends of these layers are bound by the {111} microfaces (Figure 2a). At 2 wt % Si, the cubic faces become minor faces, but they still show plane {100} regions, small octahedral pits and very large depressions formed by the {111} faces (Figure 3a,c). The octahedral faces predominate. They exhibit pronounced macrolayers with the ends built up by the {111} faces (Figure 3b). As a result, the crystals attain a polycentric structure of the {111} faces. Figure 3 also shows that the number of macrolayer generation centers can be significantly different on different faces of the same crystal. Thus, even small changes in the Si content in the system result in substantial changes in the relative growth rates of {100} and {111} faces. Obviously, the growth rate of octahedral faces decreases relative to the growth rate of cubic faces. We may suppose that this process is accompanied by the impurity blocking of some defects on the {100} faces, which leads to the formation of numerous pyramidal growth pits. Therefore, in the Mg-C system, the morphology variation in the cube-octahedron series is unambiguously determined by the influence of silicon additive. This is in contrast to the conventional transition metal catalysts, where the cube-to-octahedron transformation of the diamond morphology is primarily determined by the P-T parameters of synthesis.

4. Materials and Methods

Synthesis experiments in the Mg-Si-C system were performed using a split-sphere multi-anvil high-pressure apparatus [25]. All experiments were conducted at a pressure of 7.5 GPa with high-pressure cells in the form of a tetragonal prism with dimensions 19 × 19 × 22 mm. A PtRh6/PtRh30 thermocouple (Krastsvetmet, Krasnoyarsk, Russia) was used in each experiment for the temperature measurements. Given the data of calibration experiments [26,27], the accuracy of temperature and pressure measurements was ±40 °C and ±0.2 GPa, respectively. The high-pressure cell and sample assemblies were the same in all experiments of this series and was similar to that used in our previous studies of the Mg-C system [11,12]. A mixture of Mg and Si powders and four 0.5 mm synthetic diamond seed crystals were packed into a thick-walled graphite capsule, 6.9 mm in diameter and 6.5 mm high, which was the carbon source. The graphite capsule was enveloped from all sides with a 0.1 mm thick molybdenum foil to prevent diffusion of the high pressure cell components during experiments. A schematic of the initial sample assembly is shown in Figure 1a. In all experiments, the main parameters were kept constant: P = 7.5 GPa, T = 1800 °C, and a run time of 30 min. Only the solvent composition was changed by varying the Mg/Si ratio. The purity of starting reagents was as follows: graphite—99.99%, magnesium—99.99%, and silicon—99.99%. After experiments, the run products were dissolved in dilute hydrochloric acid. Residual graphite and newly formed diamond were weighed to determine the degree of graphite-to-diamond conversion (α) in each experiment, $\alpha = M_{Dm}/(M_{Dm} + M_{Gr}) \times 100$, where M_{Dm} is the mass of synthesized diamond, and M_{Gr} is the mass of residual graphite. The recovered diamond crystals and associated phases were studied using an Axio Imager Z2m optical microscope (Carl Zeiss Microscopy, Jena, Germany) and a Tescan MIRA3 LMU scanning electron microscope (Tescan, Brno, Czech Republic). The composition of the Mg-Si alloys formed in the experiments was analyzed by the energy dispersive spectrometry (EDS) (Oxford Instruments, Abington, UK). Spectroscopic characterization of the diamond crystals was performed by means of infrared (IR) absorption, photoluminescence (PL) and Raman scattering. IR spectra were measured using a Bruker Vertex 70 Fourier transform infrared (FTIR) spectrometer fitted with a Hyperion 2000 microscope (Bruker Optics, Ettlingen, Germany). Raman/PL spectra were measured using a Horiba J.Y. LabRAM HR800 spectrometer fitted with an Olympus BX41 microscope (Horiba Jobin Yvon S.A.S., Lonjumeau, France). A 523-nm solid state laser (Laser Quantum, Stockport, UK) was used as the excitation source. For the low temperature measurements, a Linkam FTIR600 heating/freezing stage was used (Linkam Scientific Instruments, Tadworth, UK).

5. Conclusions

We investigated the effect of the Mg/Si ratio on diamond crystallization in the Mg-Si-C system at 7.5 GPa and 1800 °C. We found that, as the Si content of the system increases from 0 to 2 wt %, the degree of the graphite-to-diamond transformation increases from 50 to 100% and remains at about this level up to 20 wt % Si. We associated this with a higher carbon solubility in the binary melts of two carbide-forming elements, Mg and Si. A further increase in the silicon content of the system leads to a decrease in the degree of graphite-to-diamond transformation and then to a complete termination of diamond synthesis at Si content greater than 50 wt %. This phenomenon is associated with the shift of the melt compositions to the silicon-enriched region, which is accompanied by a significant decrease in carbon solubility. With the Si content of up to about 15 wt %, only diamond crystallizes in the Mg-Si-C system; at Si contents in the range of 20 to 50 wt %, simultaneous crystallization of diamond and silicon carbide occurs; and only SiC crystallizes at a higher Si content in the system.

The morphology diamond crystallized in the Mg-Si-C system is determined by the relative development of {100} and {111} faces and changes from cubic to octahedral as the Si content increases.

In the Mg-C system, the stable growth form is cube, cube-octahedral crystals form at 1 wt % Si, and at ≥ 2 wt % Si the dominant growth form is octahedron. Without silicon impurity, growth of the {100} faces occurs by macrolayers, the ends of which are bound by the {111} microfaces. Upon addition of Si, macrolayers with the {111} ends form on octahedron faces, and crystals acquire the polycentric structure.

Over the entire range of the diamond producing compositions, the crystallized diamonds are nitrogen-free due to the gettering properties of magnesium and possibly silicon. As the Si content of the growth system increases, the synthesized diamonds tend to incorporate higher amounts of silicon impurities, which give rise to the localized vibrational mode at 1338 cm^{-1} observed in the IR spectra. The 1.68 eV Si-V centers dominate in photoluminescence spectra of the diamond crystals produced at Si contents ≤ 10 wt %; at higher Si contents, the 1.68 eV centers are accompanied by a system with a zero-phonon line at 1.722 eV. The established correlations between the Si content of the Mg-Si-C growth system and optical characteristics of the synthesized diamonds can be applied for developing methods of production of diamond with controlled properties.

Acknowledgments: This work was supported by the Russian Science Foundation under Grant No. 14-27-00054.

Author Contributions: Yuri Palyanov conceived and designed the experiments; Yuri Borzdov performed the experiments; Yuri Palyanov, Denis Nechaev and Yuliya Bataleva studied run products and analyzed the data; Igor Kupriyanov performed spectroscopic characterization of crystallized diamond and SiC; Yuri Palyanov and Igor Kupriyanov wrote the paper.

References

1. Bundy, F.P.; Hall, H.T.; Strong, H.M.; Wentorf, J.R. Man-made diamonds. *Nature* **1955**, *176*, 51–55. [CrossRef]
2. Bovenkerk, H.P.; Bundy, F.P.; Hall, H.T.; Strong, H.M.; Wentorf, J.R. Preparation of diamond. *Nature* **1959**, *184*, 1094–1098. [CrossRef]
3. Kanda, H. Classification of the catalysts for diamond growth. In *Advances in New Diamond Science and Technology*; Saito, S., Fujimori, N., Fukunaga, O., Kamo, M., Kobashi, K., Yoshikawa, M., Eds.; MYU: Tokyo, Japan, 1994; pp. 507–512.
4. Wedlake, R.J. Technology of diamond growth. In *The properties of diamond*; Field, J.E., Ed.; Academic Press: London, UK, 1979; pp. 501–535.
5. Burns, R.C.; Davies, G.J. Growth of synthetic diamond. In *The Properties of Natural and Synthetic Diamond*; Field, J.E., Ed.; Academic Press: London, UK, 1992; pp. 395–422.
6. Palyanov, Y.; Kupriyanov, I.; Khokhryakov, A.; Ralchenko, V. Crystal Growth of Diamond. In *Handbook of Crystal Growth*, 2nd ed.; Nishinaga, T., Rudolph, P., Eds.; Elsevier: Amsterdam, Holland, 2015; Volume 2a, pp. 671–713.
7. Ekimov, E.A.; Sidorov, V.A.; Bauer, E.D.; Mel'nik, N.N.; Curro, N.J.; Thompson, J.D.; Stishov, S.M. Superconductivity in diamond. *Nature* **2004**, *428*, 542–545. [CrossRef] [PubMed]
8. Akaishi, M.; Kanda, H.; Yamaoka, S. Phosphorous: An elemental catalyst for diamond synthesis and growth. *Science* **1993**, *259*, 1592–1593. [CrossRef] [PubMed]
9. Palyanov, Y.N.; Kupriyanov, I.N.; Sokol, A.G.; Khokhryakov, A.F.; Borzdov, Y.M. Diamond growth from a phosphorus-carbon system at HPHT conditions. *Cryst. Growth Des.* **2011**, *11*, 2599–2605. [CrossRef]
10. Palyanov, Y.N.; Kupriyanov, I.N.; Borzdov, Y.M.; Surovtsev, N.V. Germanium: A new catalyst for diamond synthesis and a new optically active impurity in diamond. *Sci. Rep.* **2015**, *5*, 14789. [CrossRef] [PubMed]
11. Palyanov, Y.N.; Borzdov, Y.M.; Kupriyanov, I.N.; Khokhryakov, A.F.; Nechaev, D.V. Diamond crystallization from an Mg-C system at high pressure high temperature conditions. *CrystEngComm* **2015**, *17*, 4928–4936. [CrossRef]
12. Palyanov, Y.N.; Kupriyanov, I.N.; Borzdov, Y.M.; Bataleva, Y.V. High-pressure synthesis and characterization of diamond from an Mg–Si–C system. *CrystEngComm* **2015**, *17*, 7323–7331. [CrossRef]

13. Palyanov, Y.N.; Kupriyanov, I.N.; Borzdov, Y.M.; Khokhryakov, A.F.; Surovtsev, N.V. High-pressure synthesis and characterization of Ge-doped single crystal diamond. *Cryst. Growth Des.* **2016**, *16*, 3510–3518. [CrossRef]

14. Khokhryakov, A.F.; Sokol, A.G.; Borzdov, Y.M.; Palyanov, Y.N. Morphology of diamond crystals grown in magnesium-based systems at high temperatures and high pressures. *J. Cryst. Growth* **2015**, *426*, 276–282. [CrossRef]

15. Khokhryakov, A.F.; Nechaev, D.V.; Palyanov, Y.N. Unusual growth macrolayers on {100} faces of diamond crystals from magnesium-based systems. *J. Cryst. Growth* **2016**, *455*, 76–82. [CrossRef]

16. Müller, T.; Hepp, C.; Pingault, B.; Neu, E.; Gsell, S.; Schreck, M.; Sternschulte, H.; Steinmüller-Nethl, D.; Becher, C.; Atatüre, M. Optical signatures of silicon-vacancy spins in diamond. *Nat. Commun.* **2014**, *5*, 3328. [CrossRef] [PubMed]

17. Iwasaki, T.; Ishibashi, F.; Miyamoto, Y.; Doi, Y.; Kobayashi, S.; Miyazaki, T.; Tahara, K.; Jahnke, K.D.; Rogers, L.J.; Naydenov, B.; et al. Germanium-vacancy single color centers in diamond. *Sci. Rep.* **2015**, *5*, 12882. [CrossRef] [PubMed]

18. *Quantum Information Processing with Diamond*; Prawer, S., Aharonovich, I., Eds.; Woodhead Publishing: Cambridge, UK, 2014; p. 330.

19. Kovalenko, T.V.; Ivakhnenko, S.A. Properties of diamonds seed-grown in the magnesium-carbon system. *J. Superhard Mater.* **2013**, *35*, 131–136. [CrossRef]

20. Collins, A.T.; Williams, A.W.S. The nature of the acceptor centre in semiconducting diamond. *J. Phys. C Solid State Phys.* **1971**, *4*, 1789–1800. [CrossRef]

21. Goss, J.P.; Briddon, P.R.; Shaw, M.J. Density functional simulations of silicon-containing point defects in diamond. *Phys. Rev. B* **2007**, *76*, 075204. [CrossRef]

22. Breeding, C.M.; Wang, W. Occurrence of the Si–V defect center in natural colorless gem diamonds. *Diam. Relat. Mater.* **2008**, *17*, 1335–1344. [CrossRef]

23. Sugiyama, S.; Togaya, M. Phase relationship between 3C-and 6H-silicon carbide at high pressure and high temperature. *J. Am. Ceram. Soc.* **2001**, *84*, 3013–3016. [CrossRef]

24. Jepps, N.W.; Page, T.F. The 6H→3C "reverse" transformation in silicon carbide compacts. *J. Am. Ceram. Soc.* **1981**, *64*, 2830–2833. [CrossRef]

25. Pal'yanov, Y.N.; Sokol, A.G.; Borzdov, Y.M.; Khokhryakov, A.F. Fluid-bearing alkaline-carbonate melts as the medium for the formation of diamonds in the Earth's mantle: an experimental study. *Lithos* **2002**, *60*, 145–159. [CrossRef]

26. Palyanov, Y.N.; Borzdov, Y.M.; Khokhryakov, A.F.; Kupriyanov, I.N.; Sokol, A.G. Effect of nitrogen impurity on diamond crystal growth processes. *Cryst. Growth Des.* **2010**, *10*, 3169–3175. [CrossRef]

27. Sokol, A.G.; Borzdov, Y.M.; Palyanov, Y.N.; Khokhryakov, A.F. High temperature calibration a multi-anvil high-pressure apparatus. *High Press. Res.* **2015**, *35*, 139–147. [CrossRef]

Article

Fabrication of Low Dislocation Density, Single-Crystalline Diamond via Two-Step Epitaxial Lateral Overgrowth

Fengnan Li, Jingwen Zhang, Xiaoliang Wang, Minghui Zhang and Hongxing Wang *

Key Laboratory for Physical Electronics and Devices of the Ministry of Education, Xi'an Jiaotong University, Xi'an 710049, China; lfn@stu.xjtu.edu.cn (F.L.); jwzhang@mail.xjtu.edu.cn (J.Z.); xlwang@semi.ac.cn (X.W.); zhangminghuicc@mail.xjtu.edu.cn (M.Z.)
* Correspondence: hxwangcn@mail.xjtu.edu.cn; Tel.: +86-29-8266-8155

Academic Editor: Yuri N. Palyanov
Received: 15 March 2017; Accepted: 17 April 2017; Published: 18 April 2017

Abstract: Continuous diamond films with low dislocation density were obtained by two-step epitaxial lateral overgrowth (ELO). Grooves were fabricated by inductively coupled plasma etching. Mo/Pd stripes sputtered in the grooves were used to inhibit the propagation of dislocations originating from the diamond substrate. Coalescent diamond films were achieved by ELO via microwave plasma-enhanced chemical vapor deposition. Etch-pits were formed intentionally to characterize the quality of the epitaxial films and distinguish different growth areas, as dislocations served as preferential sites for etching. In the window regions, a high density of dislocations, displayed as dense etch-pits, was generated. By contrast, the etch-pit density was clearly lower in the overgrowth regions. After the second ELO step, the dislocation density was further decreased. Raman spectroscopy analysis suggested that the lateral overgrowth of diamond is a promising method for achieving low dislocation density films.

Keywords: chemical vapor deposition; CVD diamond; epitaxial lateral overgrowth (ELO); dislocations; etch-pits

---

1. Introduction

Diamond exhibits attractive intrinsic properties, such as a wide band gap energy (5.47 eV), high electric breakdown field (10 MV·cm^{-1}), high carrier mobility (3800 cm^2·V^{-1}·s^{-1} for holes), high thermal conductivity (22 W·cm^{-1}), and low dielectric constant (~5.7), making it a promising material for future electronic devices [1–5]. However, extended defects, such as dislocations and stacking faults, are common in both natural and synthetic diamond. For crystals formed via chemical vapor deposition (CVD), dislocations mainly stem from extended defects related to the substrate, such as defects in the substrate surface and defects in the bulk of the substrate. Dislocations tend to thread through the CVD film almost parallel to the growth direction [6]. Epitaxial lateral overgrowth (ELO) is a useful method to suppress the dislocation density and has been used for the growth of GaN [7–10]. The ELO method has also been applied to the growth of diamond [11,12].

Although X-ray topography, cathodoluminescence and transmission electron microscopy can be used to observe and quantify dislocations in diamond, these methods are relatively difficult to apply since they require complicated sample preparation or heavy equipment [13–15]. Moreover, it is very difficult to distinguish areas of different growth quality for the purpose of fabricating a metal mask by the ELO process, since the differences can be seen only under the observation of heavy equipment. Since lattice defects, such as vacancies, dislocations, and impurities, serve as preferential sites for etching, plasma etching treatment can be used to characterize and distinguish different growth areas by dislocation density [6]. As etch-pits are formed after plasma etching treatment, it is quite convenient to assess the growth quality in different areas. In addition, the etch-pits, which can be observed under optical microscopy, can be used as markers for pattern alignment in the subsequent ELO process.

In this work, two-step ELO via microwave plasma-enhanced chemical vapor deposition (MPCVD) was used to fabricate low dislocation density, single-crystalline diamond. Grooves were fabricated by inductively coupled plasma etching. Mo/Pd stripes sputtered in the grooves were used to hinder the propagation of dislocations originating from the diamond substrate. Plasma etching was used to characterize the quality of the epitaxial films and distinguish the different growth areas. Finally, the ELO diamond was also characterized by Raman spectroscopy and field emission scanning electron microscopy (FE-SEM).

2. Results and Discussion

Figure 1 shows the schematic process diagram of the ELO method. From Step 1 to Step 2, an Al mask with 1-µm-thick stripes was patterned on the high-temperature high-pressure (HPHT) diamond substrate. The width of the Al stripes and spacing were 10 and 40 µm, respectively. Grooves with a depth of 2 µm were fabricated in the area without Al stripe coverage via inductively coupled plasma (ICP) etching. During the ICP etching process, the Al stripes protected the underlying diamond from etching. After ICP etching, the thickness of the Al stripes decreased to approximately 200 nm.

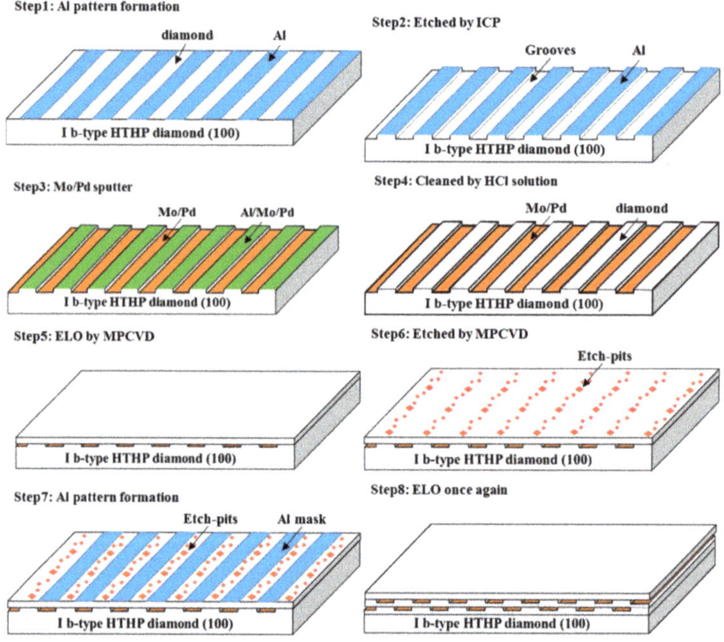

Figure 1. The schematic process diagram of the two-step epitaxial lateral overgrowth (ELO) method.

From Step 3 to Step 4, 100 nm Mo and 50 nm Pd layers were sputtered on the sample surface. The Mo layer was used to enhance the adhesion with the diamond surface, and the Pd layer was used to suppress the vertical growth of diamond. After metal sputtering, the sample was ultrasonically cleaned with an HCl solution. During this process, the Al stripes were able to lift-off the Mo/Pd layers. Thus, the areas covered with Al/Mo/Pd layers were cleaned to expose diamond windows. However, the area in the grooves covered with Mo/Pd layers were preserved after cleaning with HCl solution.

From Step 5 to Step 6, a diamond film with a thickness of 200 µm was grown by MPCVD, which produced a continuous and flat surface by lateral overgrowth. Then, the surface of the ELO layer was treated by plasma etching to form etch-pits. The stripes with high etch-pit density were grown by conventional homoepitaxial growth, and the stripes with low etch-pit density were grown by lateral overgrowth. After the first ELO step, half of the diamond surface improved in quality.

From Step 7 to Step 8, the above process was performed again for the second ELO step. During the process, the half of the diamond surface with high etch-pit density improved in quality via ELO. Since etch-pits can be observed under optical microscopy, they were used as markers for pattern alignment in the second ELO process. It should be mentioned that the Al stripes were fabricated on the lateral overgrowth area that did not show a distribution of etch-pits, thus forming diamond windows in this area during the second ELO step, as this area had a relatively lower dislocation density. Finally, the quality of the whole epitaxial film was improved.

Figure 2 shows optical images taken during the first ELO step. Figure 2a shows the image of the substrate after ultrasonic cleaning with an HCl solution. The grooves are covered with Mo/Pd layers, and the remaining areas are diamond windows. Figure 2b shows an image of the sample after 40 min of MPCVD growth. It can be observed that the diamond windows broadened from 10 µm to 25 µm by lateral overgrowth, as indicated in Area ①. Meanwhile, amorphous carbon formed on the Mo/Pd layers in the grooves, as indicated in Area ②. Figure 2c shows an image of the sample after 80 min of MPCVD growth. It can be observed that the diamond windows connected to form a continuous film. As the lateral overgrowth rate of diamond is much higher than the rate of amorphous carbon formation, the amorphous carbon was completely covered by diamond. However, the surface of the diamond film is not flat but instead has obvious stripes. Figure 2d shows the image of the sample after 240 min of MPCVD growth. It can be observed that the surface of the diamond film became smooth and flat.

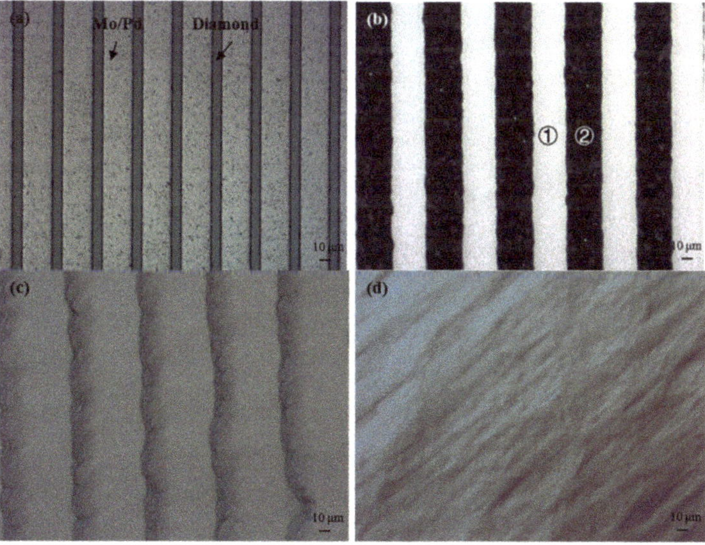

Figure 2. Optical images during the first ELO step. Panel (**a**) shows an image of the sample before MPCVD growth. Panels (**b**–**d**) show optical images of the sample after microwave plasma-enhanced chemical vapor deposition (MPCVD) growth for 40, 80 and 240 min, respectively. In Panel (**b**), Area ① and Area ② were originally diamond windows and Mo/Pd layers in the grooves, respectively.

Figure 3a,b show the Raman spectra obtained from two different areas, ① and ②, as indicated in Figure 2b. The Raman excitation wavelength was 532 nm. For Area ②, both the first-order diamond Raman line (located at 1332 cm^{-1}) and the sp^2 amorphous carbon peak (located at 1520–1580 cm^{-1}) can be observed. This suggests that amorphous carbon was first formed on the Mo/Pd stripes. In the spectrum of Area ①, the sp^2 amorphous carbon peak is no longer observed. Meanwhile, the full width at half maximum (FWHM) of the diamond Raman line is much smaller. A peak located at 1430–1470 cm^{-1} appears in the Raman spectrum, which is related to the photoluminescence of CVD diamond.

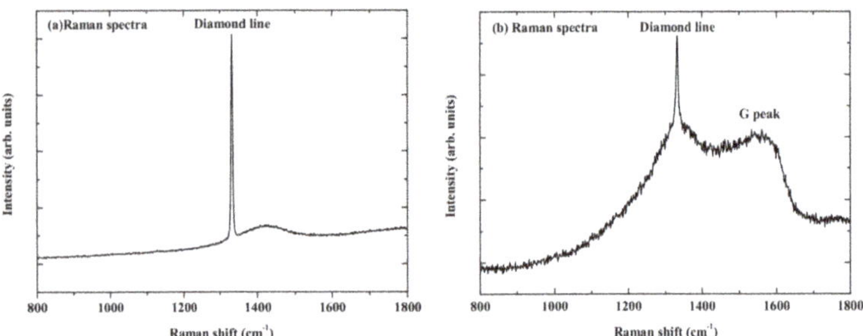

Figure 3. (**a**) Raman spectra obtained from area ① in Figure 2b. (**b**) Raman spectra obtained from area ② in Figure 2b.

Figure 4 shows images of the surface of the first ELO layer after plasma etching by MPCVD for 15 min. H$_2$/O$_2$ plasma etching treatments under relatively high pressure/power (>100 mbar, >2000 W) and with a low amount of added oxygen (<4%), usually lead to very selective etching of the (100) diamond surface [6]. Thus, defects such as dislocations located near the diamond surface typically appear as inverted pyramids with a square base after selective plasma etching. The structure of the inverted pyramids can be clearly observed in Figure 4a. The backlit optical image in Figure 4b shows that the etch-pits were concentrated in lines. Moreover, the etch-pit lines were located in the bright area. Since the window areas are transparent and the mask area blocks the backlight, this observation suggests that the lateral growth area had a low dislocation density.

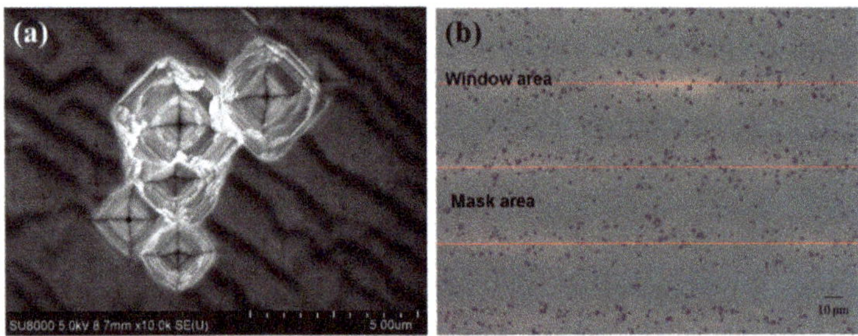

Figure 4. (**a**) FE-SEM image of etch-pits formed by plasma etching for 15 min. (**b**) Backlit optical image of the first ELO layer after plasma etching for 15 min.

Figure 5 shows FE-SEM images of the first and second ELO layer after plasma etching for 15 min. As seen in Figure 5a, the etch-pits are distributed in stripes on the surface of the first ELO layer. Dislocations originating from the substrate windows propagate to the surface, which cause dense etch-pit formation upon plasma etching. By contrast, the dislocations were effectively blocked from propagating in the lateral overgrowth area. Thus, the etch-pit density was clearly lower in these areas. As seen in Figure 5b, the etch-pits distributed in the stripes can still be observed on the surface of the second ELO layer, but the density clearly decreased. Because the dislocation density on the diamond windows decreased, fewer dislocations propagated to the surface to form etch-pits after the second ELO step. Thus, after two steps of ELO, the quality of the CVD diamond film was improved by reducing the dislocation density.

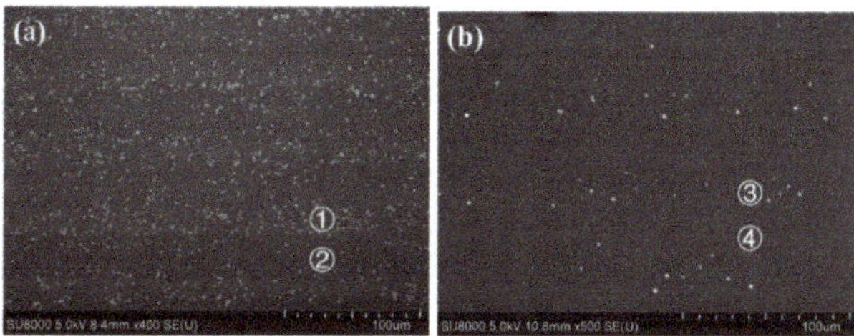

Figure 5. (**a**) FE-SEM image of the first ELO layer after plasma etching for 15 min. (**b**) FE-SEM image of the second ELO layer after plasma etching for 15 min.

Figure 6 shows FE-SEM images of the cross section of the sample after two steps of ELO. In the second ELO step, the diamond windows were located on top of the Mo/Pd stripes, leading to low dislocation density, as indicated by the dashed line. As seen in Figure 6, dark-colored triangular areas appear on the Mo/Pd stripes, which are related to amorphous carbon. A small hole and a vertical borderline can be clearly observed, which are formed when the two growth fronts of diamond coalesce. The results are in accordance with Figure 2. Initially, amorphous carbon formed on the Mo/Pd layers in the grooves. Then, the diamond windows were connected to form a continuous film by ELO.

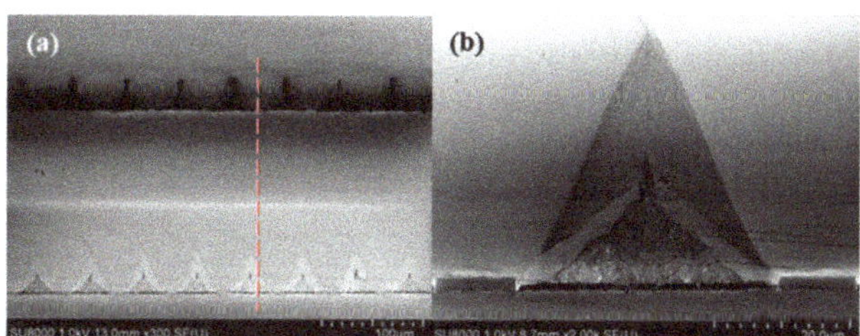

Figure 6. FE-SEM images of the cross section of the sample after two steps of ELO. Panel (**a**) shows an image of the cross section with two ELO layers. Panel (**b**) shows an image of the coalescent area.

Figure 7 shows the FWHMs of the diamond Raman lines obtained from the different surface areas after ELO, as indicated in Figure 5a,b. After the first ELO step, the FWHMs obtained from dense etch-pit Area ① ranged from 4.62 to 4.89 cm^{-1}. By contrast, the FWHMs obtained from the sparse etch-pit Area ② were all approximately 4.42 cm^{-1}. This indicates that the crystalline structure of the lateral overgrowth area was improved. After the second ELO step, the FWHMs obtained from the dense etch-pit Area ③ were all still approximately 4.42 cm^{-1}. This indicates that dislocations propagated to the surface. However, the FWHMs obtained from the sparse etch-pit Area ④ decreased to approximately 4.35 cm^{-1}. This suggests that the crystalline structure was further improved after the second ELO step.

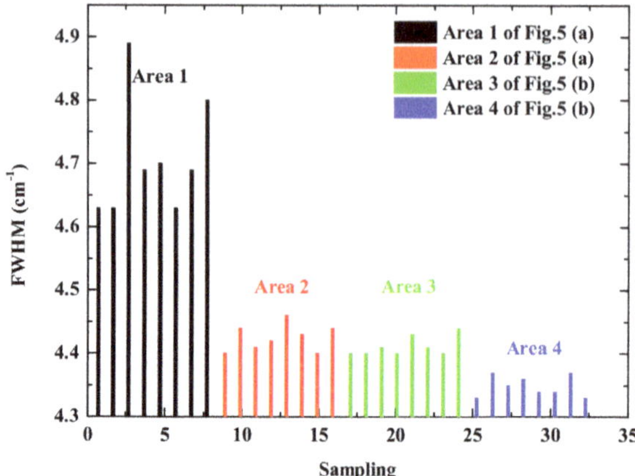

Figure 7. Full width at half maximum values (FWHMs) of the diamond Raman lines obtained from the different areas indicated in Figure 5.

3. Materials and Methods

An Ib-type high-temperature high-pressure (HPHT) diamond (100) substrate ($3 \times 3 \times 0.5$ mm^3) was used in this experiment. Before epitaxial layer growth, the substrate was cleaned in a mixed solution of nitric and sulfuric acids at 250 °C for 1 h. The growth conditions were as follows: substrate temperature: ~1150 °C; H$_2$ flow: ~500 sccm; CH$_4$ flow: ~50 sccm; gas pressure: ~120 Torr. The etch-pit formation conditions were as follows: substrate temperature: ~1020 °C; H$_2$ flow: ~400 sccm; O$_2$ flow: ~8 sccm; gas pressure: ~120 Torr. The ICP etching conditions were as follows: power: 800 W; Ar flow: ~20 sccm; O$_2$ flow: ~40 sccm; gas pressure: ~10 mTorr.

4. Conclusions

Two-step ELO on an HPHT diamond (100) substrate by MPCVD was used to fabricate a low dislocation density, single-crystalline diamond. Grooves were fabricated by ICP etching on the diamond surface, followed by sputtering Mo/Pd stripes in the grooves. Metal stripes were used to inhibit the propagation of dislocations originating from the diamond substrate by suppressing the vertical growth of diamond on the metal stripes. Plasma etching was used to characterize the quality of the epitaxial films and distinguish different growth areas.

After the first ELO step, the FE-SEM images show that the etch-pit density was clearly lower in the lateral overgrowth area. Then, the etch-pits, which can be observed under optical microscopy, were used as markers for pattern alignment in the second ELO process. Grooves and Mo/Pd stripes were fabricated again on the areas with a high density of etch-pits. After the second ELO step, the FE-SEM images show that the etch-pit density further deceased in the lateral overgrowth area. Thus, the quality of the whole epitaxial film was improved.

Acknowledgments: This work was supported by the Ministry of Science and Technology (Grant No. 2013AA03A101) of China.

Author Contributions: Fengnan Li designed and carried out the experiments. Jingwen Zhang and Xiaoliang Wang grew and etched the diamond films. Minghui Zhang and Hongxing Wang analyzed the results and reviewed the manuscript.

Conflicts of Interest: The authors declare no conflict of interest.

References

1. May, P.W. The new diamond age. *Science* **2008**, *320*, 1490–1491.
2. Isberg, J.; Hammersberg, J.; Johansson, E.; Wikstrom, T.; Twitchen, D.J.; Whitehead, A.J.; Coe, S.E.; Scarsbrook, G.A. High carrier mobility in single-crystal plasma-deposited diamond. *Science* **2002**, *297*, 1670–1672. [CrossRef] [PubMed]
3. Isberg, J.; Hammersberg, J.; Twitchen, D.J.; Whitehead, A.J. Single crystal diamond for electronic applications. *Diam. Relat. Mater.* **2004**, *13*, 320–324. [CrossRef]
4. Liao, M.Y.; Koide, Y. High-performance metal-semiconductor-metal deep-ultraviolet photodetectors based on homoepitaxial diamond thin film. *Appl. Phys. Lett.* **2006**, *89*, 113509. [CrossRef]
5. Liu, J.W.; Liao, M.Y.; Imura, M.; Tanaka, A.; Iwai, H.; Koide, Y. Low on-resistance diamond field effect transistor with high-k ZrO_2 as dielectric. *Sci. Rep.* **2014**, *4*, 06395. [CrossRef] [PubMed]
6. Naamoun, M.; Tallaire, A.; Silva, F.; Achard, J.; Doppelt, P.; Gicquel, A. Etch-pit formation mechanism induced on HPHT and CVD diamond single crystals by H_2/O_2 plasma etching treatment. *Phys. Status Solidi A* **2012**, *209*, 1715–1720. [CrossRef]
7. Nam, O.H.; Bremser, M.D.; Zheleva, T.S.; Davis, R.F. Lateral epitaxy of low defect density GaN layers via organometallic vapor phase epitaxy. *Appl. Phys. Lett.* **1997**, *71*, 2638–2640. [CrossRef]
8. Sakai, A.; Sunakawa, H.; Usui, A. Defect structure in selectively grown GaN films with low threading dislocation density. *Appl. Phys. Lett.* **1997**, *71*, 2259–2261. [CrossRef]
9. Beaumont, B.; Bousquet, V.; Vennegues, P.; Vaille, M.; Bouille, A.; Gibart, P.; Dassonneville, S.; Amokrane, A.; Sieber, B. A two-step method for epitaxial lateral overgrowth of GaN. *Phys. Status Solidi A* **1999**, *176*, 567–571. [CrossRef]
10. Hiramatsu, K.; Nishiyama, K.; Onishi, M.; Mizutani, H.; Narukawa, M.; Motogaito, A.; Miyake, H.; Iyechika, Y.; Maeda, T. Fabrication and characterization of low defect density GaN using facet-controlled epitaxial lateral overgrowth (FACELO). *J. Cryst. Growth* **2000**, *221*, 316–326. [CrossRef]
11. Bauer, T.; Schreck, M.; Stritzker, B. Epitaxial lateral overgrowth (ELO) of homoepitaxial diamond through an iridium mesh. *Diam. Relat. Mater.* **2007**, *16*, 711. [CrossRef]
12. Ando, Y.; Kamano, T.; Suzuki, K.; Sawabe, A. Epitaxial Lateral Overgrowth of Diamonds on Iridium by Patterned Nucleation and Growth Method. *Jpn. J. Appl. Phys.* **2012**, *51*, 090101. [CrossRef]
13. Kono, S.; Teraji, T.; Kodama, H.; Sawabe, A. Imaging of diamond defect sites by electron-beam-induced current. *Diam. Relat. Mater.* **2015**, *59*, 54–61. [CrossRef]
14. Umezawa, H.; Kato, Y.; Watanabe, H.; Omer, A.M.M.; Yamaguchi, H.; Shikata, S. Characterization of crystallographic defects in homoepitaxial diamond films by synchrotron X-ray topography and cathodoluminescence. *Diam. Relat. Mater.* **2011**, *20*, 523–526. [CrossRef]
15. Gaukroger, M.P.; Martineau, P.M.; Crowder, M.J.; Friel, I.; Williams, S.D.; Twitchen, D.J. X-ray topography studies of dislocations in single crystal CVD diamond. *Diam. Relat. Mater.* **2008**, *17*, 262–269. [CrossRef]

Article

Morphology of Diamond Layers Grown on Different Facets of Single Crystal Diamond Substrates by a Microwave Plasma CVD in CH_4-H_2-N_2 Gas Mixtures

Evgeny E. Ashkinazi [1], Roman A. Khmelnitskii [2,3,4], Vadim S. Sedov [1], Andrew A. Khomich [1,3], Alexander V. Khomich [1,2,3,*] and Viktor G. Ralchenko [1,5]

[1] Prokhorov General Physics Institute, Russian Academy of Sciences, Vavilova Str. 38, Moscow 119991, Russia; jane50@list.ru (E.E.A.); sedovvadim@yandex.ru (V.S.S.); antares-610@yandex.ru (A.A.K.); vg_ralchenko@mail.ru (V.G.R.)

[2] Lebedev Physical Institute, Russian Academy of Sciences, Leninskii Av. 53, Moscow 119991, Russia; roma@sci.lebedev.ru

[3] Kotelnikov Institute of Radio Engineering and Electronics, Russian Academy of Sciences, Vvedenskogo Sq. 1, Fryazino 141190, Russia

[4] Troitsk Institute for Innovation and Fusion Research, Pushkovykh Str. 12, Troitsk, Moscow 142190, Russia

[5] Harbin Institute of Technology, 92 Xidazhi Str., Harbin 150001, China

* Correspondence: alex-khomich@mail.ru

Academic Editor: Yuri N. Palyanov

Received: 30 April 2017; Accepted: 31 May 2017; Published: 6 June 2017

Abstract: Epitaxial growth of diamond films on different facets of synthetic IIa-type single crystal (SC) high-pressure high temperature (HPHT) diamond substrate by a microwave plasma CVD in CH_4-H_2-N_2 gas mixture with the high concentration (4%) of nitrogen is studied. A beveled SC diamond embraced with low-index {100}, {110}, {111}, {211}, and {311} faces was used as the substrate. Only the {100} face is found to sustain homoepitaxial growth at the present experimental parameters, while nanocrystalline diamond (NCD) films are produced on other planes. This observation is important for the choice of appropriate growth parameters, in particular, for the production of bi-layer or multilayer NCD-on-microcrystalline diamond (MCD) superhard coatings on tools when the deposition of continuous conformal NCD film on all facet is required. The development of the film morphology with growth time is examined with SEM. The structure of hillocks, with or without polycrystalline aggregates, that appear on {100} face is analyzed, and the stress field (up to 0.4 GPa) within the hillocks is evaluated based on high-resolution mapping of photoluminescence spectra of nitrogen-vacancy NV optical centers in the film.

Keywords: diamond; microwave plasma CVD; epitaxy; nanocrystalline film; photoluminescence; hillocks

---

1. Introduction

High hardness and abrasion resistance of diamond make it indispensable for machining of many materials such as composites and abrasive alloys, which leads to widespread use of diamond-based tools. Diamond coatings (DCs) produced by chemical vapor deposition (CVD) on tungsten carbide WC-Co cutting tools allow a great improvement in the tool performance by extending its lifetime, increasing the machining speed, and providing a better quality of the machined material [1]. The application of bilayered or multilayered micro- and nanocrystalline DCs further enhances the cutting tools' functional properties [2,3]. The upper nanocrystalline layer provides low roughness and low friction coefficient, as well as high bending strength [4]. The bottom layer is typically deposited in the microcrystalline diamond growth regime since it has better adhesion to the WC-Co substrate, coupled to higher hardness [5]. Moreover, compared with single-layer DCs, bilayered DCs have higher heat conductivity and increased resistance to cracking [6].

The structure of a DC deposited in microwave plasma can be controlled by varying the composition of the gas mixture during the process run. For example, the addition of nitrogen allows a transition from microcrystalline diamond (MCD) to nanocrystalline diamond (NCD) coating deposition regimes. As a result, one can obtain smooth bilayered DCs with increased wear resistance by the CVD method for an optimal nitrogen concentration [3]. The N_2 gas added to standard CH_4-H_2 mixtures strongly enhances the growth rate SC diamond seeds (typically of {100} orientation) [7,8], thus reducing the cost of CVD diamond production. In case of epitaxial growth of single crystals the growth rate and defect abundance depend on the diamond substrate face orientation, a less number of the defects forming on {100} facets [9]. Particularly, dislocation density can be significantly reduced by a lateral growth on {100} oriented substrate with a hole intentionally perforated in it as recently shown by Tallaire et al. [10].

The growth of nanocrystalline diamond on a microcrystalline diamond film has several differences from the more widely used growth on non-diamond substrates, including: (a) significant roughness of the microcrystalline film surface; (b) microcrystalline film is formed by crystallites with different faces, predominantly with {111} and {100} [11]; (c) no preliminary nucleation (seeding) is performed; and (d) there is no incubation period for growth [11]. This raises the question of possible differences of secondary nucleation and growth of nanocrystalline DCs on different faces of microcrystals of diamond films. We note that the deposition of ultrananocrystalline diamond (UNCD) films with grain typically, less than 10 nm, in a multicomponent gaseous environment, Ar-CH_4-H_2+N_2 mixtures, on polished coarse-grained polycrystalline diamond plates (the grains with random orientation and size of 60–90 μm) was reported in [12]. The UNCD growth was assumed to start epitaxially, but the high rate of secondary nucleation very rapidly transformed the film to nanocrystalline structure uniformly over the substrate area. No impact of grain orientation of the substrate on the morphology of the deposited UNCD layer was observed.

In the production of bilayer coatings of such type, it is important to deposit the NCD layer uniformly on all facets of the grains in underlying MCD film, independent on the facet orientation. The goal of the present study is the investigation of the nucleation and growth processes of DCs on different facets of a diamond single crystal as a model substrate, and examination of the growth defects on the {100} face with SEM and photoluminescence (PL) spectroscopy. To this end, we prepare faceted SC diamond substrates to deposit the diamond film simultaneously on facets of different orientations and compare the resulting morphologies for identical process parameters.

2. Experimental

The diamond coatings were synthesized in "hydrogen/methane/nitrogen" gas mixtures on SC diamond substrates using a microwave plasma CVD system ARDIS-100 (2.45 GHz, 5 kW) [13] with the following parameters: total gas flow rate of 500 sccm (H_2:460/CH_4:20/N_2:20), pressure of 130 Torr, and microwave power of 2.8 kW. The substrate temperature during deposition was about 800 °C as measured through plasma by the Mikron M770 two-color pyrometer. These deposition parameters and the nitrogen concentration in the gas mixture corresponded to deposition regime for nanocrystalline DCs on WC-Co substrates [3]. As the substrate, we used a multifaceted type IIa synthetic SC diamond produced by the high pressure—high temperature (HPHT) gradient method. The upper part of the crystal exhibited smooth {311}, {211}, {111}, and {110} growth faces, with the larger {100} face on the top. This multifaceted part was cut off along the {100} plane to remove the bottom part containing sides with negative slope. The top and bottom {100} planes were polished to roughness R_a of less than 10 nm, as measured with an optical profilometer (ZYGO NewView 5000), while the other facets remained unpolished. The roughness R_a of the as-grown faces was 90–100 nm for {111}, 120–130 nm for {211}, 160–180 nm for {311} and 100–110 nm for {110} planes. The roughness was mainly due to growth steps on those surfaces. The substrate thickness was 0.9 mm, which provided almost identical growth conditions in the CVD reactor, with a minimized temperature gradient across the substrate. At least the adjacent regions around common edges between the facets had identical temperatures during diamond deposition.

A scanning electron microscope (SEM) JSM7001F (JEOL) was used to examine the coating surface morphology. The Raman and PL spectra, excited at 473 nm wavelength, were measured at room temperature with a LabRam HR840 (Horiba Jobin-Yvon) spectrometer in a confocal configuration with the spectral resolution of 0.5 cm^{-1} and spatial resolution of ~1 μm. The spatial mapping of the PL and Raman spectra over the coating surface area was carried out by placing the sample on a motorized table (MärzhäuserWetzlar) with positioning accuracy of about 0.5 μm.

3. Results and Discussion

3.1. Diamond Deposition on Low-Index Facets of the Single Crystal Substrate

The SEM analysis shows that, under the growth conditions of nanocrystalline diamond (if a non-diamond substrate material is used), the NCD coating is formed on the {111} and {110} faces of the substrate from the very beginning. This coating is formed of grains with the size of a few tens to hundreds of nanometers. Figure 1 presents SEM images of several regions of the sample after 5 min of growth. The panoramic image in Figure 1a displays seven different faces on the HPHT diamond substrate. The original {311} and {211} faces did not contain regions with outcrops of {100} and {111} planes. However, clearly pronounced layers with {111} planes appear on those faces even at the initial growth stage, and, flat regions with {100} planes are formed, on the top of these layers. Figure 1b shows an example of surface relief formed on the {311} face. The surface morphologies of the coating grown on the {100} and {111} planes are essentially different: the {100} planes are smooth, and homoepitaxial growth in the step bunching regime [14] occurs on these planes, whereas, on the {111} planes nanocrystallites arise and develop into a rough film surrounding the {100} domains.

In the course of the deposition process, the evolution of crystallite morphology on the diamond film surface is controlled by growth rates V of the main low-index crystalline faces, first of all, {100} and {111}, as well as by the {110} and {311} faces [15]. According to the crystal growth laws the area fraction of the faces with minimum growth rate increases. The dimensionless parameter that characterizes the ratio of the growth rate of a low-index face to the growth rate of one of them, say, the {111} face is an alpha-parameter $\alpha = 3^{\frac{1}{2}} V_{100}/V_{111}$. The greater the α, the greater the fraction of the {111} faces area of the crystallite. A decrease in the substrate temperature during the CVD process leads to an increase in α and a decrease in the crystallite size [15]. An increase in the methane [CH$_4$] concentration in the gas mixture leads to an increase in the growth rate and the α parameter, a decrease in the crystallite size, and an increase in the probability of secondary nucleation, i.e., the formation of new nuclei of diamond crystallites with different crystalline orientation on the diamond surface [13]. This hydrocarbon supersaturation facilitates the growth of the nanocrystalline film. The increase in [CH$_4$] increases the concentration of hydrocarbon radicals (CH$_3$, CH$_2$, CH, etc.) and their adsorption rate on the growing surface of diamond. This increases the probability of deposition of two growth radicals side by side, including radicals with two (CH$_2$) or three (CH) free bonds, that promotes the formation of growth defects and crystalline nuclei with a different orientation. Nitrogen addition to the growth mixture also increases the diamond growth rate, among other things, due to the growth of the probability of formation of two-dimensional nuclei on the growing surface, and increases α and the probability of secondary nucleation [16]. The mechanism is similar because an atom of N embedded on the surface of diamond has two or more free bonds [15]. For large α, nanocrystallites are formed mainly by {111} faces, which are prone to twinning due to low twin formation energy [17], therefore, multiple twinning on the {111} planes determines the morphology of the nanocrystalline films.

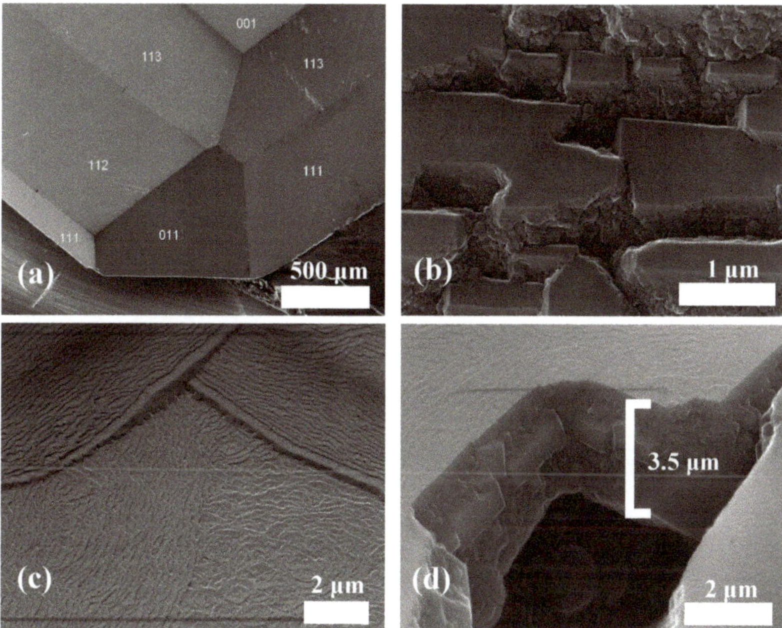

Figure 1. SEM images obtained after 5-min diamond deposition on high pressure—high temperature (HPTP) diamond substrate: (**a**) a corner part of the substrate with indicated crystallographic faces; (**b**) surface morphology of {311} face; (**c**) layer-by-layer growth on the {100} face; and (**d**) view of a lacuna at 45° angle on the face {100}.

It is believed that renucleation processes dominate under low power density or high methane concentration [11]. The observed film morphology, growth rate, and crystal size can be rationalized using a model based on the competition of H atoms, CH_3 radicals, and other CH_x (CH_2, CH, C) species to react with dangling bonds on the surface [18]. Under the CVD conditions, the concentration of CH_3 radicals in gas is much higher than the concentration of other CH_x radicals. It is generally accepted that the key role in diamond growth is played by the CH_3 radicals [18]. The CH_3 radical is relatively large and, due to steric hindrance, it is hard to be coupled to a free bond on the growing surface. This stage is assumed to limit the growth rate. The CH_3 radical coupled to the surface loses one H atom when interacting with atomic hydrogen and turns into a CH_2 group that can migrate along the growing surface [19]. This group reaches the growth step and completes the construction of the crystal. Other CH_x radicals are also coupled to free bonds. The resulting group may have a free bond. It is most probable that an atomic or molecular hydrogen is coupled to this group. However, there is a probability that another CH_x radical will be bonded. This may introduce a surface defect with a different symmetry of bonds, which gives rise to renucleation or a twin. In this model, the radical concentration ratios $[CH_3]/[H]$ and $[CH_x]/[CH_3]$ determine the probability of the renucleation event. One should take in to account the difference between the probabilities of bonding of CH_x radicals on different crystalline planes. Our experiments call into question the idea that, under the growth conditions of a nanocrystalline film, renucleation occurs on the growth surfaces of any orientation [20]. On the {100} face of SC substrate, homoepitaxial growth in the step bunching regime occurs for a long (5 min) period of time (Figure 1c,d). The concentrations of active radicals [H], $[CH_3]$, $[CH_2]$, [CH], and [C] near the {100} and {111} growth surfaces are the same, however, the growth morphologies on these surfaces are essentially different. According to the model [19], this means that the probability of incorporation of CH_3 radical in {100} plane is much higher than that in {111} plane.

Figure 1c demonstrates epitaxial character and growth steps of the {100} face. The growth steps directions may change locally (Figure 1c); however, in some places they form twisted helical structures in the form of hillocks (presumably, around screw dislocations). The steps height amounts to tens of nanometers, but sometimes they become thicker. Such case occurs when one growth layer overtakes another to form a growth layer of increased thickness (Figure 1c). There is an assumption [21] that the formation of macrosteps occurs when the growth rate of atomic steps becomes of the order of the diffusion rate of adsorbed species along the surface. Ideally, the macrosteps have atomically smooth surfaces [22]. The accumulation of thickened growth layers increases the roughness of the growth surface {100}.

Figure 1d demonstrates the growth lacunae on the face {100}, on which one can measure the thickness of a homoepitaxial film grown in 5 min. This thickness of 3.5 μm corresponds to the growth rate of 42 μm/h for the {100} face. Without the addition of N_2, the epitaxial growth rate at 4% CH_4 and 800 °C of about 5 μm/h was reported for the {100} face [23]. In other words, the addition of 4% of N_2 to the gas mixture increases the growth rate by an order of magnitude, in agreement with other publications [7,15].

At the end of the 5-min growth session, we measured the Raman spectra (Figure 2a) of the films deposited on each face of the sample [3]. The Raman spectrum on the face {100} contains only the 1st order Raman diamond line at $\nu_R = 1332.4$ cm^{-1} with full width at half magnitude (FWHM) $\Gamma_{1/2} = 2.7$ cm^{-1}. On other faces, the spectrum contains also broad bands of disordered carbon with maxima at 1355 and 1560 cm^{-1} (D and G bands), as well as trans-polyacetylene (TPA) bands with maxima at 1140 and 1490 cm^{-1}, which are characteristic of NCD.

The width of the diamond peak in the Raman spectrum increases in the series {100} → {110} → {111} → {211} → {311} from 2.7 to 4.3 cm^{-1} (Figure 2b). The diamond 1332 cm^{-1} peak broadens as a result of defects presence, and also due to a nonuniform stress. The peak shift is caused by a stress, tensile or compressive, depending on the shift sign. From the diamond Raman peak shift $\Delta\nu_R$, we estimated the stress σ in the film according to relationship σ [GPa] = $\Delta\nu_R$ [0.38 GPa/cm^{-1}] for hydrostatic stress [24], the positive or negative values of σ corresponding to compressive or tensile strength, respectively. For the {100} and {211} faces, the position of this peak corresponds to unstressed diamond. For the {111} and {110} faces, the shift of the diamond peak corresponds to the compressive stress of 0.15 GPa, while, for the {311} face the tensile stress of 0.3–0.6 GPa is assessed. The ratio of summed intensities of the TPA and G bands to the area of the diamond peak increases in the series {100} → {211} → {111} ≈ {110} → {311} according to the increase of the fraction of non-diamond phase in the film.

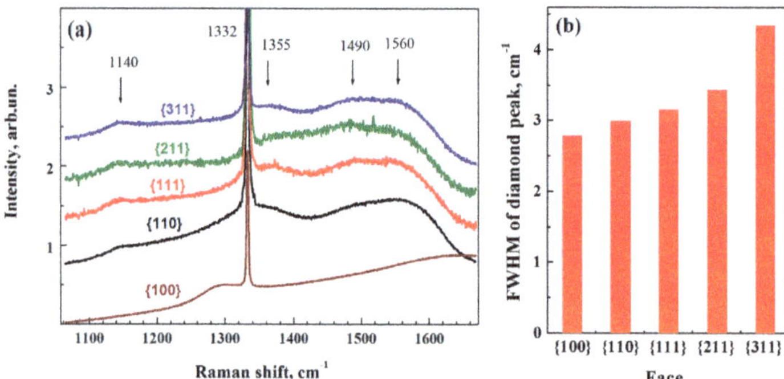

Figure 2. (a) Raman spectra on different faces after 5-min growth. The spectra are shifted along the vertical scale for clarity. Two broad bands at 1290 and 1650 cm^{-1} on {100} face are due to photoluminescence of H3 defect center [25]. The Raman spectra were normalized to the 1332.5 cm^{-1} peak intensity; (b) Diamond Raman peak width (FWHM) for CVD film on different faces.

Figure 3 demonstrates the images of the same sample after 70-min growth on the same substrate. There are no smooth epitaxial regions on {211} faces of the substrate (Figure 3a), all of them have been coated with NDC film. In contrast, on {311} faces, there still remain islands with smooth {100} oriented growth surfaces, although the area of these regions was reduced in comparison with 5-min growth sample (compare Figures 1b and 3a,b). The nanocrystallites (Figure 3b) are formed mainly by a multiple twinning mechanism both upon parallel {111} planes and by rotating the orientation of the twinned crystal at 70.5 degrees about the {110} axis. The size of the twin crystallites on {311} and {211} faces reaches hundreds of nanometers (Figure 3b). On {111} and {110} faces, the twinning manifests itself weaker, the secondary nucleation dominates [9], and the crystallite size does not exceed 100 nm.

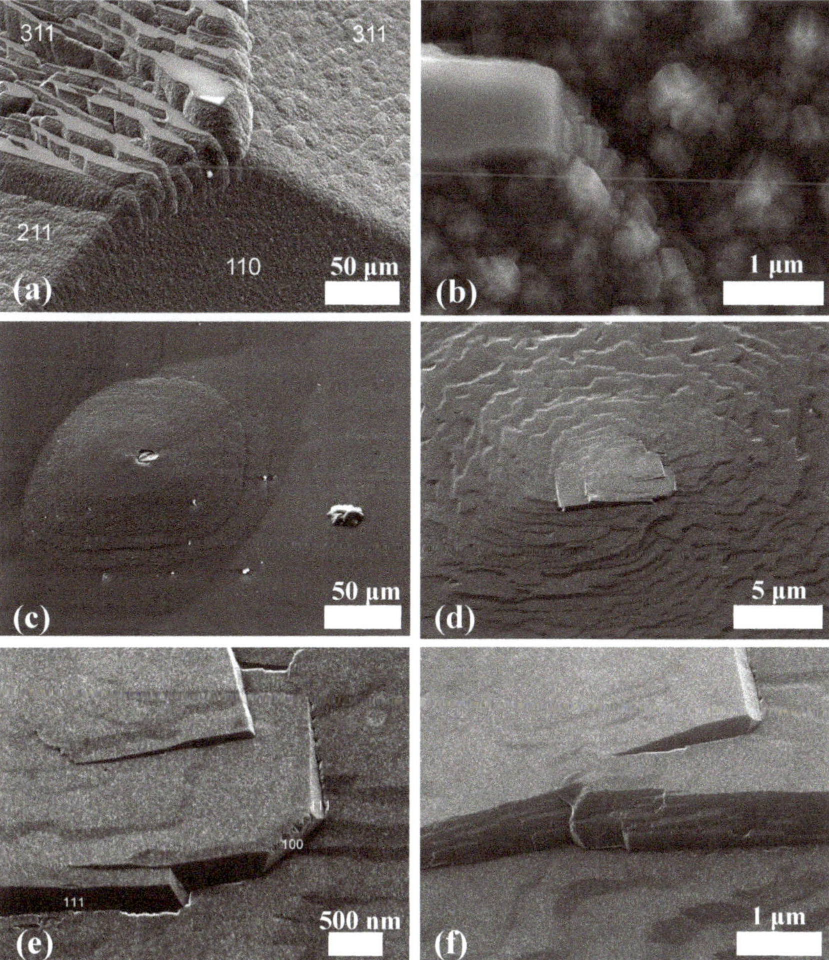

Figure 3. SEM images after 70-min growth on a single-crystal HPHT substrate: (**a**) the bevel of the substrate with indicated crystallographic faces; (**b**) {311} face at higher magnification; (**c**) stepped growth and formation of a hillock on the {100} face; (**d**) generation of a polycrystalline aggregate on the top of a hillock on the {100} face; (**e**) the same hillock, formation of low-index crystalline faces on edges of thick growth steps; (**f**) on the top of another hillock, a relief is developed on the edges of thick growth step.

3.2. Polycrystalline Aggregations at {100} Face

In contrast to other low-index faces, {100} planes continue growing predominantly by the step bunching growth mechanism, even with relatively long deposition times. The area fraction of the {100} planes decreased with the growth time, hence, the {100} faces grow with the maximum rate. This observation agrees with the assumption that the probability of CH_3 radical embedding in the {100} plane is much higher than that for other planes. The enhanced growth of layered screw structures on {100} face (Figure 3c) gives rise to hillocks [26]. As the film grows, at the top of such structures, as well as at the intersections of thickened growth steps of different directions, polycrystalline aggregations arise and develop (Figure 3c). Figure 3d–f show the initial stage of this process. One can see that growth steps become thicker and their edges become rougher. A great role in this process is played by nitrogen impurity, which is the main impurity in both natural and synthetic diamond produced by CVD and HPHT processes [27]. The incorporation efficiency of nitrogen in diamond lattice is as low as 10^{-4} [28,29]. Therefore, nitrogen is concentrated on the growing surface, contaminates and roughens the growth steps [30]. The end faces of the thickest growth macrosteps eventually acquire crystalline faceting, predominantly by {111} and {110} faces (Figure 3e). Gradually, a relief is developed on these faces (Figure 3f), on which twins readily form. The twins give rise to polycrystalline aggregates; these inclusions also may arise when lacunae (like those shown in Figure 1d) are overgrown near the defects or contaminations of the substrate.

3.3. Photoluminescence Scanning Spectroscopy on Growth Defects on {100} Face

Growth hillocks and non-epitaxial crystallites are often observed on homoepitaxial CVD diamond films. The hillock formation is believed to be caused by stacking faults or dislocations that nucleate at the interface between the surface and the epitaxial layer from lattice defects or propagate from the substrate into the CVD layer [31,32]. Those defects induce a surface roughness, facilitate incorporation of impurities, crystalline imperfections, and non-diamond phases. The hillocks could be responsible for dielectric breakdown effect and current leakage in diamond-based electronic devices which constitute the main limitations to their development and improvement of device properties. The problem of surface growth defects is not so relevant in the case of deposition of bilayer micro/nanocrystalline diamond coatings for cutting tools. Meanwhile, the features of the formation mechanism and stress state of the growth defects in CVD homoepitaxial films are still debatable. In order to examine stress distribution of such defects, we measured spatially resolved photoluminescence and Raman spectra.

Figure 4 displays the PL spectra taken at on {100} face on a hillock shown in Figure 5a, along the line A-B across the hillock, from foot to the top and further. The spectra are normalized to the integral intensity of the diamond band at 505 nm in the Raman spectrum. The PL spectra show three optical centers with zero phonon lines (ZPL) at wavelengths of 575 nm (NV^0 defect), 637 nm (NV^- defect), and 503 nm (H3 defect). The optical centers at 575 and 637 nm are formed by defects containing one nitrogen atom and one vacancy in neutral and negative charge states, respectively [25]. The NV defects are usually detected in CVD diamond films grown with the addition of nitrogen. However, the H3 optical center, formed by the $2NV^0$ defect, two nitrogen atoms and a vacancy in a neutral charge state, is less common in CVD diamond, because during growth nitrogen atoms are introduced into the crystal lattice one by one, not in pairs. In HPHT diamonds, the enhanced concentration of H3 centers is observed in less-stressed regions of the crystal and near dislocations [33].

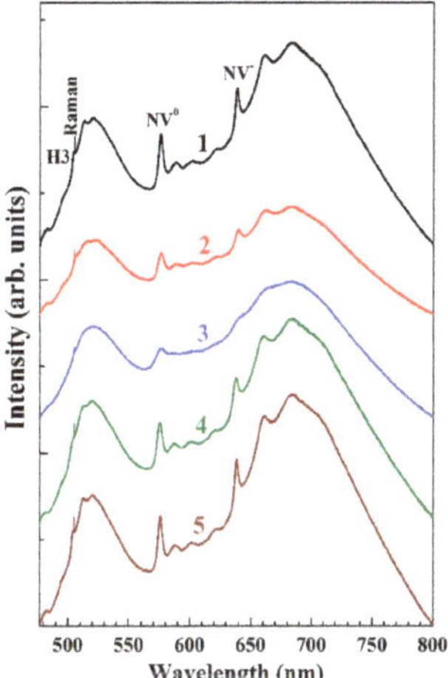

Figure 4. Photoluminescence spectra at five points (1)–(5) along the line A-B line crossing the top of the hillock, as shown in Figure 5a.

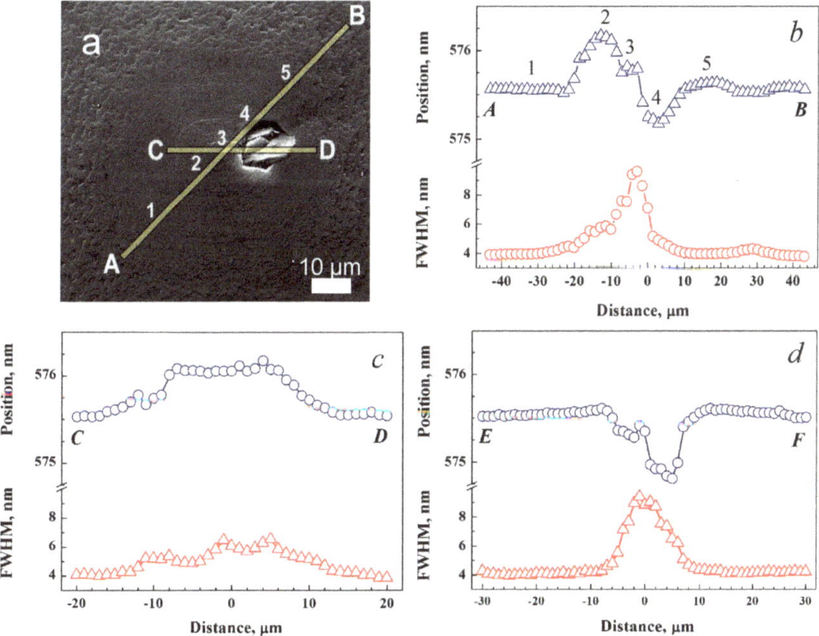

Figure 5. SEM image of a hillock with polycrystalline inclusion on the top. The lines A-B and C-D, which avoid or cross the polycrystalline inclusion, respectively, denote the direction of PL NV mapping (**a**); Variations of position (top curve) and the FWHM (bottom curve) of NV^0 ZPL at 575 nm along the lines A-B (**b**) and C-D (**c**); Variation position and width of PL at 575 nm peak along the line E-F crossing the top of another hillock without inclusions (**d**). The mapping step is 1.4 μm for A-B and 1 μm for C-D and E-F lines. The lines are guides for an eye. Locations # 1, 2 ... 5 in (**a,b**) show positions where the PL spectra presented in Figure 4 have been taken.

In CVD diamonds, the 2NV defect can be obtained by a long-term high-temperature post-growth annealing to cause aggregation of individual nitrogen atoms or NV defects via their diffusion [34,35]. Under the CVD conditions, the growth temperature and duration are considered insufficient for the diffusion of nitrogen atoms or NV defects. In the samples investigated, the 2NV defects are likely to be created immediately in process of diamond film deposition. This may occur, for example, as a result of incorporation of an N_2 molecule or a random incorporation of two nitrogen atoms next to each other. As the deposition rate of homoepitaxial diamond film increases, the incorporation efficiency of impurities into diamond also increases [36], and, according to calculations [37], the formation energy of 2NV defects in synthesized diamond substantially decreases for high nitrogen concentrations.

Owing to the presence of a vacancy, the NV optical centers are characterized by high values of strain softening, a mechanical stress leads to a significant shift in the position of its ZPL [25]. We measured the position and the FWHM of ZPLs at 575 nm and 637 nm across two hillocks. One of them contained a non-epitaxial polycrystalline aggregate on the top (Figure 5a), while the second hillock (not shown here) had no such polycrystalline inclusion. The PL spectra were measured in 1 μm or 1.4 μm steps along the lines A-B (Figure 5b) and C-D (Figure 5c), respectively, crossing the hillock near the top. The PL mapping of the second hillock was performed along the line E-F (Figure 5d), again crossing its top. The Lorentz profiles were used for fitting ZPL in the PL spectra.

The positions and widths of NV ZPL at 575 nm appreciably differed at different locations on the hillocks and revealed a complex picture along the scanning lines (Figure 5b–d). In absence of polycrystalline inclusion, the ZPL position displays a blue shift around the hillock top and the line broadening by a factor of two (Figure 5b,d). In contrast, a red shift and smaller broadening of NV ZPL is observed when the laser scan crosses the polycrystalline aggregate along the C-D path (Figure 5c). Similar profiles for ZPL position and width were also obtained for 637 nm line of NV$^-$ center (not shown in Figure 5). We attributed the observed broadening and the shift of the bands in the PL spectra to mechanical stresses in the hillock material. Spatial variations in nitrogen concentration could also cause a local stress. We estimated the stress σ from the shift Δλ of the ZPL maximum at 575 nm using the gauge factor of −0.57 GPa/nm [38]. Then, the maximum tensile stress of 0.4 GPa along the scan line A-B at point 2 (Figure 5a,b) was assessed, while the maximum compressive stress σ of 0.25 GPa corresponded to point 4 on this path.

The PL mapping along C-D line (Figure 5c) across the hillock center and non-epitaxial inclusion revealed a tensile stress in this region (the 575 nm peak shifts to longer wavelengths), the position and width of the 575 nm line correlating with the surface profile of the film. The PL mapping across the top of another hillock, having no polycrystalline inclusion, showed a different picture (Figure 5d). A stress at the hillock top causes the 575 nm peak shift by 0.75 nm to shorter wavelengths, corresponding to the compressive stress of about 0.4 GPa. A low tensile stress background is observed outside the hillock. A similar trend with PL ZPL shift and broadening was observed for NV$^-$ center at 637 nm, but not shown here. In this context, we mention the observations by Bolshakov et al. [39] the stress and variations of PL intensity of SiV centers in {100} oriented Si-doped epitaxial CVD diamond film within hillocks with polycrystalline inclusions. Using the Raman and PL spectra mapping they found a strong compressive stress on the hillock tip and an order of magnitude enhancement in the PL intensity. Our mapping results for NV PL are in line with that work.

For natural diamonds, there are several observations indicating that after the dissolution in water-containing carbonate melts the appearance of surface hillocks is typical for crystals with the high level of internal strains. Orlov [40] noted that the hillocks are often located along the intersecting striation associated with the bands of plastic deformation in diamond. Using polarized light microscopy, it was established [41] that in bulk of the crystals strongly strained regions are identified which borders coincide with the borders of the hillocks. The use of PL and Raman 3D mapping could facilitate assessment of the type and magnitude of stress within such features.

4. Conclusions

We demonstrated that the morphology of diamond films grown by a microwave plasma CVD on single crystal diamond substrates in CH_4-H_2-N_2 in presence of high concentration of nitrogen strongly depends on the substrate's crystallographic plane. The film structure on {100}, {110}, {111}, {211} and {311} faces has been examined with SEM, Raman and PL spectroscopy after a short (5 min) and prolonged (70 min) deposition time at conditions typically facilitated the formation of nanocrystalline diamond on foreign substrates. We found that only {100} face sustained homoepitaxial growth at the present experimental parameters, while only a nanocrystalline film is produced on other planes. Particularly, the NCD layer with the grain size of tens or hundreds of nanometers, often twinned, forms on {111}. This observation is important for the choice of appropriate growth parameters, in particular, for the production of bi-layer or multilayer NCD-on-MCD superhard coatings on tools, when the NCD deposition on all facets is required. The diamond epitaxy on {100} plane is possible even at high hydrocarbon and nitrogen contents in the process gas, that allows achievement of high growth rates of SC CVD diamond.

The growth of hillocks on {100} face proceeds in step-growth mode and is accompanied with the appearance of polycrystalline inclusions on the top. The early stage of the polycrystalline aggregates may include a faceting the step surface with {111} and {110} planes with further formation of the polycrystalline structure there. A tensile stress is revealed within polycrystalline inclusions on the hillocks from the shift of local PL spectra of NV optical centers, while a compressive stress is typical for the hillocks free of such inclusions. The PL spectroscopy mapping of the surface is shown to be a useful tool that allows measurement of the stress profile within growth defects in CVD diamond epitaxial films.

Acknowledgments: This work was supported by the Russian Scientific Fund (grant No. 15-19-00279).

Author Contributions: E.E.A. and R.A.K. conceived and designed the experiments; R.A.K., V.S.S. and A.A.K. performed the experiments; R.A.K. and A.V.K. analyzed the data; R.A.K., A.A.K. and V.G.R. wrote the paper.

Conflicts of Interest: The authors declare no conflict of interest.

References

1. Polini, R.; Barletta, M.; Rubino, G.; Vesco, S. Advances in the Deposition of Diamond Coatings on Co-Cemented Tungsten Carbides. *Adv. Mater. Sci. Eng.* **2012**, *2012*, 151629. [CrossRef]
2. Dumpala, R.; Chandran, M.; Rao, M.S.R. Engineered CVD Diamond Coatings for Machining and Tribological Applications. *J. Miner. Met. Mater. Soc.* **2015**, *67*, 1565–1577. [CrossRef]
3. Khomich, A.A.; Ashkinazi, E.E., Ralchenko, V.G.; Sedov, V.S.; Khmelnitskii, R.A.; Poklonskaya, O.N.; Kozlova, M.V.; Khomich, A.V. Application of Raman spectroscopy for the analysis of the structure of diamond coatings on a hard alloy. *J. Appl. Spectrosc.* **2017**, *84*, 297–304. [CrossRef]
4. Chromik, R.R.; Winfrey, A.L.; Lüning, J.; Nemanich, R.J.; Wahl, K.J. Run-in behavior of nanocrystalline diamond coatings studied by in situ tribometry. *Wear* **2008**, *265*, 477–489. [CrossRef]
5. Sun, F.; Ma, Y.; Shen, B.; Zhang, Z.; Chen, M. Fabrication and application of nano-microcrystalline composite diamond films on the interior hole surfaces of Co cemented tungsten carbide substrates. *Diam. Relat. Mater.* **2009**, *18*, 276–279. [CrossRef]
6. Shafer, L.; Hofer, M.; Kroger, R. The versatility of hot-filament activated chemical vapor deposition. *Thin Solid Films* **2006**, *515*, 1017–1024. [CrossRef]
7. Chayahara, A.; Mokuno, Y.; Horino, Y.; Takasu, Y.; Kato, H.; Yoshikawa, H.; Fujimori, N. The effect of nitrogen addition during high-rate homoepitaxial growth of diamond by microwave plasma CVD. *Diam. Relat. Mater.* **2004**, *13*, 1954–1958. [CrossRef]
8. Tallaire, A.; Achard, J.; Silva, F.; Brinza, O.; Gicquel, A. Growth of large size diamond single crystals by plasma assisted chemical vapour deposition: Recent achievements and remaining challenges. *Comptes Rendus Phys.* **2013**, *14*, 169–184. [CrossRef]
9. Lloret, F.; Fiori, A.; Araujo, D.; Eon, D.; Villar, M.P.; Bustarret, E. Stratigraphy of a diamond epitaxial three-dimensional overgrowth using doping superlattices. *Appl. Phys. Lett.* **2016**, *108*. [CrossRef]

10. Tallaire, A.; Brinza, O.; Mille, V.; William, L.; Achard, J. Reduction of dislocations in single crystal diamond by lateral growth over a macroscopic hole. *Adv. Mater.* **2017**, *29*. [CrossRef] [PubMed]

11. Koji, K. *Diamond Films: Chemical Vapor Deposition for Oriented and Heteroepitaxial Growth*; Elsevier: Amsterdam, The Netherlands, 2005; Chapter 5; pp. 31–50.

12. Ralchenko, V.; Pimenov, S.; Konov, V.; Khomich, A.; Saveliev, A.; Popovich, A.; Vlasov, I.; Zavedeev, E.; Bozhko, A.; Loubnin, N.; et al. Nitrogenatednanocrystalline diamond films: Thermal and optical properties. *Diam. Relat. Mater.* **2007**, *16*, 2067–2073. [CrossRef]

13. Bolshakov, A.P.; Ralchenko, V.G.; Yurov, V.Y.; Popovich, A.F.; Antonova, I.A.; Khomich, A.A.; Vlasov, I.I.; Ashkinazi, E.E.; Ryzhkov, S.G.; Vlasov, A.V.; et al. High-rate growth of single crystal diamond in microwave plasma in CH_4/H_2 and $CH_4/H_2/Ar$ gas mixtures in presence of intensive soot formation. *Diam. Relat. Mater.* **2016**, *62*, 49–57. [CrossRef]

14. Godbole, V.P.; Sumant, A.V.; Kshirsagar, R.B.; Dharmadhikari, C.V. Evidence for layered growth of (100) textured diamond films. *Appl. Phys. Lett.* **1997**, *71*, 2626–2628. [CrossRef]

15. Silva, F.; Bonnin, X.; Achard, J.; Brinza, O.; Michau, A.; Secroun, A.; De Corte, K.; Felton, S.; Newton, M.; Gicquel, A. Single crystal CVD diamond growth strategy by the use of a 3D geometrical model: Growth on (113) oriented substrates. *J. Cryst. Growth* **2008**, *310*, 1067–1075. [CrossRef]

16. Müller-Sebert, W.; Wörner, E.; Fuchs, F.; Wild, C.; Koidl, P. Nitrogen induced increase of growth rate in chemical vapor deposition of diamond. *Appl. Phys. Lett.* **1996**, *68*, 759–760. [CrossRef]

17. Butler, J.E.; Oleynik, I. A mechanism for crystal twinning in the growth of diamond by chemical vapour deposition. *Philos. Trans. R. Soc. A.* **2008**, *366*, 295–311. [CrossRef] [PubMed]

18. May, P.W.; Ashfold, M.N.R.; Mankelevich, Y.A. Microcrystalline, nanocrystalline, and ultrananocrystalline diamond chemical vapor deposition: Experiment and modeling of the factors controlling growth rate, nucleation, and crystal size. *J. Appl. Phys.* **2007**, *101*, 053115. [CrossRef]

19. Richley, J.C.; Harvey, J.N.; Ashfold, M.N.R. CH_2 Group Migration between H-Terminated 2×1 Reconstructed {100} and {111} Surfaces of Diamond. *J. Phys. Chem. C* **2012**, *116*, 7810–7816. [CrossRef]

20. Barbosa, D.C.; Hammer, P.; Trava-Airoldi, V.J.; Corat, E.J. The valuable role of renucleation rate in ultrananocrystalline diamond growth. *Diam. Relat. Mater.* **2012**, *23*, 112–119. [CrossRef]

21. Van der Putte, P.; Van Enckevort, W.J.P.; Giling, L.J.; Bloem, J. Surface morphology of HCl etched silicon wafers: II. Bunch formation. *J. Cryst. Growth* **1978**, *43*, 659–675. [CrossRef]

22. Hayashi, K.; Yamanaka, S.; Okushi, H.; Kajimura, K. Homoepitaxial diamond films with large terraces. *Appl. Phys. Lett.* **1996**, *68*, 1220–1222. [CrossRef]

23. Bushuev, E.V.; Yurov, V.Y.; Bolshakov, A.P.; Ralchenko, V.G.; Khomich, A.A.; Antonova, I.A.; Ashkinazi, E.E.; Shershulin, V.A.; Pashinin, V.P.; Konov, V.I. Express in situ measurement of epitaxial CVD diamond film growth kinetics. *Diam. Relat. Mater.* **2017**, *72*, 61–70. [CrossRef]

24. Tardieu, A.; Cansell, F.; Petitet, J.P. Pressure and temperature dependence of the first-order Raman mode of diamond. *J. Appl. Phys.* **1990**, *68*, 3243–3245. [CrossRef]

25. Zaitsev, A.M. *Optical Properties of Diamond: A Data Handbook*; Springer: Berlin, Germany, 2001.

26. Sunagawa, I. *Crystals: Growth, Morphology, and Perfection*; Cambridge University Press: Cambridge, UK, 2005; pp. 44–45.

27. Palyanov, Y.N.; Kupriyanov, I.N.; Khokhryakov, A.F.; Ralchenko, V.G. Crystal Growth of Diamond. In *Handbook of Crystal Growth: Bulk Crystal Growth*, 2nd ed.; Rudolph, P., Ed.; Elsevier: Amsterdam, The Netherlands, 2015.

28. Samlenski, R.; Haug, C.; Brenn, R.; Wild, C.; Locher, R.; Koidl, P. Incorporation of nitrogen in chemical vapor deposition diamond. *Appl. Phys. Lett.* **1995**, *67*, 2798–2800. [CrossRef]

29. Khomich, A.A.; Kudryavtsev, O.S.; Bolshakov, A.P.; Khomich, A.V.; Ashkinazi, E.E.; Ralchenko, V.G.; Vlasov, I.I.; Konov, V.I. Use of Optical Spectroscopy Methods to Determine the Solubility Limit for Nitrogen in Diamond Single Crystals Synthesized by Chemical Vapor Deposition. *J. Appl. Spectrosc.* **2015**, *82*, 242–247. [CrossRef]

30. Yamada, H.; Chayahara, A.; Mokuno, Y. Effects of intentionally introduced nitrogen and substrate temperature on growth of diamond bulk single crystals. *Jpn. J. Appl. Phys.* **2016**, *55*. [CrossRef]

31. Tallaire, A.; Kasu, M.; Ueda, K.; Makimoto, T. Origin of growth defects in CVD diamond epitaxial films. *Diam. Relat. Mater.* **2008**, *17*, 60–65. [CrossRef]

32. Friel, I.; Clewes, S.L.; Dhillon, H.K.; Perkins, N.; Twitchen, D.J.; Scarsbrook, G.A. Control of surface and bulk crystalline quality in single crystal diamond grown by chemical vapor deposition. *Diam. Relat. Mater.* **2009**, *18*, 808–815. [CrossRef]
33. Mccormick, T.L.; Jackson, W.E.; Nemanich, R.J. Strain and Impurity Content of Synthetic Diamond Crystals. In *Materials Research Society Symposia Proceedings*; Materials Research Society: Pittsburgh, PA, USA, 1994; Volume 349, pp. 445–450.
34. Inyushkin, A.V.; Taldenkov, A.N.; Ralchenko, V.G.; Vlasov, I.I.; Konov, V.I.; Khomich, A.V.; Khmelnitskii, R.A.; Trushin, A.S. Thermal conductivity of polycrystalline CVD diamond: Effect of annealing-induced transformations of defects and grain boundaries. *Phys. Status Solidi A* **2008**, *205*, 2226–2232. [CrossRef]
35. Jones, R.; Goss, J.P.; Pinto, H.; Palmer, D.W. Diffusion of nitrogen in diamond and the formation of A-centres. *Diam. Relat. Mater.* **2015**, *53*, 35–39. [CrossRef]
36. Okushi, H.; Watanabe, H.; Ri, S.; Yamanaka, S.; Takeuchi, D. Device-grade homoepitaxial diamond film growth. *J. Cryst. Growth* **2002**, *237–239*, 1269–1276. [CrossRef]
37. Deak, P.; Aradi, B.; Kaviani, M.; Frauenheim, T.; Gali, A. Formation of NV centers in diamond: A theoretical study based on calculated transitions and migration of nitrogen and vacancy related defects. *Phys. Rev. B* **2014**, *89*, 075203. [CrossRef]
38. Fukura, S.; Nakagawa, T.; Kagi, H. High spatial resolution photoluminescence and Raman spectroscopic measurements of a natural polycrystalline diamond, carbonado. *Diam. Relat. Mater.* **2005**, *14*, 1950–1954. [CrossRef]
39. Bolshakov, A.; Ralchenko, V.; Sedov, V.; Khomich, A.; Vlasov, I.; Khomich, A.; Trofimov, N.; Krivobok, V.; Nikolaev, S.; Khmelnitskii, R.; et al. Photoluminescence of SiV centers in single crystal CVD diamond in situ doped with Si from silane. *Phys. Status Solidi A* **2015**, *212*, 2525–2532. [CrossRef]
40. Orlov, Y.L. *The Mineralogy of Diamond*; Wiley: New York, NY, USA, 1977; p. 235.
41. Khokhryakov, A.F.; Palyanov, Y.N. Effect of crystal defects on diamond morphology during dissolution in the mantle. *Am. Mineral.* **2015**, *100*, 1528–1532. [CrossRef]

Article

Effect of Nitrogen Impurities on the Raman Line Width in Diamond, Revisited

Nikolay V. Surovtsev [1,2] and Igor N. Kupriyanov [3,*]

[1] Institute of Automation and Electrometry, Siberian Branch of the Russian Academy of Sciences, Koptyug ave., 1, 630090 Novosibirsk, Russia; lab04@iae.nsk.su

[2] Department of Physics, Novosibirsk State University, 630090 Novosibirsk, Russia

[3] Sobolev Institute of Geology and Mineralogy, Siberian Branch of the Russian Academy of Sciences, Koptyug ave., 3, 630090 Novosibirsk, Russia

[*] Correspondence: spectra@igm.nsc.ru; Tel.: +7-383-330-7501

Academic Editor: Yuri N. Palyanov

Received: 30 June 2017; Accepted: 28 July 2017; Published: 31 July 2017

Abstract: The results of a high-resolution Raman scattering study of a diamond crystal with a high content of single substitutional nitrogen impurities (550 ppm) in the temperature range from 50 to 673 K are presented and compared with the data for defect-free diamond. It is established that the increase of the nitrogen concentration in diamond leads to the temperature-independent increase of the Raman line width. Analysis of the experimental data allows us to conclude that this broadening should be attributed to the defect-induced shortening of the Raman phonon lifetime. We believe that this mechanism is responsible for the increase of the Raman line width caused by most point-like defects in diamond. No pronounced effects of the nitrogen defects on the Raman line position and phonon anharmonicity are observed.

Keywords: diamond; Raman scattering; Raman linewidth; nitrogen defects

---

1. Introduction

Raman scattering spectroscopy has proven to be an effective and powerful technique to characterize diamond and diamond-related materials synthesized under different conditions [1–10]. The first-order Raman scattering spectrum of diamond is one of the simplest of its kind and consists of a single narrow line at around 1332 cm^{-1}. Various structural imperfections present in the diamond crystal affect its Raman spectrum. The occurrence of internal strains leads to Raman line shift; fluctuations of interatomic interaction and shortening of the phonon lifetime cause Raman line broadening; the presence of defects and impurities gives rise to new lines in the spectrum. Transformation to other phases (e.g., graphite-like) also would lead to the change of the Raman spectrum. Thus, the study of Raman spectra with an emphasis on the line position and width can help in the characterization of diamonds and diamond-related materials. This message has been exploited in a prodigious number of works [11–20].

An important prerequisite for the correct interpretation of the changes in the Raman line position and width is a clear understanding of the mechanisms responsible for these changes. Despite the long prehistory of experimental and model descriptions of the Raman line in diamond, it is only recently that the ultimate experimental information has been obtained for the temperature dependence of the Raman line width in defect-free diamond [21]. It is found that, for nearly perfect diamond crystals containing negligible amounts of lattice defects and impurities, the Raman line width is described exclusively by three-phonon and four-phonon anharmonic processes. Since, in most cases, Raman spectroscopy characterization of diamond materials is performed at room temperature, the question arises of how the defect-induced contribution to the Raman line width should be described.

This question pursues whether it is due to the changes of anharmonicity, or local fluctuations of the optical phonon frequency, or the relaxation of the Raman wavevector selection rule caused by phonon confinement, or the change of phonon wavefunction as solution of the dynamic problem, or something else.

It has previously been determined that nitrogen centers in diamond lead to the increase of the Raman line width proportional to the nitrogen concentration [22,23]. However, both these works were undertaken at room temperature and it is not clear which mechanism of the Raman line broadening is responsible for this outcome. In the present work, we demonstrate that the study of the temperature dependence of the Raman line recorded with a high spectral resolution can resolve this problem.

2. Results

Due to the anharmonic effects, the Raman line width of diamond decreases with decreasing temperature. Therefore, the defect-induced broadening is better manifested in the spectra measured at low temperatures. Figure 1 shows the Raman line contours of the diamond sample with high nitrogen concentration (550 ppm) and of the defect-free diamond [21], recorded at 50 K, which was the lowest temperature in our experiments. As it can be seen, the presence of substitutional nitrogen defects leads to a significant broadening of the Raman line. At the same time, the position of the Raman line is barely, if at all, changed. The broadening induced by nitrogen defects is slightly asymmetric, but the asymmetry is low. An asymmetry factor defined as the difference between the spectral position of the maximum and the mean position at the half-maximum level is 0.07 cm^{-1} in the case of the nitrogen-containing diamond, while the width difference between the spectra of the defect-free diamond and of the diamond with nitrogen at the half-maximum level is 0.83 cm^{-1}. Thus, with this low asymmetry, we can consider the main component of the defect-induced broadening as symmetric and make use of symmetric contours for the description of the experimental Raman lines.

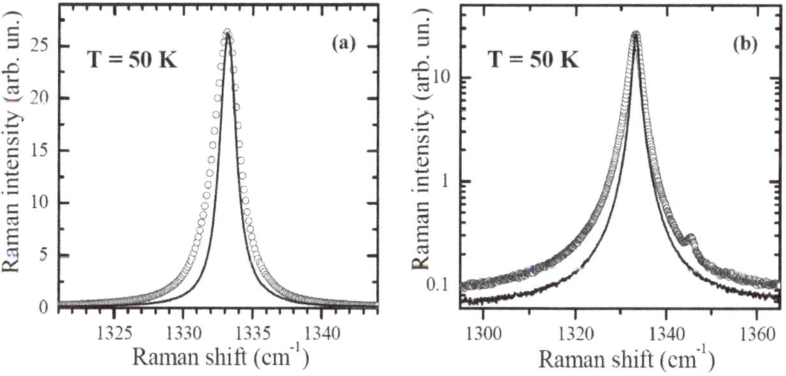

Figure 1. Raman line of the nitrogen-containing diamond (circles) and of the defect-free diamond (line) at T = 50 K: (**a**) linear scale for the intensity; (**b**) logarithmic scale for the intensity.

Another interesting feature of the Raman spectrum of the nitrogen-containing diamond is the occurrence of an additional weak peak located at 1345.5 cm^{-1} at T = 50 K. This peak is attributed to the well-known localized vibrational mode originating at substitutional nitrogen defects. This is a characteristic feature of the IR absorption spectrum of single substitutional nitrogen defects in diamond appearing at 1344 cm^{-1} at room temperature. To the best of our knowledge, this is the first time that this nitrogen-related localized mode is detected by Raman scattering. This finding opens certain prospects for the use of Raman scattering for probing the nitrogen content of micro- and nano-diamonds, for which conventional IR absorption spectroscopy cannot easily be applied. For the spectrum shown in Figure 1, it is found that the relative intensity of the 1345.5 cm^{-1} local mode peak to the diamond Raman line is about $(2.5–3.0) \times 10^{-3}$.

As we have demonstrated previously, the Raman line of diamond crystals with a negligible amount of defects can be well described by the Voigt contour, where the Gaussian component reflects the spectral resolution of the experimental setup (0.3 cm^{-1} in our experiments) and the Lorentzian component reflects the broadening caused by anharmonicity [21]. The results of the Voigt contour fitting applied to the Raman spectrum of the nitrogen-containing diamond are shown in Figure 2. It is found that fits with both the fixed Gaussian component (width set to 0.3 cm^{-1}) and with the free parameter describe the experimental line contour with good quality. However, if we suppose that the defect-induced broadening has a Gaussian-like contribution and fix the Lorentzian width, setting it to a value found for the defect-free diamond, then the quality of the fit notably decreases. Therefore, we may conclude that the substitutional nitrogen defects causes the Lorentzian-like broadening of the Raman line of diamond.

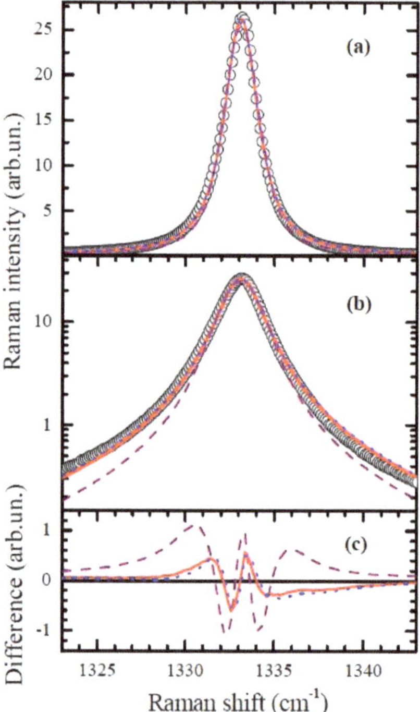

Figure 2. Raman line of the diamond with nitrogen impurities at T = 50 K (circles) and fitting curves, represented in (**a**) linear and (**b**) logarithmic scale for the intensity. The solid line is the Voight contour fit without restrictions for the fitting parameters; the dotted line is the Voight contour fit with the Gaussian width fixed to the instrumental resolution; the dashed line is the Voight contour fit with the Lorentzian width fixed to that of the Raman line of the defect-free diamond. (**c**) Differences between the experimental data and the fitting curves, employing the same designations as in (**a**,**b**).

Figure 3 shows the temperature dependence of the Raman line position and width extracted from the Voigt fit of the Raman spectra of the nitrogen-containing diamond and defect-free diamond. It can be seen that the spectral positions of the Raman line are very close for these two diamonds, and that they follow the same temperature dependence. The model description of the Raman line shift with temperature, being still a matter of research [24,25], lies beyond the scope of the present work and is not addressed here. The Lorentzian line width of the diamonds with and without nitrogen is different, but the difference is temperature independent. Three-phonon and four-phonon anharmonic processes contribute to the Lorentzian line width of the defect-free diamond, which is described by [26]:

$$\Gamma_L(T) = A(2n(\omega_0/2, T) + 1) + B(3n^2(\omega_0/3, T) + 3n(\omega_0/3, T) + 1), \tag{1}$$

where $n(\omega,T)$ is the Bose-Einstein distribution and A and B are the fitting parameters, which for the defect-free diamond are equal to 1.15 cm^{-1} and 0.2 cm^{-1}, respectively [21]. The description by Equation (1) is shown in Figure 3 for the data of the defect-free diamond. We find that the same function curve, except shifted upward by 0.87 cm^{-1}, also describes the experimental Lorentzian line width of the nitrogen-containing diamond very well. Thus, it follows that substitutional nitrogen defects in concentrations at the level of 550 ppm do not produce noticeable changes of the lattice anharmonicity, and the defect-induced broadening of the Raman line is temperature-independent.

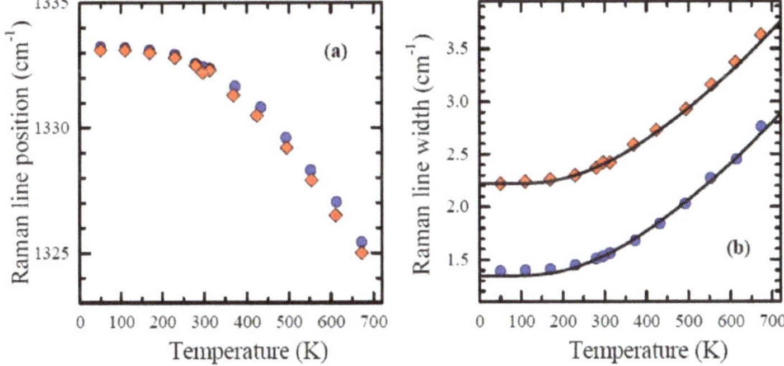

Figure 3. Temperature dependences of the Raman line parameters of the nitrogen-containing diamond (diamonds) and defect-free diamond (circles): (**a**) line position; (**b**) Lorentzian line width. The lines depict fits by Equation (1) (see text for details).

3. Discussion

The results of this study reveal that the single substitutional nitrogen defects does not change the anharmonicity of the optical phonon in diamond. In principle, one could suppose that the quasi-localized vibrational modes induced in diamond by nitrogen defects, whose local density of states extends below 500 cm^{-1} [27], may participate in the anharmonic decay of the Raman phonon. Furthermore, due to symmetry breaking, the decay of the Raman phonon into defect-induced quasi-localized modes could occur with the violation of the wavevector conservation law. Should this hold true, the occurrence of the additional channels for the anharmonic decay of the Raman phonon would modify the temperature dependence of the Raman line width. This, however, as demonstrated by our results, is not the case. With the high precision, the same anharmonic decay process, with the same parameters, describe the temperature-dependent part the Raman line width of both defect-free and nitrogen-containing diamonds. As a consequence, the defect-induced contribution to the Raman line width of a diamond crystal determined from room temperature measurements can be considered representative for the temperature-independent Raman line broadening caused by the defects, at least if the considered defect concentration is of the order of 500 ppm.

The results presented in Figures 1 and 2 demonstrate that the nitrogen concentration leads to an additional Lorentz-like broadening of the Raman line. In the case when the additional broadening would be caused by a distribution of the phonon frequency a Gaussian-like broadening is expected. Thus, the distributions of nitrogen and of defect-induce lattice distortions are uniform at the scale of the coherence volume of the optical phonon taking part in Raman scattering.

Most of the previous models dealing with the defect-induced changes of the Raman line width in diamond or similar crystals consider the relaxation of the wavevector selection rules caused by defects [28,29] as the main mechanism of the effect of the defects on the Raman spectrum. It can be described by relatively simple estimations involving the uncertainty principle [30] or by solving the equation for the phonon Green's function [31]; however, their outcomes are qualitatively similar—a shift and an accompanied broadening of the Raman line are expected. In this case, the broadened contour of defective crystals usually does not exceed the tail of the Raman contour of a defect-free crystal at either the high or low frequency side (see, for examples, figures in References [28,31]). From our results, it follows that this does not hold for the nitrogen-containing diamond, where defect-induced broadening for both sides of the Raman line is observed and the defect-induced shift of the Raman line is negligible. Furthermore, it should be noted that, as it is demonstrated by Richter et al. [30], the relaxation of the wavevector conservation law in the case of phonon confinement modifies the temperature dependence of the Raman line width. This, again, as demonstrated by our results, is not the case for nitrogen defects in diamond. While the phonon confinement model cannot describe the observed Raman line broadening correctly, we admit that the observed slight asymmetry of the Raman contour (Figure 1) could be attributed to a contribution of the phonon states in the vicinity of the Brillouin-zone center (Γ point) due to the partial breakdown of the wavevector selection rules.

A mechanism of the defect-induced shortening of the optical phonon lifetime is apparently more appropriate for the case of substitutional nitrogen defects in diamond. However, it is necessary to point out that the zone-center optical phonon has nearly zero group velocity and should be considered as not a propagating, but as an effectively standing wave. This means that the idea of the phonon mean free path evoked to describe the defect-induced scattering of acoustic phonons cannot be easily applied in the present case. To overcome this controversy, we note that the true harmonic solutions of the vibrational problem for a defective crystal are not plane waves, so the optical phonon created by the Raman process is not a vibrational eigenmode of the defective crystal. The standing wave of the optical phonon created by the Raman process undergoes elastic scattering by defects, which act as Rayleigh scatters. This leads to the attenuation of the Raman phonon, shortening its lifetime. The attenuation should be proportional to the defect concentration and result in an additional Lorentzian-like broadening, which is independent of temperature. No sharp dependence of the Raman line position on impurity concentration is expected in this case. The results of this work and the previous observations [22,23] agree well with the proposed mechanism of the defect-induced effects on the Raman line in diamonds.

In an attempt to provide a quantitative description accrued from the above model, we note that the defect-induced increase of the dissipation factor of the Raman phonon is proportional to the fractional concentration of defects, c_d, and for the defect-induced broadening, γ, one can write:

$$\gamma = F\omega_0 c_d, \tag{2}$$

where ω_0 is the phonon frequency (for diamond $\omega_0 = 1332$ cm^{-1}) and F is a dimensionless constant. For the rough estimation, the magnitude of F is an order of unity. With the nitrogen concentration of 550 ppm and F = 1, one obtains the estimation of $\gamma \approx 0.7$ cm^{-1}, which is remarkably close to the experimentally established value of 0.87 cm^{-1}. Such closeness is rather fortuitous, since F = 1 has the sense of estimation by an order of magnitude. Nevertheless, we see that the estimation by Equation (2) provides the correct order of magnitude, and it is convenient to characterize different impurities by their value of F.

In accord with previous observations [22,23], our results show that single substitutional nitrogen defects in diamond cause significant Raman line broadening, but produce very subtle effects on the Raman line position. Even for a diamond crystal with a relatively high nitrogen content (550 ppm), the shift of the Raman frequency cannot be determined with confidence as being within the accuracy of the line-center measurements (about 0.3 cm^{-1} absolute value in our experiment). On the other hand, it is well known that the presence of nitrogen impurities in diamond results in an increase in the lattice parameter [32]. This lattice dilatation should in turn manifest in the change of the Raman line position, representing the lower limit of the Raman frequency shift caused by nitrogen defects. It is therefore of interest to estimate the magnitude of this effect. Lang et al. [32] found that the dilation of the diamond lattice by single substitutional nitrogen can be described by $\Delta a/a = 0.14 \times c(N)$, where $c(N)$ is the fractional atomic concentration of substitutional nitrogen. For the diamond sample used in the present work, $c(N)$ is equal to 5.5×10^{-4} and consequently the produced lattice dilatation $\Delta a/a$ is equal to 7.7×10^{-5}. For small lattice parameter changes, we can assume the equivalence of tensile and compressive stresses and use the bulk modulus of diamond $B = 442$ GPa^{-1} and the hydrostatic pressure coefficient of the Raman line of 3.2 cm^{-1}/GPa [33]. Our calculations show that, for a diamond crystal containing 550 ppm of substitutional nitrogen, a Raman frequency shift of 0.33 cm^{-1} can be expected. This value is in reasonable agreement with our observations, so that we may speculate that the main effect of nitrogen impurities on the diamond Raman frequency is through the change of the diamond lattice parameters. Obviously, further investigations employing diamonds with even higher nitrogen content are necessary to verify this hypothesis.

4. Materials and Methods

Synthetic diamond crystal were grown by the temperature gradient growth method using a high-pressure multi-anvil apparatus of the split-sphere type [34]. A $Co_{0.7}Fe_{0.3}$ alloy was used as the solvent-catalyst. To produce diamond crystals with high nitrogen concentrations, nitrogen-containing compounds, $CaCN_2$ and Fe_3N, were added to the charge. The growth experiments were typically run at 5.5 GPa and 1400 °C for 65 h. For the purpose of the present study, a high-quality octahedral diamond crystal weighting about 2 ct and showing no metallic inclusions was selected. Using laser cutting and mechanical chipping, a plate was cleaved from the outer part of the crystal and then polished from two sides. The produced diamond plate, consisting of a single (111) growth sector, had a thickness of approximately 1 mm and linear size of 4–5 mm. The concentration of nitrogen impurities was determined using infrared (IR) absorption spectroscopy. The spectra were acquired from different locations over the sample using a Bruker Vertex 70 FTIR spectrometer fitted with a Hyperion 2000 microscope (Bruker Optics, Ettlingen, Germany). A representative IR spectrum of the diamond sample is shown in Figure 4. The concentration of nitrogen impurities in the form of single substitutional nitrogen atoms (C-centers) and nitrogen pairs (A-centers) was determined by decomposing the defect-induced absorption into the C and A components and using known conversion factors [35]. It was found that the concentration of the C form nitrogen in the sample varied within 540–560 ppm, and that of the A form nitrogen varied within 20–30 ppm. Taking into account the results of our previous study [23], we can neglect the effect of the small concentration of the A-centers, and take the average value of 550 ppm as the representative concentration of the single substitutional nitrogen defects in the studied diamond sample.

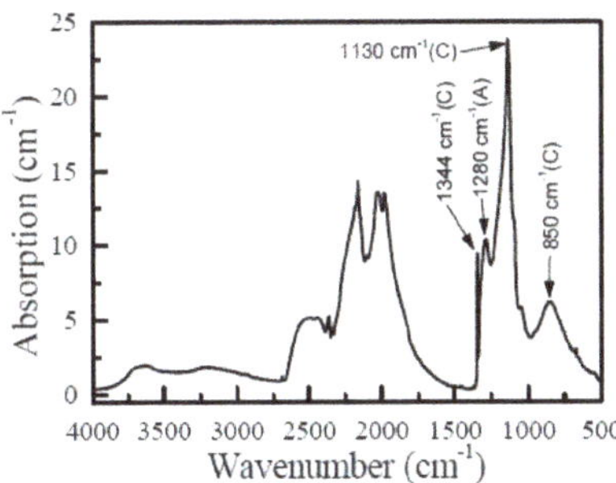

Figure 4. A representative infrared absorption spectrum of the diamond sample used in the study. The main absorption features related to the C- and A-centers are indicated.

The techniques and conditions employed for the Raman scattering experiment, as well as the spectral data processing, were the same as those used in our previous work [21]. Shortly, the Raman scattering measurements were carried out with excitation by the 514.5-nm line of an Ar-ion laser with a power of 100 mW. The spectra were recorded in a back-scattering geometry without polarization selection. An Acton TriVista 777 (Princeton Instruments, Acton, Santa Clara, CA, USA) triple spectrometer operated in the additive mode was used with 1800 groove/mm gratings. This configuration provides the pixel resolution of 0.1 cm^{-1}. The entrance slit was set to 30 microns. The instrumental spectral resolution was determined by experimental spectra of emission lines of a neon-discharge lamp, and found to be described by a Gaussian contour with a width of 0.3 cm^{-1}. The diamond sample was mounted in an optical closed-cycle cryostat or to a furnace for measurements at different temperatures. A temperature range from 50 to 700 K was covered in the Raman experiment.

5. Conclusions

From the results of the present study, the following main conclusions can be drawn: (1) The increase of the nitrogen concentration in diamond leads to additional Lorentz-like broadening of the Raman line; (2) This line width increase is independent of temperature and should be attributed to the defect-induced shortening of the Raman phonon lifetime. No pronounced effects of the nitrogen concentration on the Raman line position or anharmonicity are found. The present results establish that the increase of the Raman line width in diamonds containing point defects observed at room temperature relates to the decrease in the Raman phonon lifetime due to elastic scattering by defects.

Acknowledgments: We thank E.V. Podivilov for useful discussion and for attraction our attention to Equation (2). This work was supported by the Russian Science Foundation under Grant No. 14-27-00054. A part of the experiments was performed at the multiple-access center "High-resolution spectroscopy of gases and condensed matters" at IA&E SBRAS (Novosibirsk, Russia).

Author Contributions: The authors contributed equally to this work.

Conflicts of Interest: The authors declare no conflict of interest. The founding sponsors had no role in the design of the study; in the collection, analyses, or interpretation of data; in the writing of the manuscript, and in the decision to publish the results.

References

1. Prawer, S.; Nemanich, R.J. Raman spectroscopy of diamond and doped diamond. *Philos. Trans. R. Soc. Lond. A* **2004**, *362*, 2537–2565. [CrossRef] [PubMed]
2. Ferrari, A.C.; Robertson, J. Raman spectroscopy of amorphous, nanostructured, diamond-like carbon, and nanodiamond. *Philos. Trans. R. Soc. Lond. A* **2004**, *362*, 2477–2512. [CrossRef] [PubMed]
3. Nakashima, S.; Harima, H. Raman investigation of SiC polytypes. *Phys. Status Solidi A* **1997**, *162*, 39–64. [CrossRef]
4. Bergman, L.; Nemanich, R.J. Raman and photoluminescence analysis of stress state and impurity distribution in diamond thin films. *J. Appl. Phys.* **1995**, *78*, 6709–6719. [CrossRef]
5. Blank, V.D.; Denisov, V.N.; Kirichenko, A.N.; Kuznetsov, M.S.; Mavrin, B.N.; Nosukhin, S.A.; Terentiev, S.A. Raman scattering by defect-induced excitations in boron-doped diamond single crystals. *Diam. Relat. Mater.* **2008**, *17*, 1840–1843. [CrossRef]
6. Gouadec, G.; Colomban, P. Raman spectroscopy of nanomaterials: How spectra relate to disorder, particle size and mechanical properties. *Prog. Cryst. Growth Charact. Mater.* **2007**, *53*, 1–56. [CrossRef]
7. Yap, C.M.; Tarun, A.; Xiao, S.; Misra, D.S. MPCVD growth of ^{13}C-enriched diamond single crystals with nitrogen addition. *Diam. Relat. Mater.* **2016**, *63*, 2–11. [CrossRef]
8. Wu, G.; Chen, M.H.; Liao, J. The influence of recess depth and crystallographic orientation of seed sides on homoepitaxial growth of CVD single crystal diamonds. *Diam. Relat. Mater.* **2016**, *65*, 144–151. [CrossRef]
9. Prieske, M.; Vollertsen, F. In situ incorporation of silicon into a CVD diamond layer deposited under atmospheric conditions. *Diam. Relat. Mater.* **2016**, *65*, 47–52. [CrossRef]

10. Van Beveren, L.H.W.; Liu, R.; Bowers, H.; Ganesan, K.; Johnson, B.C.; McCallum, J.C.; Prawer, S. Optical and electronic properties of sub-surface conducting layers in diamond created by MeV B-implantation at elevated temperatures. *J. Appl. Phys.* **2016**, *119*, 223902. [CrossRef]

11. Crisci, A.; Baillet, F.; Mermoux, M.; Bogdan, G.; Nesladek, M.; Haenen, K. Residual strain around grown-in defects in CVD diamond single crystals: A 2D and 3D Raman imaging study. *Phys. Status Solidi A* **2011**, *208*, 2038–2044. [CrossRef]

12. Nasdala, L.; Hofmeister, W.; Harris, J.W.; Glinnemann, J. Growth zoning and strain patterns inside diamond crystals as revealed by Raman maps. *Am. Mineral.* **2005**, *90*, 745–748. [CrossRef]

13. Ohmagari, S.; Yamada, H.; Umezawa, H.; Chayahara, A.; Teraji, T.; Shikata, S. Characterization of free-standing single-crystal diamond prepared by hot-filament chemical vapor deposition. *Diam. Relat. Mater.* **2014**, *48*, 19–23. [CrossRef]

14. Shimizu, R.; Ogasawara, Y. Radiation damage to Kokchetav UHPM diamonds in zircon: Variations in Raman, photoluminescence, and cathodoluminescence spectra. *Lithos* **2014**, *206*, 201–213. [CrossRef]

15. Deslandes, A.; Guenette, M.C.; Belay, K.; Elliman, R.G.; Karatchevtseva, I.; Thomsen, L.; Riley, D.P.; Lumpkin, G.R. Diamond structure recovery during ion irradiation at elevated temperatures. *Nucl. Instrum. Methods Phys. Res. B* **2015**, *365*, 331–335. [CrossRef]

16. Li, Y.; Zhou, Z.X.; Guan, X.M.; Li, S.S.; Wang, Y.; Jia, X.P.; Ma, H.A. B–C bond in diamond single crystal synthesized with h-BN additive at high pressure and high temperature. *Chin. Phys. Lett.* **2016**, *33*, 028101.

17. Eaton-Magana, S.C.; Moe, K.S. Temperature effects on radiation stains in natural diamonds. *Diam. Relat. Mater.* **2016**, *64*, 130–142. [CrossRef]

18. Nasdala, L.; Steger, S.; Reissner, C. Raman study of diamond-based abrasives, and possible artefacts in detecting UHP microdiamond. *Lithos* **2016**, *265*, 317–327. [CrossRef]

19. Bensalah, H.; Stenger, I.; Sakr, G.; Barjon, J.; Bachelet, R.; Tallaire, A.; Achard, J.; Vaissiere, N.; Lee, K.H.; Saada, S.; et al. Mosaicity, dislocations and strain in heteroepitaxial diamond grown on iridium. *Diam. Relat. Mater.* **2016**, *66*, 188–195. [CrossRef]

20. Tang, Y.H.; Golding, B. Stress engineering of high-quality single crystal diamond by heteroepitaxial lateral overgrowth. *Appl. Phys. Lett.* **2016**, *108*, 052101. [CrossRef]

21. Surovtsev, N.V.; Kupriyanov, I.N. Temperature dependence of the Raman line width in diamond: Revisited. *J. Raman Spectrosc.* **2015**, *46*, 171–176. [CrossRef]

22. Hanzawa, H.; Umemura, N.; Nisida, Y.; Kanda, H.; Okada, M.; Kobayashi, M. Disorder effects of nitrogen impurities, irradiation-induced defects, and ^{13}C isotope composition on the Raman spectrum in synthetic Ib diamond. *Phys. Rev. B* **1996**, *54*, 3793–3799. [CrossRef]

23. Surovtsev, N.V.; Kupriyanov, I.N.; Malinovsky, V.K.; Gusev, V.A.; Pal'yanov, Y.N. Effect of nitrogen impurities on the Raman line width in diamonds. *J. Phys Condens. Matter* **1999**, *11*, 4767–4774. [CrossRef]

24. Lucazeau, G. Effect of pressure and temperature on Raman spectra of solids: Anharmonicity. *J. Raman Spectrosc.* **2003**, *34*, 478–496. [CrossRef]

25. Kolesov, B.A. How the vibrational frequency varies with temperature. *J. Raman Spectrosc.* **2017**, *48*, 323–326. [CrossRef]

26. Balkanski, M.; Wallis, R.F.; Haro, E. Anharmonic effects in light scattering due to optical phonons in silicon. *Phys. Rev. B* **1983**, *28*, 1928–1934. [CrossRef]

27. Briddon, P.R.; Jones, R. Theory of impurities in diamond. *Physica B* **1993**, *185*, 179–189. [CrossRef]

28. Ager, J.W., III; Veirs, D.K.; Rosenblatt, G.M. Spatially resolved Raman studies of diamond films grown by chemical vapor deposition. *Phys. Rev. B* **1991**, *43*, 6491. [CrossRef]

29. Kitajima, M. Defects in crystals studied by Raman scattering. *Crit. Rev. Solid State Mater. Sci.* **1997**, *22*, 275–349. [CrossRef]

30. Richter, H.; Wang, Z.P.; Ley, L. The one phonon Raman spectrum in microcrystalline silicon. *Solid State Commun.* **1981**, *39*, 625–629. [CrossRef]

31. Falkovsky, L.A. Width of optical phonons: Influence of defects of various geometry. *Phys. Rev. B* **2001**, *64*, 024301. [CrossRef]

32. Lang, A.R.; Moore, M.; Makepeace, A.P.W.; Wierzchowski, W.; Welbourn, C.M. On the dilatation of synthetic type Ib diamond by substitutional nitrogen impurity. *Philos. Trans. R. Soc. Lond. A* **1991**, *337*, 497–520. [CrossRef]

33. Grimsditch, M.H.; Anastassakis, E.; Cardona, M. Effect of uniaxial stress on the one-center optical phonon of diamond. *Phys. Rev. B* **1978**, *18*, 901–904. [CrossRef]
34. Palyanov, Y.N.; Borzdov, Y.M.; Khokhryakov, A.F.; Kupriyanov, I.N.; Sokol, A.G. Effect of nitrogen impurity on diamond crystal growth processes. *Cryst. Growth Des.* **2010**, *10*, 3169–3175. [CrossRef]
35. Zaitsev, A.M. *Handbook of Industrial Diamonds and Diamond Films*; Prelas, M., Popovici, G., Bigelow, L., Eds.; Marcel Dekker Inc.: New York, NY, USA, 1997; pp. 227–376.

 *crystals*

Article

Evolution in Time of Radiation Defects Induced by Negative Pions and Muons in Crystals with a Diamond Structure

Yury M. Belousov

Moscow Institute of Physics and Technology, Institutsky lane 9, 141700 Dolgoprudny, Russia; theorphys@phystech.edu; Tel.: +7-916-676-2018

Academic Editor: Yuri N. Palyanov
Received: 27 March 2017; Accepted: 9 June 2017; Published: 14 June 2017

Abstract: Evolution in time of radiation defects induced by negatively-charged pions and muons in crystals with diamond structures is considered. Negative pions and muons are captured by the nucleus and ionize an appropriate host atom, forming a positively-charged radiation defect in a lattice. As a result of an evolution in time, this radiation defect transforms into the acceptor center. An analysis of the full evolution process is considered for the first time. Formation of this acceptor center can be divided into three stages. At the first stage, the radiation defect interacts with a radiation trace and captures electrons. The radiation defect is neutralized completely in Si and Ge for a short time $t \leq 10^{-11}$ s, but in diamond, the complete neutralization time is very large $t \geq 10^{-6}$ s. At the second stage, broken chemical bonds of the radiation defect are restored. In Si and Ge, this process takes place for the neutral radiation defect, but in diamond, it goes for a positively-charged state. The characteristic time of this stage is $t < 10^{-8}$ s for Si and Ge and $t < 10^{-11}$ s for diamond. After the chemical bonds' restoration, the positively-charged, but chemically-bound radiation defect in diamond is quickly neutralized because of the electron density redistribution. The neutralization process is characterized by the lattice relaxation time. At the third stage, a neutral chemically-bound radiation defect captures an additional electron to saturate all chemical bonds and forms an ionized acceptor center. The existence of a sufficiently big electric dipolar moment leads to the electron capture. Qualitative estimates for the time of this process were obtained for diamond, silicon and germanium crystals. It was sown that this time is the shortest for diamond ($\leq 10^{-8}$ s) and the longest for silicon ($\leq 10^{-7}$ s)

Keywords: radiation defect; acceptor; chemical bond; diamond lattice; pion; muon; electron capture

1. Introduction

Radiation defects in diamond and silicon are examined actively because these semiconductors are widely used as detectors and some other devices in high energy physics. The main problem of these investigations is connected with their radiation hardness (see e.g., some recent works [1–6]). Radiation defects induced by protons, neutrons and heavier particles with kinetic energies $E \geq 100$ MeV are studied in most works. After slowing down to kinetic energies E less than ionization energies of host atoms in crystals, these impinging particles stop in the lattice, damaging it. If the recoil energy is higher than the lattice binding energy, a host atom will be displaced from its site. Numerical modeling of these processes is carried out, e.g., in [3,5]. In [7], many types of these radiation defects in diamond are well described. The second problem is implantation of ions in crystals for preparing necessary impurity atoms. Ion implantation is a commonly-used method for modifying properties of materials in the field of microelectronics. The application for diamond is represented, e.g., in [8].

For a number of reasons, radiation defects induced by light negatively-charged particles used to be out of interest in high energy physics experiments. First, these particles are secondary particles, as a rule, and, second, they cannot inflict many lattice damages compared to protons. Nevertheless, the interaction of these particles with crystals can be very important for many different applications of electronic devices.

We will consider in this article radiation defects induced by light negatively-charged particles like pions (π-mesons) and muons in crystals with a diamond structure. These particles do not destroy the crystal structure, like heavy particles, but can create specific defects in the lattice. Indeed, they are captured by a nucleus creating an impurity atom and, thus, can change the electronic properties of a crystal. Negative pions and muons are respectively long-lived particles: the lifetimes of charged pion and muon are $\tau_{\pi^{\pm}} \approx 2.6 \times 10^{-8}$ s and $\tau_{\mu} \approx 2.2 \times 10^{-6}$ s, respectively. Pions are born as usual when high energy protons are stopped in a target, and muons are born after the decay of pions:

$$\pi^{\pm} \to \mu^{\pm} + \nu_{\mu},$$

where ν_{μ} is a muon antineutrino for the negative muon and a neutrino for the positive one. This picture can be observed in cosmic rays. Negatively-charged pions and muons are stopped in a media very effectively because of the capture by nuclei.

The capture mechanism differs for pions and muons, but the result manifests in the same way in electronic properties, because they form finally the same acceptor impurity. Consider a capture of negative pions by stable nuclei of the main semiconductors: C, Si and Ge. In diamond, we have only one stable isotope C^{12}, and a capture of a negative pion gives rise to the boron acceptor:

$$C^{12} + \pi^{-} \to B^{11} + n,$$

where n is a decay neutron. The boron nucleus spin is $I = 3/2$.

In silicon isotope Si^{28} (92% in nature, see e.g., [9]) can capture π^{-} and than transform into the aluminum acceptor:

$$Si^{28} + \pi^{-} \to Al^{27} + n.$$

Processes in germanium are more complicated, because it has only two stable isotopes with an atomic number equal to 70 and 72 with 21.2% and 22% in nature, respectively [9], which can capture a negative pion and decay to an appropriate gallium isotope. Therefore, we have:

$$Ge^{70} + \pi^{-} \to Ga^{69} + n \qquad \text{and} \qquad Ge^{72} + \pi^{-} \to Ga^{71} + n.$$

Both gallium isotopes possess a spin equal to 3/2. All isotopes B^{11}, Al^{27}, Ga^{69} and Ga^{71} are stable, and the capture of the negative pions irreversibly changes the concentration of the main acceptor impurities in semiconductors.

A capture process of negatively-charged muon in crystals strongly differs from the negative pion capture process. Consider this difference in more detail. Positively- and negatively-charged muons (μ^{+} and μ^{-}) are widely used for research of condensed matter in many different areas, for the simulation of the behavior of hydrogen-like light element impurities and chemical processes with atomic hydrogen (see e.g., [10]). The application of muons for materials' investigation has become possible due to a well-developed μSR-technique based on the possibilities of supervision for a muon magnetic moment in the sample. Negatively- and positively-charged muons ($\mu^{\mp}$) are unstable leptons with spin 1/2.

The negatively-charged muon (μ^{-}) decays according to the scheme:

$$\mu^{-} \to e^{-} + \nu_{\mu} + \tilde{\nu}_{e},$$

where ν_μ and $\tilde{\nu}_e$ are muonic neutrino and electronic antineutrino, respectively. The escape probability of a decay electron depends on the angle between the electron momentum direction and the average muon spin **s**, due to what appears to be the possibility to study local fields of a target. A muon has a relatively high decay time of $\tau_\mu \approx 2.2 \times 10^{-6}$ s. The large lifetime allows investigating with a high precision the processes with a characteristic time $t < 10^{-5}$ s, which provides the opportunity for a μSR-technique for the material property studies, well comparable with the possibilities of the widely-applied methods of NMR and ESR.

The behavior of μ^+ and μ^- in a medium is radically different. From the chemical point of view, the positively-charged muon is a light element impurity modeling a light hydrogen isotope. The negatively-charged muon cascades into the ground $1s$-state forming a muonic atom (μ-atom). The mass of a muon equals 207-times the mass of an electron, and therefore, its binding energy with an atomic nucleus is 207-times larger than that of the electron. After a muon capture, much energy is released, leading to a high ionization of a target atom due to the emission of Auger electrons. Further, the target Auger electrons are captured by the positively-charged radiation-induced defect. Due to a high muon mass value, the negative muon screens a nuclear charge Z, which is effectively becoming $Z - 1$. After defect neutralization, a replacement impurity is formed, or a muonic atom, similar to an atom isotope with a nuclear charge $Z - 1$.

This fact was well known since the initial stage of the muon research (see e.g., [11]) and gave rise to the foundation of the muon method of materials research (μSR). Systematic study of impurities' formation with a nuclear charge equal to $Z - 1$ in condensed matter was carried out at the early stages of the μSR-research [12–14]. The muonic atom formed inside a semiconductor lattice models an acceptor center. For example, in diamond ($Z = 6$), the negative muon, as a result of capture by a nucleus, forms a pseudo-boron, or muonic boron, which can be designated as $_\mu$B. In Si ($Z = 14$) and Ge ($Z = 32$), the negative muon is captured by a nucleus forming the pseudo-aluminum $_\mu$Al with a nuclear charge equal to $Z = 13$ and pseudo-gallium $_\mu$Ga with a nuclear charge equal to $Z = 31$, respectively. These chemical elements are the main acceptor impurities in silicon and germanium semiconductors. Therefore, a radiation defect induced by a negative muon is unstable and disappears after its decay. However, nevertheless, this kind of defect is very interesting because it provides the possibility to study the evolution in time of the processes considered above.

The study of acceptor center properties using μ^- was suggested in [15]. The possibility to extract valuable information about the hyperfine structure and interactions with a lattice of acceptor centers in different semiconductors with the help of negative muons was shown in the works [16–19]. Recently, $\mu^- SR$ research of synthetic diamond crystals were carried out in [20–23]. We will examine evolution in time of radiation defects induced by negative muons in the following, taking in mind that they are the same for negative pions, as well.

The total process of this kind of radiation defect formation can be separated into two principally different stages. At the first stage, a center with a large positive charge appears after the stopping of a negative muon or pion. This center interacts with electrons of a trace, created by the charged particle, when it is decelerated in a crystal. As a result of this interaction, the positively-charged center partially compensates its charge or becomes neutralized if it is possible. At the second stage, the center with a compensated charge restores broken chemical bonds with a lattice and then forms an acceptor center.

Now, we will outline briefly the main results of the first stage and show the difference between diamond and other diamond structure crystals following [24]. After that, we will consider the second stage in more detail.

2. Interaction with the Trace

When a negative pion is captured by a nucleus or a negative muon is captured to the K-shell of the muonic atom, substantial energy ($E \geq 1$ keV) is released. A totally ionized positively-charged center and Auger electron ionization environment appear, creating secondary electrons. This process takes short a time, $t \sim 10^{-14}$ s. After the ionization, free electrons lose their energy; for a while, it will be of the order of the forbidden gap energy. This is a diffusion process, when ionized impurities and host atoms are neutralized, and it takes respectively a long time $t \leq 10^{-10}$ s. All of these findings were obtained as a result of a numerical modeling of a neutralization process of muonic atoms in a kinetic approximation for diamond and silicon crystals with different concentrations of impurities [24].

A different situation is observed in diamond and silicon already at this stage. The capture of a negative muon on a silicon nucleus creates the number of free charge carriers approximately at two orders more with respect to diamond. This result is connected with a difference in the number of Auger electrons, in ionization energies of impurities and a forbidden energy band for these two crystals.

Numerical calculations have shown that the recombination frequency of electrons with positively-charged ions in diamond reaches approximately 10^7 s^{-1} only at a respectively short interval of time 10^{-10} s. In silicon, the recombination frequency reaches approximately 10^{11} s^{-1} for the same interval of time. As a result, all ionized impurities in silicon including a muonic atom $_\mu$Al are neutralized for a very short time $t \approx 10^{-11}$ s. The probability of the neutralization of a muonic atom $_\mu$B is less than 10^{-3} for the interval $t \approx 10^{-10}$ s. The recombination frequency in diamond sharply falls for $t > 10^{-10}$ s, and the neutralization time in this process becomes more than both a muon lifetime and a characteristic time for chemical bonds' restoration.

Thus numerical modeling has shown that a positively-charged radiation defect, created by a negative muon in silicon and germanium, must be quickly neutralized before chemical bonds with the lattice can be restored. In diamond, we observe other behavior. Namely, the radiation defect must restore chemical bonds with the lattice to be positively charged. Therefore, we need to consider different initial states of the radiation defect at the second stage for diamond and other crystals with the diamond structure.

The second stage of the radiation defect formation consists of a few steps that lead to an acceptor center formation. A neutral defect with restored chemical bonds is not an acceptor center yet, because there exist unsaturated chemical bonds. Therefore, we need to consider at least three steps of an acceptor center formation:

(1) restoration of broken chemical bonds and neutralization of the radiation defect;
(2) capture of a missing electron and saturation of chemical bonds of a neutral radiation defect in the lattice; formation of an ionized acceptor center;
(3) formation of an acceptor center in the ground state.

The first two steps are discussed in this article.

3. Electron States of a Neutral Radiation Defect in Si and Ge

The muonic impurity atom is in an exited state just after formation because its chemical bonds with host atoms are broken. In accordance with the standard idea of quantum chemistry, only electrons with the same principal quantum number can create a chemical bond if they were on an unfilled energy level of the atom. In this case, they form hybridized states. For lattices with diamond structure electron states, ns and np are represented with equal probability, where $n = 2$, 3 and 4 relate to C, Si and Ge, respectively. Hybridized states are formed in atomic time, but chemical bonds' formation is determined by exchange interactions that are weaker than Coulomb interactions, which form the appropriate atomic configuration.

When a chemical bond is formed, a significant energy (of the order of some eV) can be emitted. In gasses and liquids, this excess energy can be transferred to the third body. This kind of energy transfer in crystal must be connected with a phonon emission. One-phonon emission with the energy of ≥ 1 eV in covalent crystal is impossible. Therefore, a transfer of this energy value could be realized in the case of a multi-phonon process. This kind of process have a very small probability. Thus, a radiation transition with a photon emission seems to us more preferable with respect of the other processes.

In this section, we consider this process for a neutralized radiation defect in Si and Ge when three electrons are in hybridized states [25]. Consider an impurity atom with the nuclear charge

$Z - 1$, which is formed as a result of a neutralization process in an atomic time and has an atomic configuration where electrons at the external shell are in the "mixed", but not in the ground, state:

$$|\psi_{in}\rangle = |ns\,np^2\rangle. \tag{1}$$

We assume that the electron configuration with the principal quantum number less then n is completely occupied, and the state of such electrons is described by the unperturbed wave function of the free atom. The initial state of the radiation defect in a silicon lattice is sketched in Figure 1a.

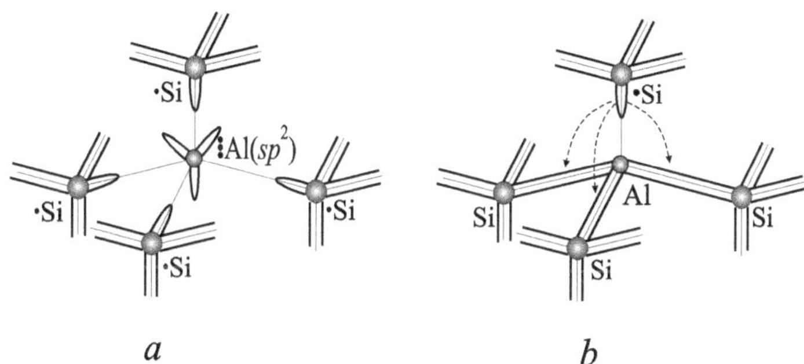

Figure 1. Radiation defect μAl state in a silicon lattice: (**a**) all bonds with host atoms are broken in the initial state, and the defect state is determined by the "mixed" function $3s3p^2$; (**b**) three electrons of the impurity form chemical bonds with host atoms in the final state, and an unsaturated (broken) bond is equiprobable for four nearest neighbors of the cluster $(AlSi_4)^0$.

The mixed state (1) forms a chemical bond, if it possesses by maximal spin $S = 3/2$, and a spacial part of its wave function one may represent in the view of equiprobable superposition of three Slater's determinants:

$$\Psi_{sp^2}(\mathbf{r}_1, \mathbf{r}_2, \mathbf{r}_3) = \frac{1}{\sqrt{3}}\left(\Psi_{+1}(\mathbf{r}_1, \mathbf{r}_2, \mathbf{r}_3) + \Psi_0(\mathbf{r}_1, \mathbf{r}_2, \mathbf{r}_3) + \Psi_{-1}(\mathbf{r}_1, \mathbf{r}_2, \mathbf{r}_3)\right), \tag{2}$$

where:

$$\Psi_M(\mathbf{r}_1, \mathbf{r}_2, \mathbf{r}_3) = \frac{1}{\sqrt{6}}\begin{pmatrix} \psi_{ns}(\mathbf{r}_1) & \psi_{np,m}(\mathbf{r}_1) & \psi_{np,m'}(\mathbf{r}_1) \\ \psi_{ns}(\mathbf{r}_2) & \psi_{np,m}(\mathbf{r}_2) & \psi_{np,m'}(\mathbf{r}_2) \\ \psi_{ns}(\mathbf{r}_3) & \psi_{np,m}(\mathbf{r}_3) & \psi_{np,m'}(\mathbf{r}_3) \end{pmatrix}. \tag{3}$$

Here, $M = m + m'$, and $m, m' = 0, \pm 1$.

A spacial part of the wave function of the defect in a final state can be represented as a superposition of three hybridized states forming the chemical bond with host atoms of the lattice:

$$\Psi_{Cr}(\mathbf{r}_1, \mathbf{r}_2, \mathbf{r}_3) = \frac{1}{\sqrt{6}}\sum_{\mathcal{P}}(-1)^{\mathcal{P}}\Psi_{\mu A}(\mathcal{P}(\mathbf{r}_1, \mathbf{r}_2, \mathbf{r}_3)). \tag{4}$$

Here, summation is carried out over all permutations $\mathcal{P}$ of the valence electrons of the impurity and:

$$\Psi_{\mu A}(\mathbf{r}_1, \mathbf{r}_2, \mathbf{r}_3) = \tfrac{1}{2}\Big(\psi_{n_1}(\mathbf{r}_1)\psi_{n_2}(\mathbf{r}_2)\psi_{n_3}(\mathbf{r}_3) + \psi_{n_2}(\mathbf{r}_1)\psi_{n_3}(\mathbf{r}_2)\psi_{n_4}(\mathbf{r}_3) \\ + \psi_{n_3}(\mathbf{r}_1)\psi_{n_4}(\mathbf{r}_2)\psi_{n_1}(\mathbf{r}_3) + \psi_{n_4}(\mathbf{r}_1)\psi_{n_1}(\mathbf{r}_2)\psi_{n_2}(\mathbf{r}_3)\Big). \tag{5}$$

The unit vectors $\mathbf{n}_a$ are directed from the impurity to nearest neighbors (along the directions of the chemical bonds). The one-particle functions $\psi_{\mathbf{n}_a}(\mathbf{r})$ are the hybridized states with directed bonds, and they can be written in a form (see e.g., [26,27]):

$$\psi_{\mathbf{n}_a}(\mathbf{r}) = \alpha_a \psi_{ns}(\mathbf{r}) + \beta_{\mathbf{n}_a} \psi_{np,\mathbf{n}_a}(\mathbf{r}) \tag{6}$$

and satisfy normalization conditions:

$$\int \psi_{\mathbf{n}_a}^*(\mathbf{r}) \psi_{\mathbf{n}_b}(\mathbf{r}) d\mathbf{r} = \delta_{ab}. \tag{7}$$

From this condition, we can obtain relations for the coefficients in the superposition (6):

$$\alpha_a^2 + \beta_{\mathbf{n}_a}^2 = 1, \quad \sum_a \alpha_a^2 = 1, \quad \sum_a \beta_{\mathbf{n}_a}^2 = 3, \quad \alpha_a \alpha_b + \beta_{\mathbf{n}_a} \beta_{\mathbf{n}_b} \cos(\widehat{\mathbf{n}_a, \mathbf{n}_b}) = 0. \tag{8}$$

4. Electron States of a Positively-Charged Radiation Defect in Diamond

A positively-charged radiation defect in diamond has the effective nuclear charge $Z = 5$. Its atomic configuration contains only two electrons in an external (unfilled) electron shell, and they are in the "mixed" state [28]:

$$|\psi_{\text{in}}\rangle = |2s\,2p\rangle. \tag{9}$$

We suppose also that the electronic configuration for the principle quantum number $n = 1$ is completely filled, and electronic states of the external atomic shell are described by unperturbed wave functions of a free atom. The initial state of the radiation defect in a diamond lattice is shown schematically in Figure 2a.

The mixed state (9) forms a chemical bond with the nearest host atoms of the lattice. We express a wave function of the initial state (9) in distinguish from the state (1) in the form of superposition with all possible spin states:

$$\Psi_{sp}(\mathbf{r}_1, \mathbf{r}_2) = A_0 \Psi_{sp}^{(0)}(\mathbf{r}_1, \mathbf{r}_2)|0,0\rangle + A_1 \Psi_{sp}^{(1)}(\mathbf{r}_1, \mathbf{r}_2) \sum_{M_S} |1, M_S\rangle, \tag{10}$$

where:

$$\Psi_{sp}^{(S)}(\mathbf{r}_1, \mathbf{r}_2) = \frac{1}{\sqrt{2}} \left(\psi_{2s}(\mathbf{r}_1)\psi_{2p}(\mathbf{r}_2) + (-1)^S \psi_{2p}(\mathbf{r}_1)\psi_{2s}(\mathbf{r}_2) \right), \tag{11}$$

$\psi_{2s}(\mathbf{r})$ is the wave function of the $2s$-state. We assume that all p-states with different projections have equal probabilities:

$$\psi_{2p}(\mathbf{r}) = \frac{1}{\sqrt{3}} \left(\psi_{21,+1}(\mathbf{r}) + \psi_{21,0}(\mathbf{r}) + \psi_{21,-1}(\mathbf{r}) \right).$$

$S = 0, 1$ are the values of the total electron spin; $|S, M_S\rangle$ is the appropriate spin-state vector. Spin states with different projections are considered as having equal probabilities; so coefficients in the superposition (10) satisfy the following condition:

$$|A_0|^2 + 3|A_1|^2 = 1.$$

The space part of the defect wave function in the final state must correspond to the determined value of the total electron spin S, and this can be represented in the form of the superposition of hybridized states providing a chemical bond with the lattice host atoms:

$$\Psi_{\text{Cr}}^{(S)}(\mathbf{r}_1, \mathbf{r}_2) = \frac{1}{\sqrt{6}} \sum_{\mathbf{n}_a, \mathbf{n}_b} \Psi_{\mathbf{n}_a \mathbf{n}_b}^{(S)}(\mathbf{r}_1, \mathbf{r}_2), \tag{12}$$

where summation is carried out over all possible directions of the chemical bond of impurity valence electrons with the nearest neighbors of the lattice,

$$\Psi^{(S)}_{\mathbf{n}_a \mathbf{n}_b}(\mathbf{r}_1, \mathbf{r}_2) = \frac{1}{\sqrt{2}} \left(\psi_{\mathbf{n}_a}(\mathbf{r}_1)\psi_{\mathbf{n}_b}(\mathbf{r}_2) + (-1)^S \psi_{\mathbf{n}_a}(\mathbf{r}_2)\psi_{\mathbf{n}_b}(\mathbf{r}_1) \right). \tag{13}$$

The unit vectors $\mathbf{n}_a$ are directed from the impurity to the neighbor atoms (along the direction of chemical bonds) like in (5). One-particle wave functions $\psi_{\mathbf{n}_a}(\mathbf{r})$ of hybridized states with directed bonds are determined by Equations (6)–(8).

The final state of the charged impurity is described by a function similar to the superposition (10) where the wave functions with a determined spin must be replaced by Expression (13).

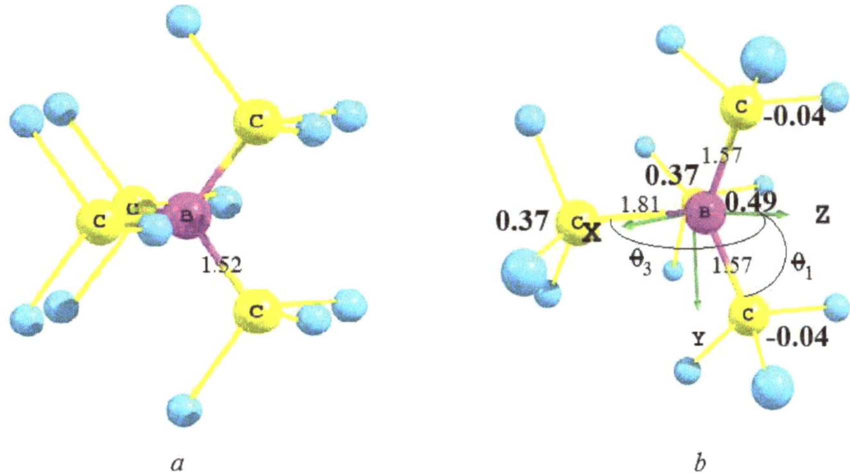

Figure 2. Structure of the radiation $(_\mu B)^+$ defect in a diamond lattice: (**a**) the $(_\mu BC_4)^+$ cluster has a T_d symmetry in the initial triplet state; (**b**) it has a lower C_{2v} symmetry with angles $\theta_1 = 65°$ and $\theta_3 = 128°$ in the final triplet state. The internuclear distances are given in Å. The spin densities on atoms are marked by bold type.

5. Formation of the Neutral Center $(_\mu A\, A_4)^0$ in Si and Ge

A lifetime of exited states (1) and (10) is determined by a rate of a radiation transition in a bond state and can be calculated by using Fermi's "golden rule":

$$dw = \frac{2\pi}{\hbar} \left| \langle \psi_{in} | \hat{V}_{\text{rad}} | \psi_{Cr} \rangle \right|^2 \delta(E_f - E_i) dv_f. \tag{14}$$

The interaction operator is:

$$\hat{V}_{\text{rad}} = \frac{e}{mc} \sum_a \hat{\mathbf{p}}_a \mathbf{A}(\mathbf{r}_a), \tag{15}$$

where $\mathbf{A}(\mathbf{r})$ is the vector-potential of the free radiation field.

Let us consider now only the term for one electron with $a = 1$ in the operator (14) to simplify the following calculations. In this case, matrix elements of the perturbation operator could be represented by the expression:

$$\langle \psi_{in} | \hat{\mathbf{p}}_1 | \psi_{Cr} \rangle_{sa} = \int \psi^*_{ns}(\mathbf{r}_1)\hat{\mathbf{p}}_1 \psi_{\mathbf{n}_a}(\mathbf{r}_1)d\mathbf{r}_1 \int \psi^*_{np,m}(\mathbf{r}_2)\psi_{\mathbf{n}_b}(\mathbf{r}_2)d\mathbf{r}_2 \int \psi^*_{np,m'}(\mathbf{r}_3)\psi_{\mathbf{n}_c}(\mathbf{r}_3)d\mathbf{r}_3 \tag{16}$$

and:

$$\langle\psi_{in}|\hat{\mathbf{p}}_1|\psi_{\mathrm{Cr}}\rangle_{pa} = \int \psi^*_{np,m}(\mathbf{r}_1)\hat{\mathbf{p}}_1\psi_{\mathbf{n}_a}(\mathbf{r}_1)d\mathbf{r}_1\int \psi^*_{np,m'}(\mathbf{r}_2)\psi_{\mathbf{n}_b}(\mathbf{r}_2)d\mathbf{r}_2\int \psi^*_{ns}(\mathbf{r}_3)\psi_{\mathbf{n}_c}(\mathbf{r}_3)d\mathbf{r}_3, \qquad (17)$$

where the indexes are a, b, $c = 1$, 2, 3, 4.

If we direct the axis $z||\mathbf{n}_1$, then the other three p-states in hybridized states (6) turn out as a result of the rotation of the state $\psi_{n1,0}(\mathbf{r})$ in the state with a rotation moment projection equal to zero on the axes $\mathbf{n}_b$. In this case, we get the opportunity to calculate easy integrals incoming in Expressions (16) and (17):

$$\int \psi^*_{ns}(\mathbf{r})\psi_{\mathbf{n}_a}(\mathbf{r})dr \;=\; \alpha_a\langle ns|ns\rangle = \alpha_a; \qquad (18)$$

$$\int \psi^*_{np,m}(\mathbf{r})\psi_{\mathbf{n}_a}(\mathbf{r})dr \;=\; \beta_a\langle np,m|\widehat{R}(\theta_a,\varphi_a)|n1,0\rangle, \qquad (19)$$

where $\widehat{R}(\theta_a,\varphi_a)$ is the rotation operator, and the matrix elements (19) are determined by the second column of the rotation matrix:

$$\langle n1,\pm 1|\widehat{R}(\theta_a,\varphi_a)|n1,0\rangle = \mp\frac{1}{\sqrt{2}}\sin\theta_a e^{-i\varphi_a}, \quad \langle n1,0|\widehat{R}(\theta_a,\varphi_a)|n1,0\rangle = \cos\theta_a. \qquad (20)$$

We put in Equation (18) the obvious expressions of the matrix elements of the rotation operator (see e.g., [29]).

Without reduction of the generality of the calculations, we consider a matrix element only for the z- projection of a momentum operator. Therefore, we have:

$$\int \psi^*_{ns}(\mathbf{r}_1)\hat{p}_{1z}\psi_{\mathbf{n}_a}(\mathbf{r}_1)d\mathbf{r}_1 \;=\; \beta_a\int \psi^*_{ns}(\mathbf{r}_1)\hat{p}_{1z}\widehat{R}(\theta_a,\varphi_a)\psi_{n1,0}(\mathbf{r}_1)d\mathbf{r}_1 = \beta_a\cos\theta_a I^{(n)}_{sp}, \qquad (21)$$

$$\int \psi^*_{np,m}(\mathbf{r}_1)\hat{p}_{1z}\psi_{\mathbf{n}_a}(\mathbf{r}_1)d\mathbf{r}_1 d\mathbf{r}_1 \;=\; \alpha_a\int \psi^*_{n1,0}(\mathbf{r}_1)\hat{p}_{1z}\psi_{ns}(\mathbf{r}_1)d\mathbf{r}_1 = \alpha_a I^{(n)*}_{sp}, \qquad (22)$$

where:

$$I^{(n)}_{sp} = \int \psi^*_{ns}(\mathbf{r})\hat{p}_z\psi_{n1,0}(\mathbf{r})d\mathbf{r}. \qquad (23)$$

For clarity, we show some intermediate calculations. The total matrix element in Expression (14) consists of 72 different items corresponding to different matrix elements between states of the superpositions (2) and (4). However, it is enough to calculate only four of them. We give them below.

$$\begin{aligned}\langle\Psi_{+1}|\hat{p}_{1z}|\Psi_{\mathbf{n}_1,\mathbf{n}_2,\mathbf{n}_3}\rangle = &-\tfrac{1}{4\sqrt{3}}\beta_1\beta_2\beta_3\left(\sin\theta_2\cos\theta_3 e^{-i\varphi_2}-\cos\theta_2\sin\theta_3 e^{-i\varphi_3}\right)I^{(n)}_{sp}\\ &+\tfrac{1}{4\sqrt{3}}\alpha_1\left(\beta_2\alpha_3\sin\theta_2 e^{-i\varphi_2}-\alpha_2\beta_3\sin\theta_3 e^{-i\varphi_3}\right)I^{(n)*}_{sp}.\end{aligned} \qquad (24)$$

The rest of the three matrix elements are determined by the other possible sets of $\mathbf{n}_a$ for the electrons with coordinates $\mathbf{r}_1$, $\mathbf{r}_2$ and $\mathbf{r}_3$:

$$\begin{aligned}\langle\Psi_{+1}|\hat{p}_{1z}|\Psi_{\mathbf{n}_2,\mathbf{n}_3,\mathbf{n}_4}\rangle = &-\tfrac{1}{4\sqrt{3}}\beta_2\beta_3\beta_4\cos\theta_2\left(\sin\theta_3\cos\theta_4 e^{-i\varphi_3}-\cos\theta_3\sin\theta_4 e^{-i\varphi_4}\right)I^{(n)}_{sp}\\ &-\tfrac{1}{4\sqrt{3}}\alpha_2\left(\alpha_3\beta_4\sin\theta_4 e^{-i\varphi_4}-\beta_3\alpha_4\sin\theta_3 e^{-i\varphi_3}\right)I^{(n)*}_{sp};\end{aligned} \qquad (25)$$

$$\langle\Psi_{+1}|\hat{p}_{1z}|\Psi_{\mathbf{n}_3,\mathbf{n}_4,\mathbf{n}_1}\rangle \;=\; -\frac{1}{4\sqrt{3}}\beta_4\left(\beta_1\beta_3\cos\theta_3 I^{(n)}_{sp}-\alpha_1\alpha_3 I^{(n)*}_{sp}\right)\sin\theta_4 e^{-i\varphi_4} \qquad (26)$$

$$\langle\Psi_{+1}|\hat{p}_{1z}|\Psi_{\mathbf{n}_4,\mathbf{n}_1,\mathbf{n}_2}\rangle \;=\; \frac{1}{4\sqrt{3}}\beta_2\left(\beta_1\beta_4\cos\theta_4 I^{(n)}_{sp}-\alpha_1\alpha_4 I^{(n)*}_{sp}\right)\sin\theta_2 e^{-i\varphi_2}. \qquad (27)$$

It is easy to see that permutations of electrons in the superposition (5) do not change expressions for the matrix elements (24)–(27). Therefore, the number of permutations in the state (5) with similar expressions reduces the total number of items by six-times.

We get a very cumbersome expression for the arbitrary values of the parameters α_a, β_a, θ_a and φ_a. However, it is necessary to take into account that the system under consideration has a symmetry at less C_{3v}. In this case, the result could be essentially simplified. We examine the simplest case at first, when the system has a tetrahedral symmetry and all parameters in the hybridized states (6) are equal to each other:

$$\alpha_a = \alpha = \frac{1}{2}, \quad \beta_a = \beta = \frac{\sqrt{3}}{2}. \tag{28}$$

If the vector $\mathbf{n}_4$ lies in the xz plane, the angles θ_a, φ_a are equal:

$$\theta_1 = 0, \quad \theta_2 = \theta_3 = \theta_4 = \theta, \quad \cos\theta = -\frac{1}{3}, \quad \varphi_2 = -\varphi_3 = \frac{2\pi}{3}, \quad \varphi_4 = 0. \tag{29}$$

In the case of symmetrical structure, we get the following expressions for the matrix elements:

$$\langle \Psi_{+1} | \hat{p}_{1z} | \Psi_{\mathbf{n}_1, \mathbf{n}_2, \mathbf{n}_3} \rangle = \frac{i}{8} \beta \left(\beta^2 \sin 2\theta I_{sp}^{(n)} - \alpha^2 \sin\theta I_{sp}^{(n)*} \right), \tag{30}$$

$$\langle \Psi_{+1} | \hat{p}_{1z} | \Psi_{\mathbf{n}_2, \mathbf{n}_3, \mathbf{n}_4} \rangle = \frac{\sqrt{3}}{8} \left(1 - \frac{1}{\sqrt{3}} \right) \beta \sin\theta \left(\beta^2 \cos^2\theta I_{sp}^{(n)} - \alpha^2 I_{sp}^{(n)*} \right), \tag{31}$$

$$\langle \Psi_{+1} | \hat{p}_{1z} | \Psi_{\mathbf{n}_3, \mathbf{n}_4, \mathbf{n}_1} \rangle = \langle \Psi_{+1} | \hat{p}_{1z} | \Psi_{\mathbf{n}_4, \mathbf{n}_1, \mathbf{n}_2} \rangle = -\frac{1}{4\sqrt{3}} \beta \sin\theta \left(\beta^2 \cos\theta I_{sp}^{(n)} - \alpha^2 I_{sp}^{(n)*} \right). \tag{32}$$

Adding up Expressions (30)–(32), we get:

$$\langle \Psi_{+1} | \hat{p}_{1z} | \Psi_{Cr} \rangle = -\frac{1 - i\sqrt{3}}{8\sqrt{3}} \beta^3 \sin 2\theta \sin^2\frac{\theta}{2} I_{sp}^{(n)}. \tag{33}$$

The matrix elements for the state with $M = -1$ are calculated by a similar way:

$$\langle \Psi_{-1} | \hat{p}_{1z} | \Psi_{Cr} \rangle = -\frac{1 + i\sqrt{3}}{8\sqrt{3}} \beta^3 \sin 2\theta \sin^2\frac{\theta}{2} I_{sp}^{(n)}. \tag{34}$$

For the state with $M = 0$:

$$\langle \Psi_0 | \hat{p}_{1z} | \Psi_{Cr} \rangle = -\frac{i}{2\sqrt{2}} \beta^3 \sin^2\theta \cos^2\frac{\theta}{2} I_{sp}^{(n)}. \tag{35}$$

After substitution of the values o Parameters (28) and (29) and adding up Expressions (33)–(35), we have:

$$\langle \Psi_{in} | \hat{p}_{1z} | \Psi_{Cr} \rangle = -\frac{\beta^3}{4\sqrt{3}} \left(\sin 2\theta \sin^2\frac{\theta}{2} + i\sqrt{2} \sin^2\theta \cos^2\frac{\theta}{2} \right) I_{sp}^{(n)} = \frac{1 - i4}{72\sqrt{2}} I_{sp}^{(n)}. \tag{36}$$

For the calculation of the integral I_{sp}, we take the appropriate wave functions of the hydrogen-like atom with an effective nuclear charge equal to $\tilde{Z}$. In accordance with Slater [30], an effective charge is determined as $\tilde{Z} = Z - \sigma$, where Z is the real nuclear charge and σ is a screening constant.

In a silicon crystal, an aluminum μ-atom $_\mu$Al is formed. It has the principle quantum number $n = 3$, and appropriate calculations for $_\mu$Al give the following results:

$$I_{sp}^{(3)} = -i\frac{10\tilde{Z}}{3\sqrt{6}}, \quad \text{in dimensional units} \quad I_{sp}^{(3)} = -i\frac{10}{3\sqrt{6}}\frac{\tilde{Z}\hbar}{a_0}, \tag{37}$$

where a_0 is the Bohr radius.

In a germanium lattice, a gallium μ-atom $_\mu$Ga with the principle quantum number $n = 4$ must be formed. The unknown value of the matrix element for $_\mu$Ga is equal to:

$$I_{sp}^{(4)} = i\tilde{Z}\sqrt{5}\left(\frac{3}{4}\right)^2, \quad \text{correspondingly} \quad I_{sp}^{(4)} = i\sqrt{5}\left(\frac{3}{4}\right)^2\frac{\tilde{Z}\hbar}{a_0}. \tag{38}$$

In the calculation of a transition probability per unit time, we will take into account that at least three electrons participate in the matrix element of the operator (19), and the number of spin states in Determinant (3) is $2S + 1 = 4$:

$$dw_{\text{hybr}} = \frac{2\pi}{\hbar^2}(2S+1)\left(\frac{e}{mc^2}\right)^2\frac{2\pi\hbar c^2}{\omega}|3\langle\Psi_{\text{in}}|\hat{p}_{1z}|\Psi_{\text{Cr}}\rangle|^2\cos^2\vartheta\,\delta(\omega_{if}-\omega)\frac{k^2dk\,d\Omega}{(2\pi)^3}. \tag{39}$$

Here, ω_{if} appropriates the transition frequency of a neutral radiation defect from the energy level of the corresponding free atom state on the energy level corresponding to a hybridized state in a lattice.

After integration over the wavevector of photons and averaging over all angles, we get:

$$w_{\text{hybr}} = \begin{cases} (17\times5^2/2^33^6)\,\tilde{Z}_{\text{Al}}^2\alpha^3\omega_{if}^{\text{Si}} \approx 3.8\times10^8s^{-1} & \text{for} \quad \text{silicon,} \\ (17\times15/2^{12})\,\tilde{Z}_{\text{Ga}}^2\alpha^3\omega_{if}^{\text{Ge}} \approx 4.9\times10^8s^{-1} & \text{for} \quad \text{germanium,} \end{cases} \tag{40}$$

where $\alpha = 1/137$ is the fine structure constant. We have substituted here the effective charge $\tilde{Z}_{\text{Al}} \approx 3.5$ and $\tilde{Z}_{\text{Ga}} \approx 5.0$ in accordance with Slater [30].

6. Formation of Neutral Center $(_\mu B\,C_4)^0$ in Diamond

Formulae (16) and (17) must be modified for diamond in accordance with Equations (11) and (13) as follows:

$$\langle\psi_{in}|\hat{\mathbf{p}}_1|\psi_{\text{Cr}}\rangle_{sa} = \int\psi_{2s}^*(\mathbf{r}_1)\hat{\mathbf{p}}_1\psi_{\mathbf{n}_a}(\mathbf{r}_1)d\mathbf{r}_1\int\psi_{2p}^*(\mathbf{r}_2)\psi_{\mathbf{n}_b}(\mathbf{r}_2)d\mathbf{r}_2 \tag{41}$$

and:

$$\langle\psi_{in}|\hat{\mathbf{p}}_1|\psi_{\text{Cr}}\rangle_{pa} = \int\psi_{2p}^*(\mathbf{r}_1)\hat{\mathbf{p}}_1\psi_{\mathbf{n}_a}(\mathbf{r}_1)d\mathbf{r}_1\int\psi_{2s}^*(\mathbf{r}_2)\psi_{\mathbf{n}_b}(\mathbf{r}_2)d\mathbf{r}_2, \tag{42}$$

To derive a common expression now, we consider that the Z-axis does not coincide with any of the bond directions $\mathbf{n}_a$. Therefore, as for (18) and (19), we have:

$$\int\psi_{2s}^*(\mathbf{r})\psi_{\mathbf{n}_b}(\mathbf{r})d\mathbf{r} = \alpha_b\langle2s|2s\rangle = \alpha_b; \tag{43}$$

$$\int\psi_{2p}^*(\mathbf{r})\psi_{\mathbf{n}_b}(\mathbf{r})d\mathbf{r} = \frac{1}{3}\beta_b(2\cos\varphi_b - i\sqrt{2}\sin\theta_b\sin\varphi_b + \cos\theta_b). \tag{44}$$

Without the reduction of the generality of the calculations, we consider a matrix element only for the z- projection of a momentum operator.

$$\int\psi_{2s}^*(\mathbf{r}_1)\hat{p}_{1z}\psi_{\mathbf{n}_a}(\mathbf{r}_1)d\mathbf{r}_1 =$$

$$\frac{\beta_a}{\sqrt{3}}\sum_m\langle2,1,0|\hat{R}(\mathbf{n}_a)|2,1,m\rangle\int\psi_{2s}^*(\mathbf{r}_1)\hat{p}_{1z}\psi_{21,0}(\mathbf{r}_1)d\mathbf{r}_1 = \frac{\beta_a}{\sqrt{3}}\cos\theta_a I_{sp}^{(2)}, \tag{45}$$

$$\int\psi_{2p}^*(\mathbf{r}_1)\hat{p}_{1z}\psi_{\mathbf{n}_a}(\mathbf{r}_1)d\mathbf{r}_1 d\mathbf{r}_1 = \frac{\alpha_a}{\sqrt{3}}\int\psi_{21,0}^*(\mathbf{r}_1)\hat{p}_{1z}\psi_{2s}(\mathbf{r}_1)d\mathbf{r}_1 = \frac{\alpha_a}{\sqrt{3}}I_{sp}^{(2)*}, \tag{46}$$

where $I_{sp}^{(2)}$ is determined by Equation (23) for $n = 2$.

The interaction operator (15) conserves the total spin, and we need to calculate only matrix elements for the superpositions (10) and (12) between states with equal total spins. For states with a total electron spin $S = 1$, the matrix elements of the interaction operator in Expression (14) are equal to zero because of the symmetry of the two-particle states (11) and (13). Therefore, contrary to calculations performed for neutral defects in Si and Ge, it is necessary to calculate matrix elements only for singlet states. Accordingly, a radiation transition in dipole approximation for triplet-states (with a maximum spin value of the $(_\mu B)^+$ defect) is forbidden. Note that singlet spin states constitute only 1/4 part of all spin states under the assumption of an equal probability of a population of all spin states. This fact leads to the reduction of a total probability transition by nine-times with respect to the probability transition between triplet states.

Taking into account a local symmetry C_{2v} of the $(_\mu B C_4)^+$ cluster, we direct axes as shown in Figure 2b and introduce the following designations: $\theta_1 = \theta_2 \equiv \theta_1, \theta_3 = \theta_4 \equiv \theta_3, \varphi_1 = \pi/2,$ $\varphi_2 = 3\pi/2, \varphi_3 = 0, \varphi_4 = \pi$. We have, respectively, $\beta_1 = \beta_2 \equiv \beta_1, \alpha_1 = \alpha_2 \equiv \alpha_1, \beta_3 = \beta_4 \equiv \beta_3,$ $\alpha_3 = \alpha_4 \equiv \alpha_3$. After summation over all bonds, one obtains the expression:

$$
\begin{aligned}
\langle \psi_{in} | \hat{p}_{1z} | \psi_{Cr} \rangle = \frac{A_0}{3\sqrt{2}} \Big\{ & \frac{1}{\sqrt{3}} I_{sp}^* \left(\beta_1^2 \cos^2 \theta_1 + 4\beta_1 \beta_3 \cos \theta_1 \cos \theta_3 + \beta_3^2 \cos^2 \theta_3 \right] \\
& + \left(\alpha_1^2 + 4\alpha_1 \alpha_3 + \alpha_3^2 \right) I_{sp} \Big\} .
\end{aligned}
\tag{47}
$$

We can determine superposition coefficients in (6). Taking into account the relations (8) as $\alpha_a^2 = 1 - \beta_a^2$ and, respectively, $\alpha_1 = \alpha_2, \beta_1 = \beta_2, \alpha_3 = \alpha_4, \beta_3 = \beta_4$. Therefore, we have:

$$
\beta_{1,3} = \frac{1}{\sqrt{2}\sin\theta_{1,3}}, \quad \alpha_{1,3} = \frac{1}{\sqrt{2}}\sqrt{1 - \cot^2 \theta_{1,3}}.
\tag{48}
$$

Here, the condition $2\alpha_1^2 + 2\alpha_3^2 = 1$ and, respectively, $\cos(\widehat{\mathbf{n}_1,\mathbf{n}_3}) = \cos\theta_1 \cos\theta_3$ were used. Finally, the matrix element (47) can be written in the form:

$$
\begin{aligned}
\langle \psi_{in} | \hat{p}_{1z} | \psi_{Cr} \rangle = \frac{A_0}{3\sqrt{2}} \Big\{ & \frac{1}{\sqrt{3}} I_{sp}^* \left(\cot^2 \theta_1 + 4 \cot \theta_1 \cot \theta_3 + \cot^2 \theta_3 \right] \\
& + \left(2 - \cot^2 \theta_1 - \cot^2 \theta_3 + 4\frac{\sqrt{\cos 2\theta_1 \cos 2\theta_3}}{\sin\theta_1 \sin\theta_3} \right) I_{sp} \Big\} .
\end{aligned}
\tag{49}
$$

For the appropriate wave functions of the hydrogen-like atom with an effective nuclear charge equal to $\tilde{Z}$, the integral $I_{sp}^{(2)}$ is equal to:

$$
I_{sp}^{(2)} = -i\tilde{Z}, \quad \text{or in dimension units} \quad I_{sp}^{(2)} = -i\frac{\tilde{Z}me^2}{\hbar} = -i\frac{\tilde{Z}\hbar}{a_0}.
\tag{50}
$$

An effective charge for a boron atom is equal to $\tilde{Z}_B \approx 2.6$ [30].

In the calculation of the probability of a transition per unit time, we shall take into account that two electrons participate in the matrix element of the operator (15). Carrying out integration over a wavevector of photons and averaging over all angles, we obtain:

$$
w_{\text{hybr}} = \frac{16}{3} |A_0|^2 f^2(\theta_1, \theta_3) \tilde{Z}^2 \alpha^3 \omega_{sp}.
\tag{51}
$$

Here, ω_{sp} corresponds to a transition frequency of the charged radiation defect from the energy level of the free ion $(_\mu B)^+$ to the energy level of the hybridized (bound) state in the lattice $(_\mu B C_4)^+$.

In accordance with the matrix element (49) and the result of the matrix element (50) calculation, a configuration factor $f(\theta_1, \theta_3)$ is equal to:

$$
f(\theta_1, \theta_3) = \frac{1+\sqrt{3}}{\sqrt{3}} \left(\cot^2 \theta_1 + \cot^2 \theta_3 \right) + \frac{4}{\sqrt{3}} \cot \theta_1 \cot \theta_3 - 4\frac{\sqrt{\cos 2\theta_1 \cos 2\theta_3}}{\sin\theta_1 \sin\theta_3} - 2.
\tag{52}
$$

A transition frequency ω_{sp} and angles for a configuration factor $f(\theta_1, \theta_3)$ were calculated numerically in [28] by the quantum-chemical methods. The crystalline chemical environment of clusters in Figure 2 has been taken into account by the procedure [31] based on a passivation of unnatural valences on a border of the cluster by hydrogen atoms. The variation of the geometrical position of the H* atoms, if it is possible, ensures the stoichiometry of the charge distribution on the carbon atoms of the model $C_5H^*_{12}$ fragment (Figure 2a, where the B atom is substituted by the central C atom).

The initial state of the radiation $(_\mu B)^+$ defect in a diamond lattice is modeled by the tetrahedral structure of Figure 2a where the central carbon atom is substituted by the B atom, and the non-equilibrium length of four B–C bonds coincides with the equilibrium length of 1.523 Å for C–C bonds found by us earlier after geometry optimization of the central structural $C_5H^*_{12}(T_d)$ fragment. As a result of a structural relaxation, the defect transits into the final hybridized state $|\psi_{Cr}\rangle$ described by the lowest in energy triplet structure of the $[BC_4H^*_{12}]^+(T, C_{2v})$ cluster (Figure 2b). The energy of such a transition is calculated from the difference between the total energies of levels $|\psi_{in}\rangle$ and $|\psi_{Cr}\rangle$ to be equal to 1.17 eV.

The angles in the cluster are equal to $\theta_1 \approx 65°$, $\theta_3 \approx 128°$. Calculations of a spin density at the center of the cluster give the value of the superposition parameter $A_0 = \sqrt{7}/4$. The effective charge for $2s$ and $2p$ states of the boron atom is $\widetilde{Z} = 2.6$. Substituting the calculated values into Formulas (51) and (52), we obtain a numerical estimate of the radiation transition rate of the impurity center $(_\mu B)^+$ into a hybridized state:

$$w_{hybr} = \tau_+^{-1} \approx 1.7 \times 10^{11} s^{-1}. \tag{53}$$

The obtained value confirms the validity of the assumption on the kinetics of a charged radiation defect $(_\mu B)^+$ thermalization reported in [24]. The hybridization rate (53) is two orders of magnitude less than the rate of a non-hybridized charged center formation.

The hybridized charged center $(_\mu B)^+$ quickly, during characteristic lattice times, transfers into a neutral state. Therefore, the neutralization time of a charged defect formed by a negative muon in a diamond lattice is determined by the value of (53), and at least two order higher than that for silicon and germanium (40).

7. Formation of an Ionized Acceptor Center

In this section, we will consider the process of an electron capture on the neutral radiation defect with totally restored chemical bonds and the formation of an acceptor center in the ionized state.

According to our cluster calculations, the neutral $[(_\mu B)C_4]^0$ defect has C_{3v} symmetry and creates the substantial electric dipole moment, which is equal to 1.08D, in a diamond lattice. Here, D is Debye, the unit of an electric dipole moment in the atomic system of units (D= 10^{-18}CGSE). The dipole moment is directed along the symmetry axis. Therefore, we can suppose that any neutral cluster of a type $[(_\mu A) A_4]^0$ in crystals with a diamond structure possesses an electric dipole moment of the order of 1D.

This electric dipole moment gives rise to an interaction necessary to capture a lattice electron and the form of an ionized acceptor center. The neutral center $_\mu A$ has unsaturated chemical bonds because a crystal lattice turns out to be deformed. This deformation is a reason to change a phonon spectrum and the local phonon mode appearance. Chemical bounds are saturated after the missing electron capture. The new cluster is an ionized acceptor center and possesses a local crystal symmetry. When the ionized acceptor center is formed, an appropriate phonon of the local mode is radiated, and crystal deformations disappear. Therefore, the problem is very similar to the problem of the thermalization of molecular ions in molecular crystals and cryocrystals of noble atoms (see e.g., [32–34]). An exact solution of the problem taking into account a crystal symmetry is scarcely possible. However, qualitative estimations can be obtained in some approximations [35,36].

7.1. Effective Hamiltonian and Interaction Operator

Detailed calculations of an ionized acceptor center formation in diamond, silicon and germanium were carried out in [35,36], and here, we will present the main results. The electric dipole moment creates a scalar potential, and the interaction energy with lattice electrons is $\mathcal{U} = e\varphi = e(\mathbf{dr})/(\varepsilon r^3)$, where ε is a dielectric penetration. Therefore, taking into account the displacement $\mathbf{u}$ $(\mathbf{r} \rightarrow \mathbf{r} + \mathbf{u})$ and neglecting the changing of the denominator of the potential, we can write an electron-phonon interaction operator:

$$\widehat{V}_{e-ph} \approx \frac{e(\mathbf{d}\hat{\mathbf{u}})}{\varepsilon r^3} = \frac{ed}{\varepsilon r^3} \cos\theta \hat{u}, \tag{54}$$

The operator of radial displacements can be determined in the approach of an isotropic elastic media. For this reason, we need to study vibrations of a sphere at the center of which is placed an electric dipolar moment $\mathbf{d}$. This dipolar moment creates an electric induction $\mathbf{D}$, and a displacement vector $\mathbf{u}$ satisfies the following equation:

$$\ddot{u}_i - c_{\parallel}^2 \Delta u_i - \frac{\varkappa}{\varepsilon^2 \rho} D_i D_k \triangle u_k = 0, \tag{55}$$

where $c_{\parallel}$ is the longitudinal sound velocity, $\varkappa$ is the dielectric susceptibility and $\varepsilon = 1 + 4\pi\varkappa$, ρ is the density of a media.

To solve Equation (55), we will take into account only radial vibrations of a deformed crystal lattice with respect to the center of the sphere. The center of this sphere coincides with our radiation defect. Therefore, we can introduce as usual for problems with central symmetry a radial displacement $\chi = ur$. In this case, we have the more simple equation for χ:

$$\ddot{\chi} - \left(c_{\parallel}^2 + \frac{4\varkappa d^2}{\varepsilon^2 \rho} \frac{1}{r^6} \right) \chi'' = 0, \tag{56}$$

where χ'' is the second derivative on the radial variable r. Making the one-dimensional Fourier transformation for the function $\chi(r,t)$:

$$\chi_{\omega,k} = 2 \int\limits_{-\infty}^{\infty} dt\, e^{i\omega t} \int\limits_{0}^{\infty} dr\, \sin(kr)\chi(r,t), \tag{57}$$

we obtain a dispersion relation in a long wavelength approximation:

$$(\omega^2 - k^2 c_{\parallel}^2 - \Omega^2)\chi_k = 0, \qquad \Omega = \frac{d}{\varepsilon R_1^4}\sqrt{\frac{\varkappa}{6\pi\rho}}, \tag{58}$$

where R_1 is the radius of the first coordination sphere. More detailed calculations are represented in Appendix A.

The numerical estimates for the frequency Ω in diamond, silicon and germanium crystals are presented in Table 1 (the parameters oft the crystals were taken from [37,38]). Since a dipole moment d is considered as an unknown parameter, the numerical estimates are presented in units of Debye.

Table 1. Physical parameters of the C, Si and Ge crystals and the estimate of the frequency Ω.

Crystal	Density ρ, g cm^{-3}	R_1, 10^{-8} cm	Dielectric Constant	Ω, 10^9 s^{-1}	Ionization Energy of the Acceptor, eV
Diamond	3.51	3.57	5.75	8.1 d	$_\mu$B, 0.37
Silicon	2.33	5.43	11.97	1.4 d	$_\mu$Al, 0.069
Germanium	5.323	5.66	16.0	0.66 d	$_\mu$Ga, 0.011

Now, we can construct an effective Hamiltonian describing the radial vibrations of a lattice and determine the operator of radial displacements $\hat{u}$ in Operator (54). Consider radial vibrations in a sphere with the radius R_D with boundary conditions $u(R_D) = 0$. Keeping a dependence on time t, we need to use discrete Fourier amplitudes χ_n instead of Equation (57):

$$\chi_n = 2 \int_0^{R_D} dr \, \sin(k_n r) \chi(r, t), \tag{59}$$

where $k_n = \pi n / R_D$.

Therefore, χ_n and $\rho \dot{\chi}_n$ are generalized coordinates and the momentum of the system under consideration, respectively. The effective Hamiltonian for a system of independent oscillators can be written as:

$$\widehat{H}_{ph} = \sum_n \frac{1}{2} \hbar \omega_n \left(\widehat{\mathcal{P}}_n^2 + \widehat{\mathcal{Q}}_n^2 \right). \tag{60}$$

Here,

$$\omega_n = \sqrt{c_\parallel^2 k_n^2 + \Omega^2}, \tag{61}$$

and the dimensionless generalized momentum and coordinate are defined as usual:

$$\widehat{\mathcal{P}}_n = \rho \widehat{\dot{\chi}}_n / p_{0n}, \quad \widehat{\mathcal{Q}}_n = \widehat{\chi}_n / q_{0n}, \tag{62}$$

while the units of the generalized momentum and coordinate are equal to:

$$q_{0n} = \sqrt{\frac{\hbar}{\omega_n \rho R_D}}, \quad p_{0n} = \sqrt{\hbar \omega_n \rho R_D}.$$

Introducing, as usual, the annihilation and creation operators:

$$\widehat{b}_n = \frac{1}{\sqrt{2}} \left(\widehat{\mathcal{Q}}_n + i \widehat{\mathcal{P}}_n \right), \quad \widehat{b}_n^\dagger = \frac{1}{\sqrt{2}} \left(\widehat{\mathcal{Q}}_n - i \widehat{\mathcal{P}}_n \right), \quad [\widehat{b}_n, \widehat{b}_{n'}^\dagger] = \delta_{nn'}, \tag{63}$$

we obtain the operator of radial displacements of a lattice, which must be substituted in Equation (54), in the form:

$$\widehat{u} = \frac{\widehat{\chi}}{r} = \frac{1}{r} \sum_n \sqrt{\frac{2\hbar}{\rho \omega(k_n) R_D}} \sin(k_n r) \left(b_n^\dagger + b_n \right). \tag{64}$$

It is easy to see that the interaction operator (54) has a strong singularity at $r \to 0$, which leads to the divergence of matrix elements. However, Operator (54) is obtained in a dipolar approximation; its expression is valid for respectively large values of the radius vector and is not applicable at $r \to 0$. In this case, as usual (see e.g., [39]), the potential is taken as constant at $r < R_0$, where R_0 is a certain characteristic distance, i.e.,

$$\widehat{V}_{e-ph} \approx \begin{cases} \left(ed/(\varepsilon r^3) \right) \cos \theta \widehat{u}, & \text{for} \quad r > R_0, \\ \left(edr/(\varepsilon R_0^4) \right) \cos \theta \widehat{u}, & \text{for} \quad r < R_0. \end{cases} \tag{65}$$

In our case, we can take $R_0 \geq r_0$, where r_0 is the length of chemical bonds in the lattice.

7.2. Electron Capture Rate

The probability of electron capture per unit time can be calculated with the well-known Fermi golden rule:

$$dw_{if} \equiv dw_{\text{capt}} = \frac{2\pi}{\hbar} \left| \langle f | \widehat{V}_{e-ph} | i \rangle \right|^2 \delta(E_f - E_i) dv_f \frac{d\mathbf{k}}{(2\pi)^3}. \tag{66}$$

We will consider a case of low temperature, and the initial state $|i\rangle = |\mathbf{k}\rangle |0_{ph}\rangle$ corresponds to the free electron in the valence band with the wave-vector $\mathbf{k}$ and the absence of exited radial phonons.

The final state is determined by the captured electron to the $|ns\rangle$ or $|np\rangle$ hybridized state of the cluster and by the excitation of the radial phonon with the wavevector $\mathbf{k}_{\mathrm{ph}}$:

$$|f\rangle = \left(\begin{array}{c} |ns\rangle \\ |np\rangle \end{array} \right) |\mathbf{k}_{\mathrm{ph}}\rangle.$$

Here, $n = 2, 3$ and 4 for C, Si and Ge, respectively. Equation (66) takes into account that all electrons having wavevectors in the interval from $\mathbf{k}$ to $\mathbf{k} + d\mathbf{k}$ can be captured by the cluster.

The volume element of the final states in the case of one-dimensional motion is equal to:

$$dv_f = R_D \frac{dk_{\mathrm{ph}}}{2\pi} do,$$

where do is the element of the solid angle into which a phonon is emitted. The energies of the initial and final states are:

$$E_i = E_0 + \frac{\hbar^2 k^2}{2m^*}, \quad E_f = E_- + \hbar\omega(k_{\mathrm{ph}}), \quad E_0 - E_- = \hbar\triangle,$$

m^* is the effective mass of the electron. To obtain numerical results, the ionization energy of the cluster (acceptor) is taken for $\hbar\triangle$.

Integration of (66) over k gives:

$$dw_{\mathrm{capt}} = \frac{m^*}{(2\pi\hbar)^2} \frac{2}{3\rho} \left(\frac{ed}{\varepsilon} \right)^2 \left| e\langle f \left| F(r)e^{ik_{\mathrm{ph}}r} \right| \mathbf{k}_0\rangle \right|^2 \frac{k_0}{\omega(k_{\mathrm{ph}})} dk_{\mathrm{ph}}d\Omega, \tag{67}$$

where a vector of the final state $|f\rangle_e$ takes into account electron states only, and according to Equation (65),

$$F(r) = \begin{cases} r^{-4}, & \text{for} \quad r > R_0, \\ R_0^{-4}, & \text{for} \quad r < R_0. \end{cases} \tag{68}$$

Here:

$$k_0^2 = \frac{2m^*}{\hbar}(\triangle - \omega_{\mathrm{ph}}). \tag{69}$$

For the further analysis, it is convenient to introduce the dimensionless parameters and variables:

$$x = \frac{\widetilde{Z}}{na_0} r, \quad \widetilde{k}_{\mathrm{ph}} = \frac{na_0}{\widetilde{Z}} k_{\mathrm{ph}}, \quad \widetilde{k}_0 = \frac{na_0}{\widetilde{Z}} k_0, \tag{70}$$

where $n = 2, 3$ and 4 for C, Si and Ge, respectively. The estimated characteristic parameters for considered crystals are summarized in Table 2.

Table 2. Characteristic parameters estimated for C, Si and Ge.

Crystal	r_0, 10^{-8} cm	$\widetilde{Z}$	$\widetilde{r}_0$	ω_0, 10^{14}c^{-1}	$k_{0,max}$, 10^7 cm^{-1}	$\widetilde{k}_{0,max}\widetilde{r}_0$	$k_{ph,max}$, 10^8 cm^{-1}	$\widetilde{k}_{ph,max}\widetilde{r}_0$
C	1.54	2.6	3.78	3.32	3.1	0.48	4.2	15.9
Si	2.34	3.5	5.15	1.05	1.34	0.31	1.18	2.78
Ge	2.44	5.0	5.75	1.24	0.54	0.13	0.32	0.77

All considered crystals are anisotropic, and the sound velocity $c_{\parallel}$ in the dispersion relation (58) depends on the direction of a phonon propagation. To take this into account, we will use, as usual, the average value of the longitudinal velocity of sound $\langle c_{\parallel}\rangle$, neglecting the effects of anisotropy.

We define a characteristic frequency:

$$\omega_0 = \langle c_\parallel \rangle \frac{\tilde{Z}}{na_0} \tag{71}$$

and corresponding dimensionless frequencies $\tilde{\omega} = \omega_{\text{ph}}/\omega_0$, $\tilde{\triangle} = \triangle/\omega_0$ and $\tilde{\Omega} = \Omega/\omega_0$. Taking into account the dispersion relation of the radial phonons (58), we can write the electron capture rate in the form:

$$w_{\text{capt}} = G_n \int \frac{d\Omega}{4\pi} \int\limits_{\tilde{\Omega}}^{\tilde{\triangle}} \frac{\left| \tilde{k}_0 \mathcal{A}(\tilde{k}_{\text{ph}}, \tilde{k}_0) \right|^2}{\sqrt{(\tilde{\triangle} - \tilde{\omega})(\tilde{\omega}^2 - \tilde{\Omega}^2)}} d\tilde{\omega}, \tag{72}$$

where:

$$G_n = \frac{1}{6\pi\rho m^*} \left(\frac{2m^*}{\hbar \langle c_\parallel \rangle} \right)^{3/2} \left(\frac{ed}{\varepsilon} \right)^2 \left(\frac{\tilde{Z}}{na_0} \right)^{13/2}. \tag{73}$$

The dimensionless matrix element $\mathcal{A}$ is determined by Function (68):

$$\mathcal{A}(\tilde{k}_{\text{ph}}, \tilde{k}_0) = {}_e\langle f \left| F(x)e^{i\tilde{k}_{\text{ph}}x} \right| \tilde{k}_0 \rangle. \tag{74}$$

Equation (72) shows that the electron can be captured both in s- and p-states of the hybridized state of the cluster. As was shown in [36], the capture rate to the p-state w^p is approximately two orders less than the capture rate to s-state w^s. Therefore, in the following, we can consider the electron capture to the s-state. In this case, the matrix element (74) is given by the expression:

$$\mathcal{A}_n^s(\tilde{k}_{\text{ph}}, \tilde{k}_0) = \frac{1}{\sqrt{2\pi}\tilde{k}_0} \int\limits_0^\infty F(x)R_{n0}(x)e^{i\tilde{k}_{\text{ph}}x} \sin(\tilde{k}_0 x)x\,dx = \tilde{k}_0^{-1}(I_1^{(n)}(\tilde{\omega}) - iI_2^{(n)}(\tilde{\omega})), \tag{75}$$

where $I_1(\tilde{\omega})$ and $I_2(\tilde{\omega})$ are the real and imaginary parts of the matrix element, respectively. Finally, the capture rate (72) has the form:

$$w_{\text{capt}} \approx w^s = G_n \int\limits_{\tilde{\Omega}}^{\tilde{\triangle}} \frac{(I_1^{(n)}(\tilde{\omega}))^2 + (I_2^{(n)}(\tilde{\omega}))^2}{\sqrt{(\tilde{\triangle} - \tilde{\omega})(\tilde{\omega}^2 - \tilde{\Omega}^2)}} d\tilde{\omega}, \tag{76}$$

Obtained Formula (76) determines the formation rate of the ionized acceptor center through the capture of the electron of the medium on the neutral radiation defect induced by the negative muon or pion in crystals. Analytical expressions depend on well-known characteristic parameters of the medium except for two parameters of the cluster, namely, the electric dipolar moment d and the parameter R_0 in Equation (65). The dependence on d of the integral in Equation (75) is weak and can assume that $w_{\text{capt}} \propto d^2$. Unfortunately, the dependence of the results on the parameter R_0 is more critical, because the matrix elements exponentially depend on the lower integration limit. Nevertheless, the parameter R_0 cannot be smaller than the length of the chemical bound in the lattice. We can determine the upper limit also $R_0 \leqslant R_1$, the radius of the first coordination sphere.

Numerical calculation were performed for several different values of the uncertain parameters d and R_0. The results are summarized in Table 3.

According to the obtained results in the considered range of the uncertain parameters R_0 and d, the spread of the estimates of the capture rates is about two orders of magnitude and seems at first unsatisfactory. However, it is not surprising because both the dipole approximation and the dispersion relation for the radial phonons (58) are quite rough for $r \sim R_1$. Nevertheless, the considered interaction mechanism is fairly justified and describes the process of the electron capture on the neutral defect of the lattice with the formation of the ionized state of the acceptor center.

Table 3. Estimate of the formation rate of ionized acceptor centers $(\mu A)^-$ in the C, Si and Ge crystals.

d	x_0, C	w_{capt}, C$\cdot 10^8$ s^{-1}	x_0, Si	w_{capt}, Si $\cdot 10^8$ s^{-1}	x_0, Ge	w_{capt}, Ge $\cdot 10^8$ s^{-1}
1.0	3.78	22.0	5.12	0.71	5.76	14.3
0.5	3.78	5.9	5.12	0.19	5.76	3.8
0.2	3.78	0.98	5.12	0.034	5.76	0.65
1.0	5.5	2.2	7.5	0.095	7.5	2.4
0.5	5.5	0.58	7.5	0.024	7.5	0.64
0.2	5.5	0.09	7.5	$2.6 \cdot 10^{-3}$	7.5	0.11

8. Discussion

We have considered the total process of an acceptor center formation in crystals with the diamond structure, which appears as a radiation defect induced by negative pions or muons. It was shown that the evolution of this kind of radiation defects can be divided into two physically different stages. At the first stage, the negatively-charged particle is stopped in the crystal and captured by a nucleus with the charge Z (in the case of π^-) or at the K-shell of a muonic atom (in the case of μ^-). Both negative pions and muons create a host nucleus with an effective charge $Z - 1$. This strongly-charged center interacts with trace electrons and captures them. This stage of radiation defect neutralization exists because of Coulomb interaction. Numerical calculations show that this stage of neutralization strongly differs in diamond and other crystals. Namely, the radiation defect is completely neutralized in Si and Ge for a relatively short time $\tau_n \leq 10^{-11}$ s. In diamond, this radiation defect can be completely neutralized for a long time $\tau_n > 10^{-6}$ s.

The second stage of evolution of this radiation defect is connected with restoration of chemical bonds with the lattice as the first step and formation of an appropriate acceptor center as the final step. Chemical bonds in Si and Ge are restored for neutral radiation defects and in C chemical bonds can be formed for a single-fold charged center. Formation of a chemically-bound radiation defect is accompanied with sufficiently large energy emission. Therefore, the process of chemical bonds' formation can be described with the help of radiation transition. Our estimates gave rather a long time for this step: $\tau_{hybr} \approx 2.6 \times 10^{-9}$ s for Si and $\tau_{hybr} \approx 2.0 \times 10^{-9}$ s for Ge. Charged radiation defect in C forms chemical bonds very quickly: $\tau_{hybr} \approx 0.6 \times 10^{-11}$ s. We can see that this time is many orders shorter than τ_n. Therefore, the radiation defect in diamond is neutralized in the chemically-bound state. This time cannot be strictly estimated, but we can suppose that it is determined by characteristic electronic times in the lattice and must be of the order of 10^{-10} s. We can conclude that the first step of the formation of a chemically-bound neutral radiation defect is approximately two orders shorter in diamond with respect to silicon and germanium.

The second step finishes the formation of an acceptor center in the ionized state. This step is similar in all of the above-mentioned crystals, but very complicated for obtaining good quantitative results. We can see that the chemically-bound neutral radiation defect is a cluster with unsaturated chemical bonds. This unsaturated chemical bond can be saturated if an electron of a valence band of the crystal is captured by the cluster. What kind of interaction gives an opportunity to capture a missing electron? The neutral cluster possesses a relatively large electric dipole moment, which interacts with lattice electrons. This interaction can be qualitatively described in the dipolar approximation. Our analytical and numerical calculations show that:

$$0.5 \times 10^{-10}\,\text{s} \leq \tau_{capt} \leq 2 \times 10^{-8}\,\text{s} \quad \text{for diamond,}$$
$$1.4 \times 10^{-8}\,\text{s} \leq \tau_{capt} \leq 2 \times 10^{-7}\,\text{s} \quad \text{for silicon,}$$
$$0.5 \times 10^{-9}\,\text{s} \leq \tau_{capt} \leq 1.7 \times 10^{-8}\,\text{s} \quad \text{for germanium.}$$

The total time for the acceptor center formation in the ionized state as a result of a radiation defect induced by negative pions and muons is the sum of times for all steps:

$$\tau_{ac^-} = \tau_n + \tau_{hybr} + \tau_{capt},$$

and is determined by the longest item. The final step is slower with respect to the first two steps, and we can conclude that the formation time of the ionized acceptor center is the shortest for diamond ($\leq 2 \times 10^{-8}$ s) and the longest for silicon ($\leq 2 \times 10^{-7}$ s). These values are comparable with characteristic times in semiconductor devices.

The ionized acceptor center is neutralized through the mechanism of Coulomb capture of the hole from the valence band. This process is well studied in many articles (see e.g., [39–41]), and we will not concern ourselves with this problem here.

9. Conclusions

The obtained results will be useful both for μSR experiments and research of different radiation defects in semiconductors. The considered approach can be applied to crystals with the sphalerite-type structure ($A_{III}B_V$ semiconductors, e.g., GaAs, InSb, CdS), which are widely used in electronic devices. Unfortunately, these crystals are more complicated for analysis because of the large variety of possible impurity centers. In addition, the model of chemical bonds must be modified for some calculations.

Acknowledgments: I acknowledge my colleagues Leonid E. Fedichkin and Leonid P. Sukhanov for fruitful discussions. This work is supported by the Ministry of Education and Science of the Russian Federation 16.7162.2017/8.9.

Conflicts of Interest: The authors declare no conflict of interest.

Appendix A

The function $\chi(r,t)$ can be represented in the form of the "one-dimensional" Fourier-transform:

$$\chi(r,t) = 2 \iint\limits_{-\infty}^{\infty} \frac{d\omega dk}{(2\pi)^2} \chi_{\omega,k} e^{-i\omega t} \sin(kr), \tag{A1}$$

$$\chi_{\omega,k} = 2 \int\limits_{-\infty}^{\infty} dt\, e^{i\omega t} \int\limits_{0}^{\infty} dr\, \sin(kr)\chi(r,t). \tag{A2}$$

An integral equation can be obtained for the Fourier amplitude given by Equation (A1) taking into account that the expression for the force in Equation (54) is valid for $r > R_1$, where R_1 is the radius of the first coordination sphere, i.e.,

$$(\omega^2 - k^2 c_{\parallel}^2)\chi_k = \frac{4\varkappa d^2}{\varepsilon^2 \rho} \int\limits_{R_1}^{\infty} dr\, e^{ikr} \frac{1}{r^6} \chi''(r,t). \tag{A3}$$

After passage of the Fourier amplitude on the right-hand side of Equation (A3), this equation is reduced to the form:

$$(\omega^2 - k^2 c_{\parallel}^2)\chi_k = \frac{4\varkappa d^2}{\pi \varepsilon^2 \rho} \times \int\limits_{0}^{\infty} dk' k'^2 \chi_{k'} \int\limits_{R_1}^{\infty} \frac{dr}{r^6} \left(\cos(k-k')r - \cos(k+k')r \right). \tag{A4}$$

Another serious simplification should be done in Equation (A4). The maximum contribution to the second integral comes from wavevectors $k' \approx k$. For this reason, the term $\cos(k+k')r$ can be neglected, and the Fourier amplitude $\chi_{k'}$ can be taken at the point k. In particular, this means that that the main contribution to the integral comes from the values $k' \leq r^{-1}$. Correspondingly, in the accepted approximation, the integral on the right-hand side can be modified to the form:

$$\int\limits_{0}^{\infty} dk' k'^2 \chi_{k'} \int\limits_{R_1}^{\infty} \frac{dr}{r^6} \cos(k-k')r \approx \chi_k \int\limits_{R_1}^{\infty} \frac{dr}{r^6} \int\limits_{0}^{r^{-1}} k'^2 dk' = \frac{\chi_k}{24R_1^8}.$$

After that, the simple dispersion relation (58) is obtained.

References

1. Honniger, F.; Fretwurst, E.; Lindstrom, G. DLTS measurements of radiation induced defects in epitaxial an MCz silicon detectors. *Nucl. Instrum. Methods Phys. Res. Sect. A* **2007**, *583*, 104–108.

2. Grilj, V.; Skukan, N.; Jakšić, M. Irradiation if thin diamond detectors and radiation hardness tests using MeV protons. *Nucl. Instrum. Methods Phys. Res. Sect. B* **2013**, *306*, 191–194.

3. Guthoff, M.; de Boer, W.; Muller, S. Simulation of beam induced lattice defects of diamond detectors using FLUKA. *Nucl. Instrum. Methods Phys. Res. Sect. A* **2014**, *735*, 223–228.

4. Sato, Yu.; Shimaoka, T.; Kaneko, J.H. Radiation hardness of a single crystal CVD diamond detector for MeV energy protons. *Nucl. Instrum. Methods Phys. Res. Sect. A* **2015**, *784*, 147–150.

5. Grilj, V.; Skukan, N.; Jakšić, M. The evaluation of radiation damage parameter for CVD diamond. *Nucl. Instrum. Methods Phys. Res. Sect. B* **2016**, *372*, 161–164.

6. Makgato, T.H.; Sideras-Haddad, E.; Ramos, M.A. Magnetic properties of point defects in proton irradiated diamond. *J. Magn. Magn. Mater.* **2016**, *413*, 76–80.

7. Vins, V.G. New radiation induced defects in HPHT synthetic diamonds. *Diam. Relat. Mater.* **2005**, *14*, 364–368.

8. Dresselhaus, M.S.; Kalish, R. *Ion implantation in Diamond, Graphite and Related Materials*; Springer: Berlin, Germany, 1992.

9. Vonsovsky, S.V. *Magnetizm of Microparticles*; Nauka: Moscow, Russia, 1973; pp. 1–197. (In Russian)

10. Smilga, V.P.; Belousov, Y.M. *The Muon Method in Science*; Nova Science: New York, NY, USA, 1994; pp. 1–420.

11. Waisenberg, A.O. *Mu-Meson*; Nauka: Moscow, Russia, 1964; pp. 1–400. (In Russian)

12. Dzhuraev, A.A.; Evseev, V.S. Mechanism of mu-meson depolarization in molecular condensed media. *Sov. Phys. JETP* **1972**, *35*, 615–819.

13. Dzhuraev, A.A.; Evseev, V.S.; Myasishcheva, G.G. Depolarization of negative muons in solids. *Sov. Phys. JETP* **1972**, *35*, 748–752.

14. Dzhuraev, A.A.; Evseev, V.S. Depolarization of negative muons in condensed molecular media. *Sov. Phys. JETP* **1974**, *39*, 207–211.

15. Gorelkin, V.N.; Smilga, V.P. On the theory of precession of the polarization vector of μ^--mesons in mu-nucleonic atoms. *Sov. Phys. JETP* **1974**, *66*, 586–591.

16. Gorelkin, V.N.; Grebinnik, V.G.; Gritsaj, K.I. μSR-study of the behavior of impurity atoms in silicon. *Phys. At. Nucl.* **1993**, *56*, 1316–1319.

17. Mamedov, T.N.; Duginov, V.N.; Grebinnik, V.G. Investigation of the behavior of the impurity atoms in Si by μSR-method. *Hyperfine Interact.* **1994**, *86*, 717–722.

18. Mamedov, T.N.; Chaplygin, I.L.; Duginov, V.N. Anomalous frequency shift of negative muon spin precession in *n*-type silicon. *Hyperfine Interact.* **1997**, *105*, 345–349.

19. Mamedov, T.N.; Gritsaj, K.I.; Stoykov, A.V. μ^-SR investigations in silicon. *Phys. B* **2000**, *289–290*, 574–577.

20. Mamedov, T.N. *Study of Boron Acseptor Center in Synthetic Diamond by μSR-Technique*; JINR Preprint P14-2007-12; JINR: Dubna, Russia, 2007. (In Russian)

21. Mamedov, T.N.; Baturin, A.S.; Blank, V.D. Non-equilibrium charge carrier dynamics in synthetic diamond studied by μSR-method. *Diam. Relat. Mater.* **2008**, *17*, 1221–1214.

22. Mamedov, T.N.; Baturin, A.S.; Gritsaj, K.I. Muonic atom as an acceptor centre in diamond. *J. Phys. Conf. Ser.* **2014**, *551*, 012046.

23. Baturin, A.S.; Gorelkin, V.N.; Mamedov, T.N.; Soloviev, V.R. Absolute negative mobility of charge carriers in diamond and interpretation of μSR experiments. *Phys. B* **2006**, *374–375*, 340–346.

24. Antipov, S.A.; Belousov, Y.M.; Soloviev, V.R. Dynamics of charge carriers at the place of the formation of a muonic atom in diamond and silicon. *JETP* **2012**, *115*, 866–875.

25. Belousov, Y.M. Formation of radiation defect induced by negative muons in crystals with diamond structure. *J. Phys. Conf. Ser.* **2012**, *343*, 012013.

26. Pauling, L. *Natureof the Chemical Bond*, 3rd ed.; Cornell University Press: Ithaca, NY, USA, 1960.

27. Coulson, C.A. *Valence*; Oxford University Press: Oxford, UK, 1961.

28. Belousov, Y.M.; Sukhanov, L.P. Formation of chemically bound positively charged radiation defect induced by negative muon in diamond crystals. *Diam. Relat. Mater.* **2015**, *58*, 10–15.

29. Landau, L.D.; Lifshitz, E.M. *Quantum Mechanics. Nonrelativistic Theory*, 3rd ed.; Pergamon: Oxford, UK; New York, NY, USA, 1977.

30. Slater, J.C. Atomic Shielding Constants. *Phys. Rev.* **1930**, *36*, 57–64.

31. Groshev, G.E.; Sukhanov, L.P. Quantum-chemical cluster models of the $Si(111)/SiO_2$ interface. *Theor. Exp. Chem.* **1990**, *26*, 268–275. (In Russian)

32. Belousov, Y.M.; Smilga, V.P. Theory of muon spin relaxation in condensed phases of hydrogen isotopes. *JETP* **1994**, *79*, 811–818.

33. Belousov, Y.M. Theoretical description of the muon spin depolarization in the crystalline phase of ^{3}He. *JETP* **2007**, *104*, 215–229.

34. Belousov, Y.M.; Smilga, V.P. Theory of muon spin relaxation in condensed phases of hydrogen isotopes. *Hyperfine Interact.* **1997**, *106*, 19–25.

35. Belousov, Y.M. Process of an acceptor center $_\mu B^-$ formation in diamond crystals. *J. Phys. Conf. Ser.* **2014**, *551*, 012047.

36. Belousov, Y.M. Process of negative-muon-induced formation of an ionized acceptor center $(_\mu A)^-$ in crystals with diamond structure. *JETP* **2016**, *123*, 1160–1169.

37. Madelung, O. *Semiconductors—Basic Data.*, 3rd ed.; Springer: Berlin/Heidelberg, Germany, 2004.

38. Adachi, S. *Properties of Group-IV, III-V and II-VI Semiconductors. Wiley Series in Materials for Electronic & Optoelectronic Applications*; John Wiley & Sons: Chichester, UK, 2005.

39. Abakumov, V.N.; Perel, V.I.; Yasievich, I.N. *Nonradiative Recombination in Semiconductors*; SPb.Institute of Nuclear Physics RAS: Amsterdam, The Netherlands, 1997.

40. Lax, M. Cascade capture of electron in solids. *Phys. Rev.* **1960**, *119*, 1502–1523.

41. Abakumov, V.N.; Yasievich, I.N. Cross section for recombination of an electron with a positively charged center in a semiconductor. *Sov. Phys. JETP* **1976**, *44*, 345–349.

Article

Thermochemical Wear of Single Crystal Diamond Catalyzed by Ferrous Materials at Elevated Temperature

Lai Zou [1,*], Yun Huang [1], Ming Zhou [2] and Guijian Xiao [1]

[1]   State Key Laboratory of Mechanical Transmission, Chongqing University, Chongqing 400044, China; yunhuang@samhida.com (Y.H.); xiaoguijian@cqu.edu.cn (G.X.)

[2]   School of Mechanical and Electrical Engineering, Harbin Institute of Technology, Harbin 150001, China; zhouming@hit.edu.cn

*   Correspondence: zoulai@cqu.edu.cn; Tel.: +86-23-6766-9883; Fax: +86-23-6766-9663

Academic Editor: Yuri N. Palyanov
Received: 25 March 2017; Accepted: 17 April 2017; Published: 19 April 2017

Abstract: Single crystal diamond has been recognized as the optimal tool material in ultra-precision machining. However, the excessive tool wear prevents it from cutting ferrous materials. This paper conducts a series of thermal analysis tests under the conditions of different gas atmospheres, heating temperatures, crystallographic planes and workpiece materials, in order to clarify the details of thermochemical wear of diamond catalyzed by iron at elevated temperature. Raman scattering analysis was performed to identify the transformation of diamond crystal structure. Energy dispersive X-ray analysis was used to detect the change in chemical composition of the work material. X-ray photoelectron spectroscopy was adopted to confirm the resultants of interfacial thermochemical reactions. The experimental results revealed that the diamond wear included the graphitization, diffusion and oxidation. Temperature was considered as the key factor affecting these wear mechanisms. The initial graphitization temperatures of diamond catalyzed by iron under different conditions were obtained, and the graphitized degree relied heavily on the crystallographic plane while being insensitive to the workpiece material. The diffusion wear rule was preliminarily achieved by the established prediction model of the carbon atoms diffusing into the iron lattice, and the types and resultants of interfacial chemical reactions were deduced.

Keywords: diamond; thermochemical wear; graphitization; diffusion; oxidation; ferrous materials; chemical reaction; thermal analysis

---

1. Introduction

Natural single crystal diamond has been widely used in ultra-precision machining of nonferrous metals owing to its outstanding mechanical and physicochemical properties [1,2]. However, cutting of ferrous metals with diamond tools has not been successful in application to date as the tool wear rate is severe [3,4]. Meanwhile, with the requirement for high precision and complex moulds to be made of ferrous materials increasing greatly in recent years, it is difficult to guarantee the demands for machining efficiency and machining accuracy using the conventional processes [5]. Therefore, it is necessary to investigate the tool wear of single crystal diamond in the cutting operation of ferrous metals.

Several technological methods have been proposed for overcoming this serious tool wear issue: turning of medium carbon steels in different carbon-saturated atmospheres [6,7]; turning of stainless steels in a low temperature environment [8]; adopting the ultrasonic vibration assisted turning technique to decrease the cutting force and temperature for controlling tool wear and obtaining optical-quality surface [9,10]; restraining the chemical reactions between carbon atoms and iron atoms at high temperature by surface modification of the diamond tool and workpiece materials [11]; and proposing an ion-shot coolant method for micro-cutting of ferrous materials [12].

Although these auxiliary processing methods mentioned above could decrease the diamond tool wear to various degrees, they have not met the requirements of industrial application, as rather less attention has been paid to the wear mechanisms of a single crystal diamond tool. Thornton et al. [13] performed experimental investigations on turning medium carbon steels under the conditions of different gas atmospheres and cutting speeds. The results revealed that diamond graphitization was the main cause for tool wear, which was attributed to the local high temperature, the continuous clean surface with high chemical activity produced by cutting and the oxygen molecules in the air. In addition, it was pointed out that there was no diamond diffusion wear in the cutting process. Paul et al. [14] found that ferrous metals were classified as non-diamond turnable materials ascribed to the unpaired numbers of d-shell electrons of the iron atoms, and the wear rate was proportional to the number of unpaired d electrons. Shimada et al. [15,16] suggested that the tool wear had a strong correlation with thermochemical reactions between diamond and ferrous metals based on the results of static erosion tests. However, there was not much detailed information about the functional mechanism of each wear form. Narulkar et al. [17] utilized molecular dynamics and realistic interaction potentials as the first direct evidence that the structure of diamond at the cutting edge initially transformed from diamond cubic into hexagonal graphite in the presence of iron. Subsequently, the graphitic carbon diffused into the iron surface. Zou et al. [18] chiefly investigated the graphitization wear of single crystal diamond against ferrous metals by frictional wear experiments. The results showed that the effect of mechanical force on tool wear should not be ignored and the diamond graphitized degree was proportional to mechanical force and sliding velocity. There was, as yet, no consistent explanation of the diamond tool wear mechanism, which was a limitation on the development of the machining process.

In view of this, a series of thermal analysis experiments simulating the wear process of single crystal diamond were proposed in this work, in order to investigate the thermochemical wear of diamond surface catalyzed by iron at elevated temperature under different gas atmospheres, for thoroughly understanding the individual and intensive diamond tool wear mechanism in machining of ferrous materials.

2. Materials and Methods

2.1. Sample Preparation

Similar sized natural single crystal diamonds were used in these tests, which were polished with crystallographic plane (110) and (100), respectively. Four kinds of stainless steels with different carbon content were selected as the workpiece materials, which were widely used in mechanical and electronic fields. Table 1 presents the chemical compositions of each work material. These ferrous materials were cut by wire electrical discharge machining to the diameter and thickness of 4, and 0.2 mm, respectively. It is worth mentioning that the diamonds and steels were cleaned carefully in an ultrasonic vessel with acetone before each thermal analysis experiment.

Table 1. Chemical compositions of the work samples used in tests.

Workpiece	Chemical Compositions (w%, Balance Fe)								
	C	Mn	Si	S	P	Ni	Cr	Ti	Mo
0Cr18Ni9	0.045	1.13	0.35	0.003	0.033	8.04	17.10	–	0.03
1Cr18Ni9Ti	0.07	1.21	0.37	0.01	0.03	8.07	17.99	0.40	–
2Cr13	0.27	0.39	0.56	0.01	0.02	0.14	12.40	–	–
3Cr13	0.29	0.21	0.89	0.004	0.021	0.21	12.47	–	–

Subsequently, the diamond polished with the designated crystallographic plane was set at the bottom surface of an aluminum oxide crucible in contact with the ferrous specimen. Figure 1 shows the schematic model of the thermal corrosion between diamond and steel specimen.

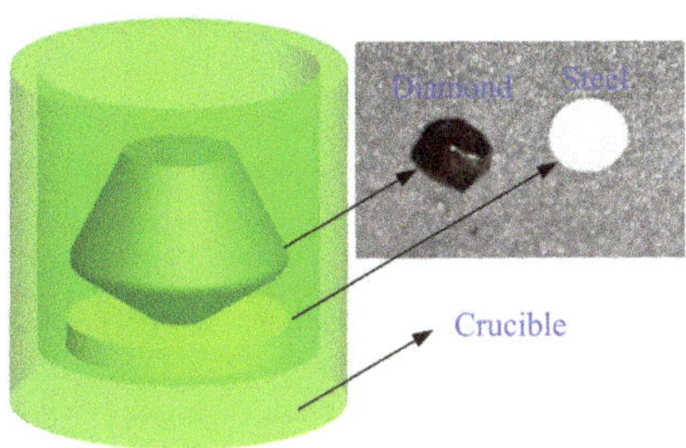

Figure 1. Schematic model of thermal analysis between diamond and steel specimen.

2.2. Characterization

2.2.1. Experimental Procedure

In order to accurately describe the thermochemical wear process of the single crystal diamond catalyzed by ferrous materials at elevated temperature, thermal analysis experiments were performed using a TGA1600 simultaneous thermal analyzer with two functions, including differential thermal analysis (DTA) and thermal gravimetric (TG). Figure 2 displays the working principle of this thermal analysis technique. In this work, the samples that consisted of diamond and steel were heated in a designated temperature range, and the heating rate was set to 10 K/min. Moreover, the argon or air continuously penetrated into the crucible with a flow rate of 40 mL/min during the heating process. The total mass of the samples was approximately 100 mg. Table 2 presents the parameters of the thermal analysis experiments.

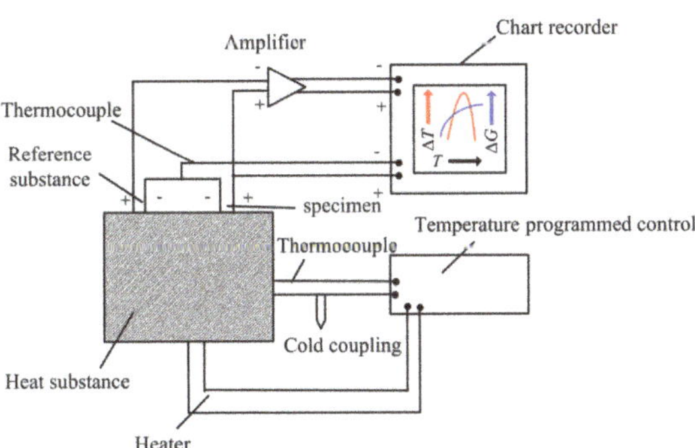

Figure 2. Working principle of the thermal analysis technique.

Table 2. Parameters of the thermal analysis experiments.

Parameters	Value
Heating range (K)	293~1273/1573
Heating rate (K/min)	10
Gas flow rate (mL/min)	40
Gas atmosphere	argon/air
Polished plane of diamond	(100)/(110)
Workpiece materials	0Cr18Ni9/1Cr18Ni9Ti/2Cr13/3Cr13

2.2.2. Analytic Technique

Raman spectroscopy (RS) was considered as the most effective tool to distinguish the structural changes of the carbon cluster [19]. The crystal structure transforming from tetrahedral diamond to regular hexagonal graphite was identified by the BWS435-532 Raman spectrometer (B&W Tek, Newark, NJ, USA) in this work. The 532 nm wavelength of an Ar ion laser can be focused onto the sample surface down to a spot size of 1 μm^2 with the system resolution of 5 cm^{-1}, and these parameters were kept unanimous for detecting the diamond worn surfaces in all the tests.

The thermal corrosion surface morphology of the samples was examined using the FEI Quanta200 scanning electron microscope (SEM) (FEI, Hillsboro, OR, USA). The changing chemical composition of the ferrous specimen before and after the test could be surveyed by the energy dispersive X-ray (EDX) analysis, in order to study the influence of correlative factors on the diffusion wear of the carbon atoms in single crystal diamond continuously diffused onto the corrosion surface of the ferrous materials.

ESCALAB250 X-ray photoelectron spectroscopy (XPS) (Thermo Fisher Scientific, Waltham, MA, USA) was used to detect the chemical compounds generated at the thermal corrosion interface between diamond and steel specimen for the purpose of confirming the type of thermochemical reactions. It was the fundamental research for further studying the diamond oxidation wear. The energy scanning range of this instrument was 0~5000 eV, continuously adjustable in the range of 1~400 eV with the step size of 1 eV, and the beam spot area was in the range of 900~20 μm with the step size of 50 μm.

3. Results and Discussion

3.1. Thermal Analysis Results

Initially, the thermochemical changes of the experimental samples during the heating process were qualitatively analyzed in accordance with the characteristics of the DTA curves, and then the existence and corresponding temperature range of these changes were further concluded in combination with the features shown in the TG curves. Ultimately, the thermal corrosion process of the single crystal diamond catalyzed by ferrous materials at high temperature could be precisely described using the related theory and analytic technique.

Figure 3 shows the DTA and TG curves of the specimens heated in different gas environments from room temperature to 1573 K and 1373 K. It can be found that an obvious exothermic peak emerged corresponding to the temperature in the vicinity of 1383 K from the DTA curve in Figure 3a. As the transformation of diamond to graphite was an exothermic process in view of the thermodynamics theory, it can be excluded that this phenomenon was caused by oxidation of the experimental specimens in argon atmosphere. Therefore, it could be initially regarded that the emergence of this exothermic peak was due to the phase transition from the diamond structure to the graphite structure. Besides, it was clear that the overall quality of the samples basically did not change, as shown from the TG curve in Figure 3a because no gas escape would be produced by the chemical reaction in this atmosphere.

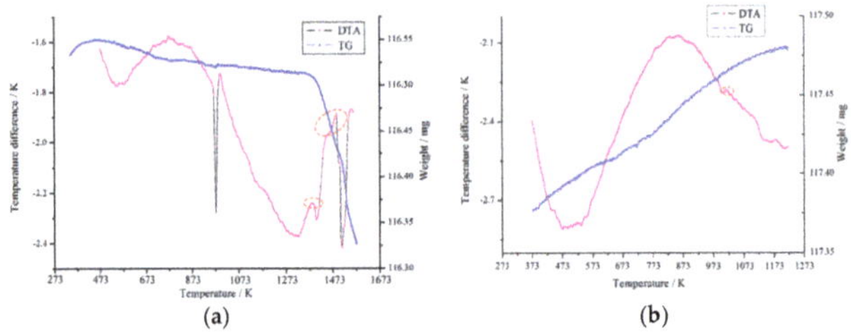

Figure 3. Differential thermal analysis (DTA) and thermal gravimetric (TG) curves of the specimens heated in different gas environments; the workpiece material was 2Cr13; the diamond crystallographic plane was (100); the temperature range was from 293 K to 1573 K and 1273 K. (**a**) DTA–TG curves in argon; (**b**) DTA–TG curves in air.

Moreover, the thermal analysis curves shown in Figure 3b were analyzed in accordance with the same method. The emergence of the tiny exothermic peak could be primarily attributed to the diamond graphitization while the temperature was approximately 1013 K. As a series of complex oxidation-reduction reactions might occur at the corrosion interface between diamond and steel in air condition, it would result in an obvious exothermic process and the total weight of the specimens increased.

Figure 4 presents the worn morphology of the single crystal diamonds under the same experimental conditions as shown in Figure 3. It can be clearly seen that the corrosion degree of diamond in argon was far lower than that in air, and part of the steel specimen was attached on the diamond surface under the effect of high temperature. The detailed wear process will be discussed in the following part about the thermochemical wear mechanisms of diamond catalyzed by the ferrous materials.

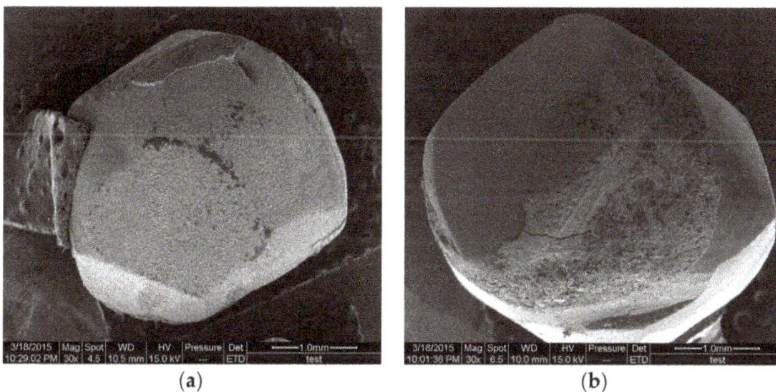

(**a**)   (**b**)

Figure 4. Worn morphology of the single crystal diamonds corroded by the ferrous materials; the workpiece material was 2Cr13; the diamond crystallographic plane was (100); the temperature range was from 293 K to 1573 K and 1273 K. (**a**) Worn morphology in argon; (**b**) Worn morphology in air.

3.2. Thermochemical Wear Mechanisms

The thermal corrosion interface between diamond and steel specimen was further analyzed using advanced testing methods mentioned above based on the results of the thermal analysis experiments, and the thermochemical wear mechanisms of single crystal diamond will be studied and revealed in depth from the several subsequent aspects.

3.2.1. Graphitization Wear

Figure 5 shows the Raman spectra curves of the diamond corrosion surface. The peaks corresponding to the Raman shift at approximately 1582 cm^{-1} were attributed to the sp^2 hybridized orbit of graphite, while others corresponding to the Raman shift at approximately 1334 cm^{-1} were attributed to the sp^3 hybridized orbit of diamond [20]. In view of the occurrence of the graphite structure, it can be confirmed that the exothermic process arose due to diamond graphitization. Therefore, the initial temperature of diamond transformed into graphite was found to be 1383 K and 1013 K in different gas environments based on iron catalysis. The results showed that the initial temperature of graphitization was much lower than that with no catalysis in inert gas. It was deduced that the catalysis of steel specimens could significantly reduce the graphitization conditions and accelerate the graphitized degree of diamond.

Consequently, graphitization should be considered as a major factor for tool wear in diamond cutting of ferrous metals. In addition, this result can also be taken into account for a new research direction in diamond tool ultra-precision machining of steels in inert gas. Meanwhile, the lower onset temperature of graphitization in air condition was attributed to the existence of oxygen which would cause a large number of oxidation-reduction reactions between the diamond and iron.

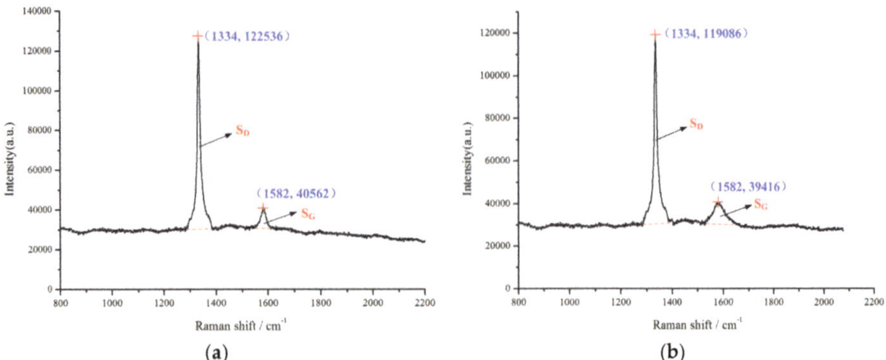

Figure 5. Raman spectra curves of the diamond surface corroded by the ferrous materials; the workpiece material was 2Cr13; the diamond crystallographic plane was (100); the temperature range was from 293 K to 1573 K and 1273 K. (**a**) Raman curves in argon; (**b**) Raman curves in air.

Equation (1) was proposed based on the principle of Gibbs free energy [21].

$$\Delta G_T^{\ominus} \leq -1100 - 4.64T \tag{1}$$

where $\Delta G_T^{\ominus}$ and T were Gibbs free energy and heating temperature, respectively. It can obviously be seen that Gibbs free energy would decrease as the temperature was increasing, which results in the diamond graphitized degree being aggravated. This could be used to explicate the appearance of the large exothermic peak when exceeding the onset temperature of diamond graphitization, as shown in Figure 3.

Moreover, Raman scattering intensity, reflected by the area of Raman peak, was proportional to the analyte concentration as indicated in the previous research [19]. Therefore, the graphitized degree of single crystal diamond could be evaluated by the ratio of the graphite peak area S_G and the diamond peak area S_D. Table 3 presents the graphitized degree of diamonds under different experimental conditions. The results revealed that the diamond crystallographic plane had an important influence on the graphitization wear, and this can be used as selection criteria of the diamond tool in the actual cutting process. However, the graphitized degree was insensitive to the chemical compositions of the work materials.

Table 3. Graphitized degree of diamonds under different experimental conditions.

S_G/S_D (%)		0Cr18Ni9	1Cr18Ni9Ti	2Cr13	3Cr13
Argon	(100)	24.6	24.3	23.8	24.1
	(110)	18.9	17.7	18.5	17.8
Air	(100)	45.8	45.1	44.2	44.5
	(110)	36.5	37.4	36.8	36.2

3.2.2. Diffusion Wear

Figure 6 presents the EDX analysis of sample surfaces corroded under the same experimental conditions mentioned above. It can be seen that the carbon content of the steel surface increased obviously after the thermal analysis experiments. It was certain that the diffusion wear occurred at the interface between ferrous material and diamond specimen, especially in argon environment. The minimal diamond diffusion wear that occurred in air condition might be due to the lower maximum heating temperature; the complex chemical reactions also had a certain influence on it.

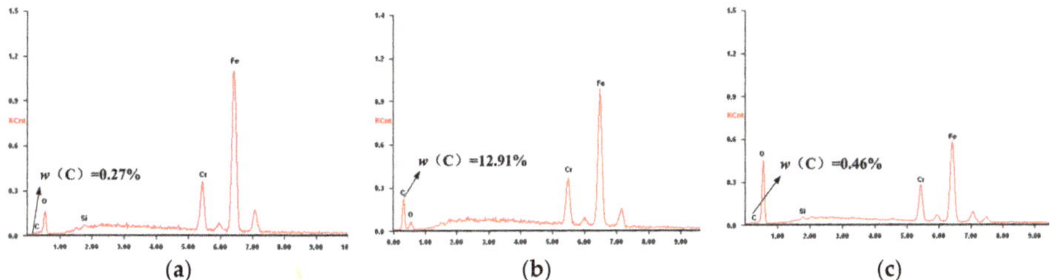

Figure 6. Energy dispersive X-ray (EDX) analysis curves of the sample surface heated in different gas environments; the workpiece material was 2Cr13; the diamond crystallographic plane was (100); the temperature range was from 293 K to 1573 K and 1273 K. (**a**) Before test; (**b**) Tested in argon; (**c**) Tested in air.

Because the iron atoms in ferrous metals had a stronger ability to dissolve carbon than other metal atoms, the carbon atoms in the steel sample could form a certain concentration gradient with diamond. This would cause carbon atoms to become dissociated from the diamond surface and to diffuse into the workpiece material at elevated temperature. However, the experimental results showed that the carbon content of the steel and diamond crystallographic plane had little effect on diffusion wear.

Furthermore, the penetration of carbon atoms into the workpiece surface was primarily considered as a unidimensional diffusion process. In this case, Fick's second law was properly proposed for modeling, as follows, with the hypothesis that the diffusion coefficient was independent of the carbon concentration.

$$\frac{\partial C(x,\tau)}{\partial \tau} = D\frac{\partial^2 C(x,\tau)}{\partial x^2} \tag{2}$$

$C(x, \tau)$ was the concentration function of carbon atoms at the diffusion distance x and the diffusion time τ, which needed to be solved to establish the diffusion model. Moreover, the diffusion coefficient D could be decided by the following equation.

$$D = D_0 e^{\ Q/RT} \tag{3}$$

where D_0, Q, R and T were the diffusion constant, the diffusion activation energy, the gas constant and the thermodynamic temperature, respectively. It can be seen from this formula that the diffusion coefficient D would generally increase exponentially with the increasing temperature as other parameters remain unchanged. The ultimate diffusion model of carbon atoms that penetrated into the iron crystal lattice was obtained as follows by authors based on relevant assumptions [22].

$$C(x,\tau) = C_0 + (C_p - C_0)\,\text{erfc}\left(\frac{x}{2\sqrt{D\tau}}\right) \tag{4}$$

where erfc(x) was the complementary error function used for solving this diffusion model. C_0 was the solubility of carbon atoms in the ferrous metal at ambient temperature, and C_p was the saturated carbon concentration at the specific temperature. It can be seen from this model that the diffusion coefficient and the diffusion time were two crucial factors affecting the distribution laws of carbon concentration, different from the initial carbon concentration of the sample. This tendency was consistent with the results mentioned above. Consequently, the diffusion wear mainly affected by high temperature should be considered as an important wear mechanism of single crystal diamond in machining of ferrous materials, and more attention should be paid to reducing temperature and designing a diffusion barrier at the tool–chip interface to suppress it. The current analytic results could be regarded as the basis for further investigation of diamond diffusion wear in combination with the reasonable experimental studies.

3.2.3. Oxidation Wear

Figure 7 shows the micro-morphology and EDX analysis of the sample surfaces produced by thermal corrosion under the same experimental conditions. As seen in Figure 7a,b, the result of energy spectrum analysis corresponding to the white block material in micro topography indicated that the carbon atoms diffused from diamond would form a certain carbide with iron atoms in the workpiece material, as a result of no other chemical reactions existing in argon atmosphere. However, the iron–carbon compound was easily decomposed by the effect of elevated temperature. It could be deduced that the possibility of diamond reacting directly with iron to cause tool wear in the cutting operation of ferrous materials was relatively small by combining with the results of thermal analysis experiments.

Unlike the above studies, it can be inferred from the other detection result in Figure 7 that the oxides appeared on the steel surface although at lower temperature in air condition. In order to further verify the existence and types of the iron oxides, the corrosion surface was analyzed by the XPS technique and the results are shown in Figure 8. It can be clearly seen from Figure 8a that the original spectrum curve of O1s coincided well with the fitting line, which could be proof that the other two spectrum curves were reliable. The electron binding energies of the two peaks were 530.14 eV and 531.42 eV, respectively. It can be determined that these two curves corresponded to the spectral lines of Fe_xO_y and the C–O bond on the basis of relevant analysis data [23], and this result definitely indicated the existence of iron oxides. It was thus that the presence of oxygen would lead to the occurrence of a large number of redox reactions at the interface between experimental samples, resulting in the diamond oxidation wear.

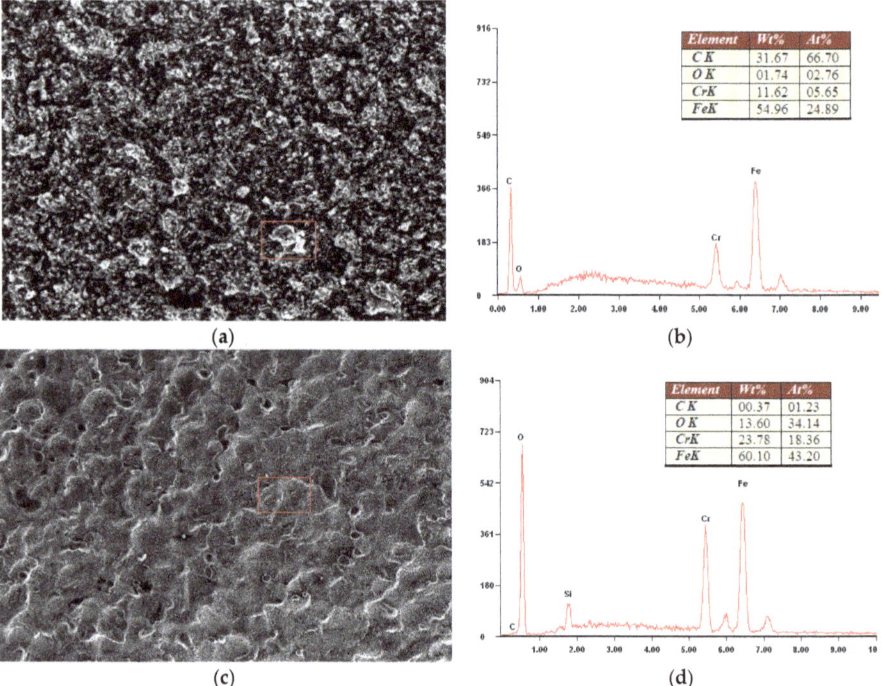

Figure 7. Micro-morphology and EDX analysis of the sample surfaces corroded under the same experimental conditions. (**a**) SEM image in argon; (**b**) EDX analysis in argon; (**c**) SEM image in air; (**d**) EDX analysis in air.

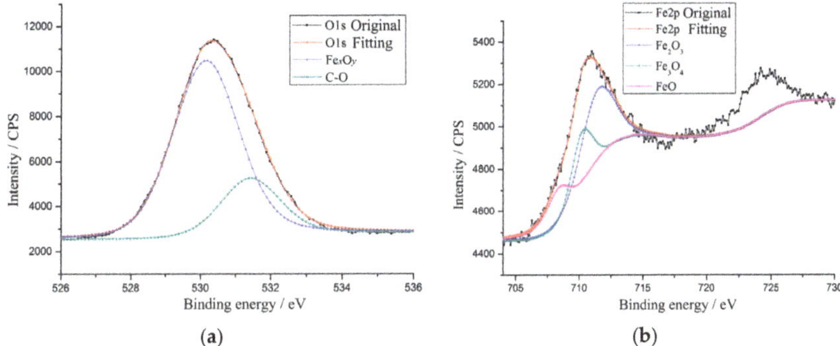

Figure 8. Electronic spectra curves of O1s and Fe2p characteristic peaks on the sample surface corroded in air condition. (**a**) Characteristic peaks of O1s; (**b**) Characteristic peaks of Fe2p.

In addition, the electron binding energies of these three peak curves shown in Figure 8b were in the sequence of 708.41 eV, 710.17 eV and 711.38 eV from small to large, and the corresponding oxides were FeO, Fe_3O_4 and Fe_2O_3, respectively. It was certain that a series of complicated thermal chemical reactions had been produced at the interface during the heating process, which could be used to explain the unclear exothermic phenomenon as shown in Figure 3b. Additionally, the peak intensity of the individual spectral line directly represented the oxide content at the corrosion interface. It can be used to further determine the type of interfacial chemical reactions and study the oxidation wear process of single crystal diamond in depth.

4. Conclusions

In this work, a series of thermal analysis experiments between diamond specimens and steels under different conditions are performed to investigate single crystal diamond wear. The main conclusions obtained are summarized as follows:

(1)   Diamond thermochemical wear includes the graphitization, diffusion and oxidation, and graphitization is the main wear mechanism. Temperature should be considered as the key factor influencing these wear mechanisms.

(2)   The initial graphitization temperatures of diamond catalyzed by ferrous materials are obtained. The graphitized degree is observably intensified with the rising temperature, and it relies heavily on the diamond crystallographic plane while being insensitive to the work material.

(3)   The diffusion wear rule is preliminarily achieved by the established prediction model, and this can be used as fundamental research for further investigation of diamond diffusion wear in combination with the reasonable experimental studies. Additionally, more attention should be paid to reducing the temperature and designing a diffusion barrier at the interface to suppress it.

(4)   It is confirmed that the possibility of diamond reacting directly with iron to cause tool wear is relatively small by comparing with thermal analysis curves in different atmospheres. However, the presence of oxygen will lead to the occurrence of a large number of redox reactions at the interface, resulting in diamond oxidation wear.

This research can be used as the basis for expanding the application areas of ultra-precision machining with a single crystal diamond tool.

Acknowledgments: This work was supported by the National Natural Science Foundation of China (Grant No. 51605056) and the Research Funds for Basic Science and Advanced Technology of Chongqing (Grant No. cstc2016jcyjA0066).

Author Contributions: Lai Zou and Ming Zhou conceived and designed the experiments; Guijian Xiao performed the experiments; Lai Zou and Yun Huang analyzed the data; Yun Huang contributed the materials; Lai Zou wrote the paper.

Conflicts of Interest: The authors declare no conflict of interest.

References

1. Cheung, C.F.; Lee, W.B. Characterisation of nanosurface generation in single-point diamond turning. *Int. J. Mach. Tools Manuf.* **2001**, *41*, 851–875. [CrossRef]
2. Fang, F.Z.; Venkatesh, V.C. Diamond turning of soft semiconductors to obtain nanometric mirror surfaces. *Int. J. Adv. Manuf. Technol.* **2002**, *19*, 637–641. [CrossRef]
3. Ramanujachar, K.; Subramanian, S.V. Micromechanisms of tool wear in machining free cutting steels. *Wear* **1996**, *197*, 45–55. [CrossRef]
4. Lane, B.M.; Dow, T.A.; Scattergood, R. Thermo-chemical wear model and worn tool shapes for single-crystal diamond tools cutting steel. *Wear* **2013**, *300*, 216–224. [CrossRef]
5. Venkatesh, V.; Swain, N.; Srinivas, G.; Kumar, P.; Barshilia, H.C. Review on the machining characteristics and research prospects of conventional microscale machining operations. *Mater. Manuf. Process.* **2016**. [CrossRef]
6. Casstevens, J.M. Diamond turning of steel in carbon-saturated atmospheres. *Precis. Eng.* **1983**, *5*, 9–15. [CrossRef]
7. Hitchiner, M.P. Factors affecting chemical wear during machining. *Wear* **1984**, *93*, 63–80. [CrossRef]
8. Evans, C. Cryogenic diamond turning of stainless steel. *CIRP Ann. Manuf. Technol.* **1991**, *40*, 571–575. [CrossRef]
9. Song, Y.C.; Nezu, K.; Park, C.H.; Moriwaki, T. Tool wear control in single-crystal diamond cutting of steel by using the ultra-intermittent cutting method. *Int. J. Mach. Tools Manuf.* **2009**, *49*, 339–343. [CrossRef]
10. Shamoto, E.; Suzuki, N. Ultrasonic vibration diamond cutting and ultrasonic elliptical vibration cutting. *Compr. Mater. Process.* **2014**, *11*, 405–454.
11. Brinksmeier, E.; Gläbe, R.; Osmer, J. Ultra-precision diamond cutting of steel molds. *CIRP Ann. Manuf. Technol.* **2006**, *55*, 551–554. [CrossRef]
12. Inada, A.; Min, S.; Ohmori, H. Micro cutting of ferrous materials using diamond tool under ionized coolant with carbon particles. *CIRP Ann. Manuf. Technol.* **2011**, *60*, 97–100. [CrossRef]
13. Thornton, A.G.; Wilks, J. The wear of diamond tools turning mild steel. *Wear* **1980**, *65*, 67–74. [CrossRef]
14. Paul, E.; Evans, C.J. Chemical aspects of tool wear in single point diamond turning. *Precis. Eng.* **1996**, *18*, 4–19. [CrossRef]
15. Tanaka, H.; Shimada, S.; Ikawa, N.; Yoshinaga, M. Wear mechanism of diamond cutting tool in machining of steel. *Key Eng. Mater.* **2001**, *196*, 69–78. [CrossRef]
16. Shimada, S.; Tanaka, H.; Higuchi, M. Thermo-chemical wear mechanism of diamond tool in machining of ferrous metals. *CIRP Ann. Manuf. Technol.* **2004**, *53*, 57–60. [CrossRef]
17. Narulkar, R.; Bukkapatnam, S.; Raff, L.M.; Komanduri, R. Graphitization as a precursor to wear of diamond in machining pure iron: A molecular dynamics investigation. *Comput. Mater. Sci.* **2009**, *45*, 358–366. [CrossRef]
18. Zou, L.; Dong, G.J.; Zhou, M. Investigation on frictional wear of single crystal diamond against ferrous metals. *Int. J. Refract. Met. Hard Mater.* **2013**, *41*, 174–179. [CrossRef]
19. Jen, F.L.; Jia, W.L.; Pal, J.W. Thermal analysis for graphitization and ablation depths of diamond films. *Diam. Relat. Mater.* **2006**, *15*, 1–9.
20. Chen, Z.C.; Subhash, G.; Tulenko, J.S. Raman spectroscopic investigation of graphitization of diamond during spark plasma sintering UO_2-diamond composite nuclear fuel. *J. Nucl. Mater.* **2016**, *475*, 1–5. [CrossRef]
21. Berman, R. The diamond-graphite equilibrium calculation: The influence of a recent determination of the Gibbs energy difference. *Solid State Commun.* **1996**, *99*, 35–37. [CrossRef]
22. Zou, L.; Zhou, M. Experimental investigation and numerical simulation on interfacial carbon diffusion of diamond tool and ferrous metals. *J. Wuhan Univ. Technol.-Mater. Sci. Ed.* **2016**, *31*, 307–314. [CrossRef]
23. Yamashita, T.; Hayes, P. Analysis of XPS spectra of Fe^{2+} and Fe^{3+} ions in oxide materials. *Appl. Surf. Sci.* **2008**, *254*, 2441–2449. [CrossRef]

MDPI
St. Alban-Anlage 66
4052 Basel
Switzerland
Tel. +41 61 683 77 34
Fax +41 61 302 89 18
www.mdpi.com

Crystals Editorial Office
E-mail: crystals@mdpi.com
www.mdpi.com/journal/crystals